The Enterprise of Science in Islam

DIBNER
INSTITUTE
FOR THE HISTORY
OF SCIENCE AND
TECHNOLOGY

Dibner Institute Studies in the History of Science and Technology
Jed Buchwald, general editor, Evelyn Simha, governor

Jed Z. Buchwald and Andrew Warwick, editors, *Histories of the Electron: The Birth of Microphysics*

Jed Z. Buchwald and I. Bernard Cohen, editors, *Isaac Newton's Natural Philosophy*

Anthony Grafton and Nancy Siraisi, editors, *Natural Particulars: Nature and the Disciplines in Renaissance Europe*

Frederic L. Holmes and Trevor H. Levere, editors, *Instruments and Experimentation in the History of Chemistry*

Agatha C. Hughes and Thomas P. Hughes, editors, *Systems, Experts, and Computers: The Systems Approach in Management and Engineering, World War II and After*

Jan P. Hogendijk and Abdelhamid I. Sabra, editors, *The Enterprise of Science in Islam: New Perspectives*

N. L. Swerdlow, editor, *Ancient Astronomy and Celestial Divination*

The Enterprise of Science in Islam

New Perspectives

---

edited by

Jan P. Hogendijk and Abdelhamid I. Sabra

The MIT Press
Cambridge, Massachusetts
London, England

This book was set in Bembo and Times New Roman by The MIT Press.

Printed and bound in the United States of America.

Library of Congress Cataloging-in-Publication Data

The enterprise of science in Islam : new perspectives / edited by Jan P. Hogendijk and Abdelhamid I. Sabra.
p. cm. – (Dibner Institute studies in the history of science and technology)
Includes bibliographical references and index.
ISBN 0-262-19482-1 (hc. : alk. paper)
1. Science, Medieval. 2. Science–Islamic countries–History. I. Hogendijk, J. P. II. Sabra, A. I. III. Series.

Q124.97 .E57 2003
509.17′671–dc21

2002029522

Contents

Introduction

Between AD 800 and 1450, the most important centers for the study of what we now call "the exact sciences" were located in the vast multinational Islamic world. The sciences denoted by this name included the mathematical sciences of arithmetic, geometry and trigonometry, and their applications in various fields such as astronomy, astrology, geography, cartography, and optics, to mention only some of the more prominent examples. During the eighth and ninth centuries, the bulk of Greek science, medicine and philosophy, and much of Indian and pre-Islamic Persian science, were appropriated by Islamic civilization through a complex process of translations from Pahlavi, Sanskrit, Greek, and Syriac, in the course of which Arabic became the language of a rich and active scientific and philosophical tradition for many centuries. In the eleventh and twelfth centuries, many Arabic scientific works and Arabic versions of Greek scientific and philosophical texts were translated into Latin, and in turn were appropriated into the Latin medieval culture. These translations were crucial for the rise of the "renaissance of the twelfth century" in Europe and they later played an important part in the development of the exact sciences during the Renaissance of the sixteenth century. However, only a small part of the total Islamic accomplishments in science was transmitted to medieval Europe. The scientific endeavors in the Islamic world of course remain as an important subject to be investigated in its own right as a distinctive aspect of Islamic culture.

The Islamic tradition in the exact sciences continued well into the nineteenth century, and abundant source material is available in the form of unpublished manuscripts in Arabic, Persian, and other languages in libraries all over the world. In the last decades, many researchers have worked on the Islamic scientific tradition, and our views of this tradition are rapidly changing as a result of recent discoveries. This process will, hopefully, continue, because important sources have not yet been identified and studied. Hence the time is not yet ripe for a reliable survey of the entire field.

The twelve chapters in this book discuss some of the new perspectives on the Islamic scientific tradition emerging from recent historical research. The chapters are revised and updated versions of contributions by experts in the history of Islamic science at a conference on New Perspectives in Islamic Science held at the Dibner Institute in November 1998. The emphasis of the conference was on the mathematical sciences, and the chapters in this volume represent a cross-section of the field. Not all important topics could be included, and the reader will notice the absence of contributions on, for example, Islamic astronomical instruments. Nevertheless, we believe that the present volume will transmit to the reader a view of an exciting and rapidly developing field of historical research.

The editors have divided the twelve chapters into six groups of two, under the headings: Cross-cultural Transmission; Transformations of Greek Optics; Mathematics: Philosophy and Practice; Numbers, Geometry and Architecture; Seventeenth-century Transmission of Astronomy; Science and Medicine in the Maghrib and al-Andalus. This subdivision does not do full justice to the multiple ways in which the twelve chapters are connected. Most of the chapters are, in one way or another, related to transmission of scientific knowledge, either from one culture to another (Kunitzsch, Burnett, Kheirandish, Sabra, Endress, Berggren, Pingree), or within the medieval Islamic world itself (Kunitzsch, Samsó, Djebbar). Three chapters discuss mainly astronomy (Burnett, Pingree, Samsó), five chapters are entirely or mainly on mathematics (Kunitzsch, Berggren, Sesiano, Dold, Djebbar), two chapters are on optics (Sabra, Kheirandish), while the chapter by Endress concerns the philosophy of the mathematical sciences. Several chapters are concerned with the relations between the exact sciences and other fields, such as natural philosophy (Kheirandish, Endress), architecture (Dold), medicine (Langermann). Some chapters concern outsider's views on mathematics and its use (Endress, Langermann), and medieval debates on scientific methodology (Berggren, Sabra). Two chapters discuss the individual mathematicians al-Kūhī (Berggren) and al-Kāshī (Dold), whose styles and attitudes turn out to be very different. The chapters by Djebbar, Langermann, and Samsó concern geographical areas, which will be indicated by their medieval Islamic names: al-Andalus is Islamic Spain and Portugal, that part of the Iberian peninsula that belonged to the medieval Islamic world, and the Maghrib is Northwestern Africa, that is, modern Morocco, Algeria, and Tunisia.

The chapters in this volume are mainly based on the analysis of (often unpublished) sources in Arabic and other languages. These documents are usually scientific, but may also include literary sources and biographies (Kunitzsch, Djebbar). In some of the chapters, the authors use special methods to argue their new insights, ranging from the philological analysis of technical terms in different languages (Burnett, Kheirandish), via the analysis of different systems of numerals (Kunitzsch, Burnett), to computer analysis of numerical tables (Samsó).

Several chapters summarize many years of research work by their authors, as the following three examples show. Berggren's characterization of al-Kūhī as a mathematician is the outcome of his project to publish a complete edition of al-Kūhī's works; Sabra's analysis of the argument of Books I–VII of Ibn al-Haytham's *Optics* is based on his Arabic edition and English translation of this large and fundamental work; Djebbar's survey of mathematics and astronomy in the Maghrib and Andalusia includes his own research and that of his pupils. Several chapters deal with striking examples of originality of the Islamic tradition (Sabra, Dold, Sesiano).

Not surprisingly, the chapters in this volume support the view that the Islamic scientific tradition was even richer, more profound, and with more complex relations to other cultures than had been thought hitherto. As more sources become gradually available, our picture also becomes more detailed. Thus, Berggren is now able to characterize Abū Sahl al-Kūhī as a conservative, "old-guard" mathematician among his contemporaries, Djebbar is able to list differences between the algebraic traditions in al-Andalus and the Maghrib, and so on. Most of the chapters also confront the reader with unsolved historical problems. Thus, Kunitzsch reminds us that it is still unknown when and where the Hindu-Arabic number symbols evolved from the Eastern Islamic forms (now used in the middle East) to the Western Islamic forms, which are virtually identical to the modern forms 1,2,3,4,5,6,7,8,9,0. In spite of all the progress, much about science in the Islamic tradition is still unknown.

Since some of the authors have different preferences for systems of referencing and on the precise ways to transcribe or write Arabic names, the editors have decided not to interfere with these preferences. Thus the same name may appear in different chapters in slightly different forms, for example, Abu'l-Rayḥān, Abū al-Rayḥān, and Abū ar-Rayḥān; al-Kūhī, al-Qūhī.

The "new perspectives" in each chapter will now be briefly outlined in the rest of this introduction, under the six headings mentioned above.

Cross-cultural Transmission

In his chapter, "The Transmission of Hindu-Arabic Numerals Reconsidered," Paul Kunitzsch strives to distinguish between, on the one hand, what can be accepted as fairly established fact in regard to this widely discussed subject, and, on the other, what still remains uncertain and in need of further study. Kunitzsch accepts the usually cited evidence presented by the Arabic and Syriac sources in support of the thesis that the nine numerals plus a symbol for an empty place initially came to the Arabs from India in the eighth century. He notes that variable forms of the nine numerals continued to be used well into the eleventh century, but rejects the hypothesis, once proposed by S. Gandz, that the dust board utilized by the Arabic mathematicians for performing their calculations later became the model for the Latin abacus.

Kunitzsch argues in the second part of his chapter that the term dust numerals (Arabic: *ḥurūf al-ghubār*), rather referred to the written figures, as opposed to the non-written numerals signified in processes known as mental reckoning or finger reckoning. The majority of the numerals current in the Arabic/Islamic West (the Maghrib and al-Andalus) could be obtained, so Kunitzsch remarks, from their Eastern counterparts. Thus, while the general line of borrowing of the numerals appears quite clear, Kunitzsch emphasizes the need for further research to establish the existence of the Hindu-Arabic numerals in the Western-Islamic sources prior to AD 1300. This should help to clear up the process of early Latin borrowing by way of the twelfth-century Arabic-Latin translations.

In his chapter, "The Transmission of Arabic Astronomy via Antioch and Pisa in the Second Quarter of the Twelfth Century," Charles Burnett studies a hitherto neglected channel of transmission of the exact sciences from the Eastern Islamic world to Christian Europe before Toledo emerged as a center of the translation industry, and the first translation of Ptolemy's *Almagest* was produced in Sicily from Greek. Burnett's first source is a medieval Latin manuscript of Book I–IV of the *Almagest* of Ptolemy (ca. 150 AD), now in Dresden. By a meticulous analysis of the technical terminology and the numerals that were used, Burnett shows that this "Dresden Almagest" was translated from an Arabic version around 1120 in Antiochia (now in Turkey), and that the translator wanted to give the impression that he had translated the work from Greek. Burnett then introduces a second source, a cosmological work in Latin entitled *Liber Mamonis*, which is written in the same technical terminology as the Dresden Almagest. Burnett shows that this work was inspired by Arabic sources, and he suggests that the title is derived from one such source, the *Verified Astronomical Tables* (Al-Zīj al-Mumtaḥan) by the astronomers around Caliph al-Maʾmūn in the early ninth century. Burnett identifies the author of the *Liber Mamonis* as Stephanus the philosopher, who originated from Pisa but who worked in Antiochia around 1120. Burnett argues that the "Dresden Almagest" was written in the same milieu. In the last part of his chapter, he studies a third work, the *Tables of Pisa*, which is based on the lost astronomical tables of the Iranian astronomer al-Ṣūfī (d. 986), and which has often been related to the Jewish astronomer Abraham ibn Ezra (d. ca. 1160). By an analysis of the systems of numeration used in the *Tables of Pisa*, Burnett establishes a clear connection with the *Liber Mamonis*, and hence with Antiochia. Thus, Antiochia is shown to be an important center for the early transmission of astronomy from Arabic into Latin.

Transformations of Greek Optics

The science of optics (Arabic: *ʿilm al-manāẓir*, Greek: *hē optikē technē*) is perhaps unique as a Greek mathematical discipline that received in the Islamic middle ages a radical transformation which ultimately succeeded in launching it on an entirely new course. As a mode of inquiry consisting of a combined mathematical and experimental approach to visual perception, optics was represented in Greek antiquity by works of Euclid (ca. 300 BC), Ptolemy (second century AD), and Theon of Alexandria (fourth century AD), all of which carried the title *Optika*, and at least two of which, namely those of Euclid and Ptolemy, came to be known in Arabic translations, some made as early as the ninth century, and given the title *al-Manāẓir*. The radical transformation took place in the first half of the eleventh century, in the large seven-part work of al-Ḥasan ibn al-Haytham, *Kitāb al-Manāẓir*. This work had the good fortune of finding its way to Muslim Spain where a Latin translation was made of it probably in the late twelfth century, thereby securing an entry into the main stream of European scientific, mathematical, and philosophical thought where it is known to have exerted considerable influence that lasted all the way up to the seventeenth century.

In her chapter, "The Many Aspects of Appearances," Elaheh Kheirandish focuses on transformations occurring already through the ninth-century Arabic assimilation of central Greek optical terms, beginning with the Greek and Arabic names for the discipline itself. She takes as her starting point a chapter in al-Fārābī's *Catalogue of the Sciences* (first half of the tenth century), a work later also widely known in Europe in a medieval Latin translation. Al-Fārābī was a philosopher, and not himself one of those who contributed to the mathematical science of vision. But his chapter offers a valuable picture of the discipline and a set of operative terms in it as understood by an intelligent witness from the century immediately preceding that of Ibn al-Haytham. Kheirandish further elaborates her analysis by critical comparisons with terms and phrases found in the writings of the ninth-century Arabic writers and their Greek sources.

Following the order of exposition in al-Fārābī's chapter, Kheirandish organizes her discussion in sections under five headings: I. Veracity and Accuracy of Vision, in which she notes a particular emphasis on two aspects of appearances, namely veracity and accuracy, and a greater emphasis in al-Fārābī on the latter than on the former; II. Justifiability and Variety of Demonstration, noting the concern of optics with explanation understood as a demonstrative procedure, and the connected quantification of visual clarity, by which she does not of course mean to imply any kind of measurement; III. Versatility and

Fallibility of Applications, where she regrets the fact that al-Fārābī's recognition of various applications of optics does not lead him to a discussion of mirrors and the principle of reflection, a neglect which in addition to the omission of surveying she finds surprising, and one of the problematic cases of transmission; and, finally, IV. Elements and Mechanisms of Vision, and V. Modes and Mediums of Operation, in which Kheirandish goes into the widespread and, sometimes, substantive confusion between optical reflection and refraction of visual rays. This confusion strangely persisted in the works of Naṣīr al-Dīn al-Ṭūsī and Quṭb al-Dīn al-Shīrāzī in the thirteenth century, and until Ibn al-Haytham's work became generally known to Arabic readers, thanks to the comprehensive *Commentary* made by al-Shīrāzī's gifted student, Abu 'l-Ḥasan Kamāl al-Dīn al-Fārīsī more than two hundred years after Ibn al-Haytham died (ca. 1041), and probably more than a hundred years after the latter's *Kitāb al-Manāẓir* began to be utilized by readers of Latin.

In the following chapter by A. I. Sabra, "Ibn al-Haytham's Revolutionary Project in Optics: The Achievement and the Obstacle," the author claims to use "revolution" in the strict sense of a conscious and radical transformation of a widely practiced and accepted approach to a whole scientific discipline, a transformation that goes to the heart of the basic assumptions of the traditional system. Sabra does not believe that Ibn al-Haytham's departure from tradition was a *creatio ex nihilo*, and accordingly he stresses elements borrowed from earlier methods and doctrines, especially those derived from Ptolemy's *Optics* and Aristotelian physics; but Sabra tries to show that these elements now serve different functions in a new system. The chapter is actually an outline of the single, continuous argument which, according to Sabra, runs through all the seven books that make up the *Optics* of Ibn al-Haytham: Having totally rejected, on the basis of empirical evidence, the visual-ray hypothesis as the foundation of previous mathematical theories of vision, and aligning himself (again on the basis of experience) with the Peripatetic view of vision as the reception of forms of light and color, Ibn al-Haytham was led to accord psychology a new, inevitable and fundamental role never realized earlier in the works of the Greek mathematicians and their Arabic successors. The new approach further leads Ibn al-Haytham to reformulate the existing explanations of rectilinear transmission, reflection and refraction of light in ways that may conform with the older geometry, but that also generate new problems posed by his own treatment of specular images, as Sabra notes.

The obstacle referred to in the title of Sabra's chapter has to do with the doctrine, inherited from Galen, locating the sensitivity of the eye in the crystalline humor. This was a fateful error which forced Ibn al-Haytham to fall back on psychology in a last attempt (in Book VII) to explain crucial experiments that he was the first to describe in the history of optics. The erroneous doctrine was itself again adopted as an empirically established fact of ocular physiology, but this time Ibn al-Haytham put his trust in what was widely reported and accepted by the medical tradition as empirical proof. It was from here, as Sabra remarks but does not discuss, that Kepler started his own researches that led to his well-known breakthrough in the *Paralipomena ad Vitellionem*, published in 1604.

MATHEMATICS: PHILOSOPHY AND PRACTICE

Gerhard Endress begins his survey chapter, "Mathematics and Philosophy in Medieval Islam," with a summary of the views of the two most important ancient Greek philosophers on mathematics and astronomy. Plato believed that number and mathematics were related to an eternal world of ideas, which is the essence and source of the changing world in which we live. Aristotle, on the other hand, believed that mathematics was abstracted from reality and not directly related to the essence of the real world. However, he believed that the properties of the real world could be deduced from a few basic principles of natural philosophy, in a way somewhat similar to the deductive reasoning in geometry. Unfortunately, Aristotle drew up his cosmological principle of uniform rotation around the center of the earth at a time when astronomy was still in a qualitative state. When Ptolemy worked out his planetary theory in the second century AD, he could only make his models correspond to reality by introducing epicycles and equant points. The resulting contradiction with the Aristotelian principle of uniform motion around the center of the earth was never resolved in antiquity.

Endress then turns to the attitude of the most important Islamic philosophers with respect to the mathematical sciences. The early philosopher al-Kindī (ca. 830) and his followers considered mathematics, in the Platonic vein, as an intermediary between philosophy and the science of nature. Al-Fārābī (d. 950) accepted the Aristotelian view of mathematics and natural science, and the same is basically true for Ibn Sīnā (Avicenna, d. 1037). These philosophers were not astronomers, but Ibn al-Haytham (d. ca. 1040) was a competent astronomer who tried to bridge the gap between Aristotelian physics and Ptolemaic astronomy. He finally came to the conclusion that astronomical models had to be deduced by demonstrative reasoning (in the way of Aristotle) from basic principles different from those enunciated by Aristotle, and that the ad-hoc devices used by Ptolemy were unsatisfactory. However, Ibn al-Haytham did not work out a new astronomical system of his own.

Endress then turns to the Andalusian astronomers and philosophers, who distinguished between the natural philosophy of Aristotle, dealing with essences and causes, and the geometrical models of the astronomers. These Andalusian scholars believed that these mathematical models dealt only with accidental properties, abstracted from reality, and thus one could not use them to penetrate into reality in its own right. They believed that Aristotle had reached human perfection, and they made a few hopeless attempts to propose Aristotelian alternatives to Ptolemy's models in order to (re?)construct a true Aristotelian cosmos. Endress presents an interesting quotation from the famous Ibn Rushd/Averroes (d. 1198), expressing frustration and unhappiness with the results.

In the final section, Endress discusses the Jewish philosopher Maimonides (d. 1204) and late Islamic thinkers, who decided that "natural philosophy" is domain of God, and that Aristotelian natural philosophy ought to be discarded as contrary to religion. They argued that the astronomers were free to imagine models that correspond to observation and make possible predictions, an attitude that may strike us as quite modern.

Whereas Endress's chapter is devoted to the views of relative outsiders (astronomers and philosophers) on the nature and use of mathematics, J. L. Berggren studies an insider's view in his chapter "Tenth-Century Mathematics through the Eyes of Abū Sahl al-Kūhī." As many other research mathematicians in his time, al-Kūhī was fascinated by geometry in the style of the Hellenistic geometers, and some of his tenth-century contemporaries regarded him as the "master of his age in the art of geometry." Berggren suggests, plausibly, that al-Kūhī was fascinated by the certainty of geometrical knowledge. He preferred to work on problems discussed in or inspired by the ancient Greek geometry of Archimedes and Apollonius, such as geometrical constructions of the regular heptagon and other figures by means of conic sections, and the determination of the volumes and centers of gravity of solids including the paraboloid. Berggren states that among tenth-century geometers, al-Kūhī was unique in finding and (generally) solving geometrical problems of some real depth.

Al-Kūhī believed in the progress of science and he says that "the science of geometry will endure and will continue to grow, in contrast to man's life span, which comes to an end." He contributed to methodological debates on analysis and synthesis, the mathematical concept of a "known ratio" in connection with the quadrature of the circle, and the question to what extent it was legitimate to use motion in geometry. Unlike most of his colleague-mathematicians, al-Kūhī was uninterested in arithmetic, algebra, and trigonometry. He even regarded mathematics based on numerical approximations as "bad" mathematics because it is not based on exact and demonstrative methods. Berggren concludes that al-Kūhī was rather conservative in his outlook, and that he may have been the last mathematician to look on mathematics with the eyes of the great Hellenistic geometers.

NUMBERS, GEOMETRY, AND ARCHITECTURE

The Islamic mathematicians contributed not only to geometry but also to the theory of numbers, and they studied a combination of the two in what they called "harmonious dispositions of numbers" (Arabic: *aʿdād al-wafq*), nowadays called magic squares. A magic square is a square array of integer numbers, with the property that the sums of all elements in each column, row, or diagonal are equal. In modern terms, the number of elements in each row, column, or diagonal is called the order of the magic square, and the sum of the elements in each row, column, or diagonal is called the magical constant.[1] Although magic squares are probably of pre-Islamic (Persian?) origin, there is no evidence that they were seriously studied before the Islamic tradition. The title of Jacques Sesiano's chapter, "Quadratus Mirabilis," is the name he has given to the most complex type of magic squares that has hitherto been found in the entire medieval Islamic tradition. The construction of this type of magic square is explained in a text by the late tenth-century mathematician al-Anṭākī, but Sesiano points out that it may have been discovered earlier. For each integer n, al-Anṭākī constructs a magic square of order n, composed of the numbers 1, 2, . . . n^2, with the following two properties:

1. The odd numbers 1, 3, . . . are placed in a rhombus in the middle of the square and the even numbers 2, 4 . . . in four triangular corners of the square;
2. The magic square is "bordered," that is to say that if all $4n - 4$ numbers on the outside are taken away, the remaining square is also a new magic square of order $n - 2$, and if all $4n - 12$ numbers on the outside of this new square are taken away, the remaining square is also a magic square, and so on, until one is left with a magic square of order 4 or 3.

Sesiano gives a transcription of the construction of this "Quadratus Mirabilis" in modern notation as well as a literal translation of the Arabic text of al-Anṭākī. His chapter shows that the theory of magic squares reached a much higher level in the tenth century than had been thought before. The term "magic square" is European, and as Sesiano points out, magic squares were mostly investigated in the Islamic tradition because of their mathematical as well as recreational interest. Most magical "applications" of these squares are relatively late and only utilize the simplest forms.

The relation between geometry, numerical computation and architecture is explored in the chapter by Yvonne Dold-Samplonius, which is entitled "Calculating Surface Areas and Volumes in Islamic Architecture." She begins with a concise introduction to Islamic architecture. Not much is known about the relations between architecture and mathematics in the early Islamic tradition, but the available evidence suggests that sophisticated mathematics entered

Islamic architecture relatively late in the tradition. (Of course, future research may show this impression to be incorrect.) Dold lists a number of rough calculations of volumes and surfaces of domes from various practical mathematical works until the thirteenth century, and she compares these calculations with the sophisticated approach in the *Key to Arithmetic* of al-Kāshī, who died around 1429 in Samarkand. Al-Kāshī computed various coefficients which enabled craftsmen to easily find surfaces and volumes of various types of domes found in Central Asia. His computations are unrelated to earlier Islamic studies on the volume of parabolic domes (paraboloids) of a more theoretical nature, which Dold characterizes as "highschool mathematics." One of the authors of these more theoretical studies was al-Kūhī, the subject of Berggren's chapter in this volume. Dold finishes her chapter with a brief look at al-Kāshī's discussion of the stalactite vaults called "muqarnas," which are characteristic of Islamic architecture.

Al-Kāshī had a talent for finding user-friendly and highly accurate approximative solutions of difficult practical problems which cannot be solved exactly, and this is why Dold-Samplonius calls him "the first modern mathematician." Her view is supported by further mathematical analysis of al-Kāshī's calculations. Thus, al-Kāshī and al-Kūhī represent two radically different types of Islamic mathematicians.

Seventeenth-century Transmission of Astronomy

The chapters in this section by David Pingree and Julio Samsó discuss concrete examples of transmission of Islamic science after the medieval period. Both chapters are related to the tradition of the famous astronomical handbook (*Zīj*) of Ulūgh Beg (1393–1449), the ruler of Samarkand who founded an astronomical observatory and appointed al-Kāshī as its director. As Pingree explains in his chapter, "The Sarvasiddhāntarāja of Nityānanda," the *Zīj* of Ulūgh Beg was revised in the early seventeenth century at the court of the Moghul emperor Shah Jahān at Delhi, and the reworking was then translated into Sanskrit by Nityānanda. Because the translation failed to find favor with the traditional Hindu astronomers at Delhi, Nityānanda decided to write a Sanskrit apology for Islamic astronomy, entitled Sarvasiddhāntarāja. Pingree presents a detailed analysis of chapters 2 and 3 of this work, on the computation of the mean and true longitudes of the planets. Nityānanda rephrased Islamic astronomy in a language which Hindu astronomers could understand, and thus he expressed the planetary mean motions in integer revolutions in a Kalpa of 4,320,000,000 years. He also tried to show that the differences between traditional Hindu astronomy and what he called the "Roman Zīj," that is, the new Islamic system, were only small. Nityānanda gave only numerical algorithms and he did not discuss the geometrical background of the Islamic (essentially Ptolemaic) models of planetary motion; since he also failed to define many of his new technical terms, his colleagues must have found it hard to understand his work. Pingree points out that some of the new vocabulary in the *Sarvasiddhāntarāja* was nevertheless used in other translations made at Delhi a few years later, for example in the Sanskrit prose version of the *Tabulae Astronomicae* of Philippe de la Hire (1640–1718). Thus, Pingree's chapter sheds light on one of the most neglected areas in the history of the transmission of Islamic astronomy, namely its development and influence in South Asia.

Julio Samsó's chapter, "On the Lunar Tables in Sanjaq Dār's *Zīj al-Sharīf*," is related to the complex transmission of astronomical knowledge from the Eastern to the Western Islamic world. From the ninth through the eleventh centuries, there existed a special type of Islamic astronomy in al-Andalus. One of the characteristics of this Andalusian astronomy is the existence of special theories to explain the supposed phenomenon of "trepidation," that is the oscillation in the ecliptic longitudes of all fixed stars with respect to the vernal point. These theories were also studied in the Maghrib. Eastern Islamic astronomers, such as Muḥyī al-Dīn al-Maghribī and Ibn al-Shāṭir, rejected trepidation, and from the late 14th century on, their theories were also transmitted to the Maghrib. From that time on, the astrologers in the Maghrib continued to use the older Andalusian astronomy, while the more sophisticated Eastern Islamic astronomy was used by the astronomers and *muwaqqits* (official time-keepers in mosques). The *Zīj* of Ulūgh Beg was apparently transmitted to the Maghrib in the 17th century, and Julio Samsó studies the influence of that work on the 17th-century *Zīj al-Sharīf* ("Noble Astronomical handbook") by Sanjaq Dār of Tunis. Samsó briefly summarizes the complex (essentially Ptolemaic) astronomical contents of this work, and using computer programs developed by Benno van Dalen, he analyzes the lunar tables, to establish the dependency of *Zīj al-Sharīf* on the *Zīj* of Ulūgh Beg in detail. The technical terminology and the concrete examples of numerical tables in Samsó's chapter may give the reader some idea of the knowledge and skills which a medieval Islamic astronomer needed in order to compile an astronomical handbook (*Zīj*) for his own city and time.

Science and Medicine in the Maghrib and al-Andalus

Ahmed Djebbar's contribution to this volume, entitled "A Panorama of Research on the History of Mathematics in al-Andalus and the Maghrib between the Ninth and Sixteenth Centuries," is a unique survey of the research done by modern historians between 1834 and 1980 on the history of medieval mathematics and astronomy in those geographical areas. Djebbar treats not only

Western literature but also the research by Arabic scholars in the nineteenth and twentieth centuries, and he comments on the historical and political motivations of these Arabic researchers. For the period after 1980, Djebbar summarizes much of the work of his own research group at the École Normale Supérieure in Algiers, which is now the most important center for the study of medieval Maghribi mathematics and astronomy. The survey shows the progress that has been made in the last decades in our understanding of medieval Maghribi science, as well as a number of its more puzzling aspects, which will have to be clarified by future research. Examples are the discontinuities in the transmission from the Eastern Islamic to the Western Islamic world, the fact that mathematics in the Maghrib was limited to arithmetic from the late 13th century onwards, the transmission of Euclid's *Elements* and other geometrical works, and the silence of the bio-bibliographical works on the algebraic tradition in al-Andalus and the Maghrib. Libraries in the Maghrib contain massive amounts of unpublished Arabic manuscripts and it is expected that further study of these sources will not only shed light on these open problems, but also reveal many more details about the mathematical and astronomical traditions in specific periods and specific areas of the Maghrib.

In his chapter "Another Andalusian Revolt? Ibn Rushd's Critique of al-Kindī's Pharmacological Computus," Tzvi Langermann touches upon the question which some have raised with regard to intellectual developments in North Africa and Muslim Spain under the rule of the ideologically driven Almohads/al-Muwaḥḥidūn (524–667/1130–1269), namely whether certain trends in science, medicine and philosophy, as well as in Islamic law and Arabic grammar, could be interpreted as manifestations of a general revisionist attitude toward the established authorities in those fields in the Eastern Islamic world. Evidence for the revisionist thesis in a scientific field has previously been expounded with reference to the conscious and reasoned rejection of the Eastern decision in favor of Ptolemaic, as opposed to Aristotelian astronomy, by Andalusian scholars, including especially the famous Ibn Rushd/Averroes, who referred to the Ptolemaic system of eccentrics and epicycles as "the astronomy of our time." Langermann is concerned with another episode: Ibn Rushd's vehement attack, in his medical textbook, *Kitāb al-Kulliyyāt*, on the work of the Eastern ninth century polymath, al-Kindī, in which the latter proposed a new, non-Galenic computus for calculating the right quantities of simple drugs, in order to produce the desired degree of their compounded elemental qualities: heat, cold, dry, and moist. Averroes objected, among other things, that al-Kindī was causing confusion in a medical subject by straying too far beyond the boundaries and the rules or laws/*qawānīn* proper to a natural inquiry, such as the art of medicine, into considerations of numbers and music. It is known that Averroes, in a commentary on Aristotle's *Meteorology*, raised an exactly parallel objection against the role of mathematics in the "physical" work of another Easterner, Ibn al-Haytham.

Langermann here refrains from directly answering the general, cultural question: hence the question mark in his title. And, following a good maxim, truth is in the details, he offers in the first part of his chapter a lucid and enlightening analysis of many of the details involved in Averroes's arguments, with due emphasis on their immediate medical context in the writings of authorities from Galen to medical authors in the Arabic tradition, including members of the distinguished Andalusian Ibn Zuhr family. In the second part of his chapter, devoted to the question of Contexts (thus in the plural), Langermann observes the lack of general interest in Kindī's computus, but refers to a lost treatise by the influential Abu 'l-ʿAlāʾ ibn Zuhr (d. 1130), directly addressed to al-Kindī's book. Why [then] was there so little interest in al-Kindī's book? Langermann asks, and delivers the most plausible answer: while pharmacologists (in the East and the West) were spurred by practical applications, Ibn Rushd, as the confirmed Aristotelian philosopher, was (doggedly?) concerned for what he took to be Aristotelian methodology. Langermann concludes with the advice that full attention should be given to the persistent interest in pharmacology in its own right. His final paragraph refers to the great Abū Marwān ibn Zuhr (Avenzoar, d. 1161), son of Abu 'l-ʿAlāʾ and close friend to Ibn Rushd, who had served as *wazīr* and physician to the founder of the Almohad dynasty, ʿAbd al-Muʾmin (1130–1163), and who had written his medical work *al-Taysīr* on the advice of Ibn Rushd. This reference calls up again the ghost of an underlying Andalusian ideological tendency whose many ramifications for the intellectual history of Muslim Spain and the Maghrib still need to be defined and adequately explored.

The Editors

Note

1. If the order of the magic square is n and the consecutive numbers are $1 \ldots n^2$, the magical constant is $\frac{1}{n}(1^2 + 2^2 + \ldots + n^2) = \frac{1}{2}n(n^2 + 1)$.

Contributors

J. Lennart Berggren is professor of mathematics at Simon Fraser University. His research interests center on the history of the mathematical sciences in ancient Greece and medieval Islam, in particular geometry and geometrical methods in geography and astronomy. He is translating and annotating the extant works of Abū Sahl al-Kūhī, and has authored or co-authored a number of articles and books, including *Episodes in the Mathematics of Medieval Islam* (1986), *Euclid's "Phaenomena"* (with R. Thomas, 1996), and *Ptolemy's Geography: An Annotated Translation of the Theoretical Chapters* (with A. Jones, 2000).

Charles Burnett is professor of the history of Islamic influences in Europe at the Warburg Institute, University of London. He has written extensively on the translation of scientific and philosophical texts from Arabic into Latin in the Middle Ages. His recent works include *The Introduction of Arabic Learning into England* (1997), *Scientific Weather Forecasting in the Middle Ages: The Writings of al-Kindī* (with Gerrit Bos, 2000), and *Abū Maʿshar on Historical Astrology: The Book of Religions and Dynasties (On the Great Conjunctions)*, 2 volumes (with K. Yamamoto, 2000).

Ahmed Djebbar is Professor of History of Mathematics at Lille University. His main research interests are in the history of Arabic mathematics in North Africa and Muslim Spain. He has been secretary to the International Commission for the History of Mathematics in Africa since 1986. Recent publications include *Une histoire de la science arabe* (2001) and *La vie et l'œuvre d'Ibn al-Bannā: un essai biobibliographique* (2001).

Yvonne Dold-Samplonius is an associate member of the Institute for Scientific Computing of the University of Heidelberg, where she directed the video "Qubba for al-Kashī," showing al-Kashī's geometrical constructions for determining the volumes of domes and arches. With her team she is now working on a three-year Muqarnas (stalactite vaults) project. With international cooperation they want to develop a computer method to create virtual muqarnas based on old plans, such as the Topkapı Scroll in Istanbul.

Gerhard Endress is professor of Arabic and Islamic Studies at the University of Bochum. His research focuses on the Arabic translation and transmission of the Greek sciences, and the history of philosophical thought in Islam. He has published textual editions and studies of the Arabic translations of Aristotle and late Hellenistic Neoplatonism, and of the Arabic philosophers. His publications include *An Introduction to Islam* (English ed. 1988, rev. 2001); *A Greek and Arabic Lexicon: Materials for a Dictionary of the Mediaeval Translations from Greek into Arabic* (with D. Gutas, since 1992); and *Averroes and the Aristotelian Tradition* (with J. A. Aertsen, 1999).

Jan P. Hogendijk is a member of the Department of Mathematics of the University of Utrecht. His research interest is the history of the mathematical sciences in Greek antiquity and medieval Islamic civilization. He has recently published an edition of the mathematical works of al-Jurjānī (10th c.) and an analysis of the works of Abū Naṣr ibn ʿIrāq (ca. 1000) on sundial theory. He is now working on an edition and English translation of a ninth-century manual of geometrical problems by Nuʿaim ibn Mūsā.

Elaheh Kheirandish is a senior resident fellow at The Dibner Institute for the History of Science and Technology, Massachusetts Institute of Technology. Her publications include *The Arabic Version of Euclid's Optics: Kitāb Uqlīdis fī Ikhtilāf al-manāẓir* (1999), and articles on the history of mathematical sciences in the Islamic world. Her current work includes two projects in the history of optics and mechanics involving the application of the electronic medium (the IOTA project at the Dibner Institute, and the Archimedes Project at Harvard's Classics Department).

Paul Kunitzsch was professor of Arabic Studies at the University of Munich from 1977 to 1995. His main fields of research are the transmission of the sciences, especially astronomy and astrology, from Antiquity to the Arabs and from the Arabs to medieval Europe; Arabic knowledge of the stars and constellations; and Oriental influences in medieval European literature. His recent publications include *The Melon-Shaped Astrolabe in Arabic Astronomy* (1999) and *Claudius Ptolemäus, Der Sternkatalog des Almagest* (1986–1991).

Y. Tzvi Langermann is associate professor of Arabic at Bar Ilan University. His research interests cover a broad range of topics in medieval science and philosophy, with special emphasis on the thought of Moses Maimonides and the intellectual history of Yemenite Jewry. Recent publications include *The Jews and the Sciences in the Middle Ages* (1999); "Studies in Medieval Hebrew Pythagoreanism"; and "Criticism of Authority in Moses Maimonides and Fakhr al-Dīn al-Razī."

David Pingree is a university professor at Brown University and chair of the Department of the History of Mathematics. His interests include the history of astronomy, astrology, and astral magic in ancient Mesopotamia, Greece, the Roman Empire, India, the Islamic world, Byzantium, and medieval Europe, as well as the transmission of these sciences from one culture to another. Recent publications include *Arabic Astrology in Sanskrit* (2001) and *Astral Sciences in Mesopotamia* (1999).

Abdelhamid I. Sabra is emeritus professor of the history of Arabic science in Harvard's Department of the History of Science. His work in the history of science first focused on seventeenth-century Europe, with a book on *Theories of Light from Descartes to Newton* (1967, 1981). He later (after 1955) turned his attention to Arabic/Islamic science, with studies, editions of primary texts, and translations. His edition of *Ibn al-Haytham's Optics, Books IV–V* (complementing his edition and translation of *Books I–III*, 1983, 1989) is in press.

Julio Samsó is professor of Arabic and Islamic studies at the University of Barcelona. He is a member of the International Academy of History of Science and former president of the International Commission on Science in Islamic Civilization. His main research interests are in the history of astronomy in Muslim Spain and in medieval North Africa. His publications include *Las ciencias de los antiguos en al-Andalus* (1992) and *Islamic Astronomy and Medieval Spain* (1994).

Jacques Sesiano is a lecturer in the history of mathematics at the École Polytechnique Fédérale de Lausanne. His publications include *Un traité médiéval sur les carrés magiques* (1996) and *Une introduction à l'histoire de l'algèbre* (1999).

I

Cross-Cultural Transmission

1

The Transmission of Hindu-Arabic Numerals Reconsidered

Paul Kunitzsch

For the last two hundred years the history of the so-called Hindu-Arabic numerals has been the object of endless discussions and theories, from Michel Chasles and Alexander von Humboldt to Richard Lemay in our times. But I shall not here review and discuss all those theories. Moreover I shall discuss several items connected with the problem and present documentary evidence that sheds light—or raises more questions—on the matter.

At the outset I confess that I believe the general tradition, which has it that the nine numerals used in decimal position and using zero for an empty position were received by the Arabs from India. All the oriental testimonies speak in favor of this line of transmission, beginning from Severus Sēbōkht in 662[1] through the Arabic-Islamic arithmeticians themselves and to Muslim historians and other writers. I do not touch here the problem whether the Indian system itself was influenced, or instigated, by earlier Greek material; at least, this seems improbable in view of what we know about Greek number notation.

The time of the first Arabic contact with the Hindu numerical system cannot safely be fixed. For Sēbōkht (who is known to have translated portions of Aristotle's *Organon* from Persian) Fuat Sezgin[2] assumes possible Persian mediation. The same may hold for the Arabs, in the eighth century. Another possibility is the Indian embassy to the caliph's court in the early 770s, which supposedly brought along an Indian astronomical work, which was soon translated into Arabic. Such Indian astronomical handbooks usually contain chapters on calculation[3] (for the practical use of the parameters contained in the accompanying astronomical tables), which may have conveyed to the Arabs the Indian system. In the following there developed a genre of Arabic writings on Hindu reckoning (*fī l-ḥisāb al-hindī*, in Latin *de numero Indorum*), which propagated the new system and the operations to be made with it. The oldest known text of this kind is the book of al-Khwārizmī (about 820, i.e., around fifty years or more after the first contact), whose Arabic text seems to be lost, but which can very well be reconstructed from the surviving Latin adaptations of a Latin translation made in Spain in the twelfth century. Similar texts by al-Uqlīdisī (written in 952/3), Kūshyār ibn Labbān (2nd half of the 10th century) and ʿAbd al-Qāhir al-Baghdādī (died 1037) have survived and have been edited.[4] All these writings follow the same pattern: they start with a description of the nine Hindu numerals (called *aḥruf*, plural of *ḥarf*; Latin *litterae*), of their forms (of which it is often said that some of them may be written differently), and of zero. Then follow the chapters on the various operations. Beside these many more writings of the same kind were produced,[5] and in later centuries this tradition was amply continued, both in the Arabic East and West. All these writings trace the system back to the Indians.

The knowledge of the new system of notation and calculation spread beyond the circles of the professional mathematicians. The historian al-Yaʿqūbī describes it in his *Tārīkh* (written 889)—he also mentions zero, *ṣifr*, as a small circle (*dāʾira ṣaghīra*).[6] This was repeated, in short form, by al-Masʿūdī in his *Murūj*.[7] In the following century the encyclopaedist Muḥammad ibn Aḥmad al-Khwārizmī gave a description of it in his *Mafātīḥ al-ʿulūm* (around 980); also he knows the signs for zero (*aṣfār*, plural) in the form of small circles (*dawāʾir ṣighār*).[8] That the meaning of *ṣifr* is really "empty, void" has been nicely proved by August Fischer,[9] who presents a number of verses from old Arabic poetry, where the word occurs in this sense. It may thus be regarded as beyond doubt that *ṣifr*, in arithmetic, indeed renders the Indian *śūnya*, indicating a decimal place void of any of the nine numerals. Exceptional is the case of the *Fihrist* of Ibn al-Nadīm (around 987, that is, contemporaneous with the encyclopaedist al-Khwārizmī). This otherwise well-informed author apparently did not recognize the true character of the nine signs as numerals; he treats them as if they were letters of the Indian alphabet.[10] He juxtaposes the nine signs to the nine first letters of the Arabic *abjad* series and says that, if one dot is placed under each of the nine signs, this corresponds to the following (*abjad*) letters *yāʾ* to *ṣād*, and with two dots underneath to the remaining (*abjad*) letters *qāf* to *ẓāʾ* (with some defect in the manuscript transmission). This sounds as if he understood the nine signs and their amplification with the dots as letters of the Indian alphabet. Even a Koranic scholar, Abū ʿAmr ʿUthmān al-Dānī (in Muslim Spain, died 1053), knows the zero, *ṣifr*, and compares it to the common Arabic orthographic element *sukūn*.[11] (For all these authors it must be kept in mind that the manuscripts in which we have received their texts date from more recent times and therefore may not reproduce the forms of the figures in the original shape once known and written down by the authors.)

Of some interest in this connection are, further, two quotations recorded by Charles Pellat: the polymath al-Jāḥiẓ (died 868) in his *Kitāb al-muʿallimīn* advised schoolmasters to teach finger reckoning (*ḥisāb al-ʿaqd*) instead of *ḥisāb al-hind*, a method needing "neither spoken word nor writing"; and the historian and literate Muḥammad ibn Yaḥyā al-Ṣūlī (died 946) wrote in his

Adab al-kuttāb: "The scribes in the administration refrain, however, from using these [Indian] numerals because they require the use of materials [writing-tablets or paper?] and they think that a system which calls for no materials and which a man can use without any instrument apart from one of his limbs is more appropriate in ensuring secrecy and more in keeping with their dignity; this system is computation with the joints (*ʿaqd* or *ʿuqad*) and tips of the fingers (*banān*), to which they restrict themselves."[12]

The oldest specimens of written numerals in the Arabic East known to me are the year number 260 Hijra (873/4) in an Egyptian papyrus and the numerals in MS Paris, BNF ar. 2457, written by the mathematician and astronomer al-Sijzī in Shīrāz between 969 and 972. The number in the papyrus (figure 1.1)[13] may indicate the year, but this is not absolutely certain.[14] For an example of the numerals in the Sijzī manuscript, see figure 1.2. It is to be noted that here "2" appears in three different forms, one form as common and used in the Arabic East until today, another form resembling the "2" in some Latin manuscripts of the 12th century, and a form apparently simplified from the latter; also "3" appears in two different forms, one form as common in the East and used in that shape until today, and another form again resembling the "3" in some Latin manuscripts of the 12th century.

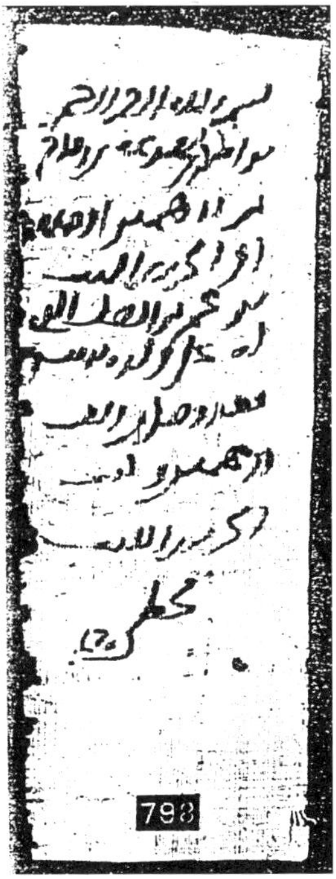

Figure 1.1
Papyrus PERF 789.
Reproduced from Grohmann, Pl. LXV, 12

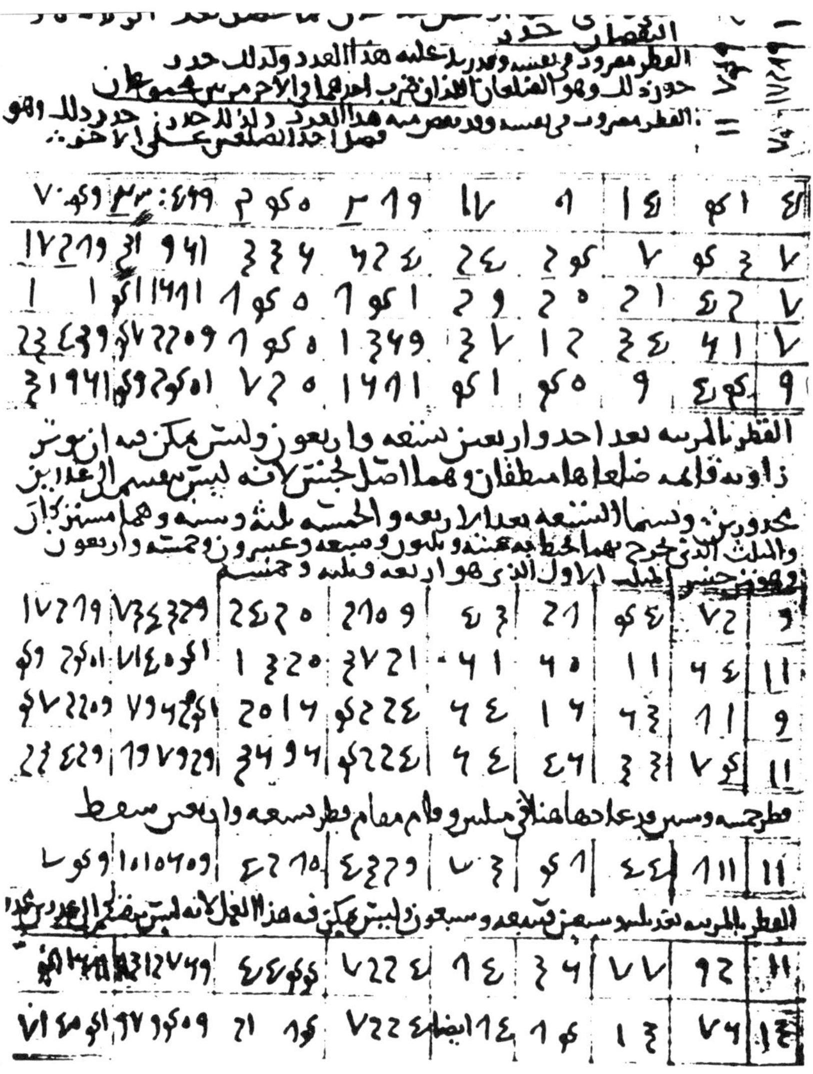

Figure 1.2
MS Paris, B. N. ar 2457, fol 85v. Copied by al-Sijzī, Shīrāz, 969–972

This leads to the question of the shape of the nine numerals. Still after the year 1000 al-Bīrūnī reports that the numerals used in India had a variety of shapes and that the Arabs chose among them what appeared to them most useful.[15] And al-Nasawī (early eleventh century) in his *al-Muqniʿ fī l-ḥisāb al-hindī* writes at the beginning, when describing the forms of the nine signs, "Les personnes qui se sont occupées de la science du calcul n'ont pas été d'accord sur une partie des formes de ces neuf signes; mais la plupart d'entre elles sont convenues de les former comme il suit"[16] (then follow the common Eastern Arabic forms of the numerals).

Among the early arithmetical writings that are edited al-Baghdādī mentions that for 2, 3, and 8 the Iraqis would use different forms.[17] This seems to be corroborated by the situation in the Sijzī manuscript. Further, the Latin adaptation of al-Khwārizmī's book says that 5, 6, 7, and 8 may be written differently. If this sentence belongs to al-Khwārizmī's original text, that would be astonishing. Rather one would be inclined to assume that this is a later addition made either by Spanish-Muslim redactors of the Arabic text or by the Latin translator or one of the adapters of the Latin translation, because it is in these four signs (or rather, in three of them) that the Western Arabic numerals differ from the Eastern Arabic ones.[18]

Another point of interest connected with Hindu reckoning and the use of the nine symbols is: how these were used and in what form the operations were made. Here the problem of the calculation board is addressed. It was especially Solomon Gandz who studied this problem in great detail and who arrived at the result that the Arabs knew the abacus and that the term *ghubār* commonly used in Western Arabic writings on arithmetic renders the Latin *abacus*.[19] As evidence for his theory he also cites from Ibn al-Nadīm's *Fihrist* several Eastern Arabic book titles such as *Kitāb al-ḥisāb al-hindī bi-l-takht* (to which is sometimes added *wa-bi-l-mīl*), "Book on Hindu Reckoning with the Board (and the Stylus)." I cannot follow Gandz in his argumentation. It is clear, on the one side, that all the aforementioned eastern texts on arithmetic, from al-Khwārizmī through al-Baghdādī, mention the *takht* (in Latin: *tabula*) and that on it numbers were written and—in the course of the operations—were erased (*maḥw*, Latin: *delere*). It seems that this board was covered with dust (*ghubār*, *turāb*) and that marks were made on it with a stylus (*mīl*). But can this sort of board, the *takht* (later also *lawḥ*, Latin *tabula*), be compared with the abacus known and used in Christian Spain in the late tenth to the twelfth centuries? In my opinion, definitely not. The abacus was a board on which a system of vertical lines defined the decimal places and on which calculations were made by placing counters in the columns required, counters that were inscribed with *caracteres*, that is, the nine numerals (in the Western Arabic style) indicating the number value. The action of *maḥw*, *delere*, erasing, cannot be connected with the technique of handling the counters. On the other side, the use of the *takht* is unequivocally connected with writing down (and in case of need, erasing) the numerals; the *takht* had no decimal divisions like the abacus, it was a board (covered with fine dust) on which numbers could be freely put down (Ibn al-Yāsamīn speaks of *naqasha*) and eventually erased (*maḥw*, *delere*). Thus it appears that the Arabic *takht* and the operations on it are quite different from the Latin *abacus*. Apart from the theoretical descriptions in the arithmetical texts we have an example where an astronomer describes the use of the *takht* in practice: al-Sijzī mentions, in his treatise *Fī kayfīyat ṣanʿat jamīʿ al-asṭurlābāt*, how values are to be collected from a table and to be added, or subtracted, on the *takht*.[20] Furthermore it is worth mentioning that al-Uqlīdisī adds to his arithmetical work a Book IV on calculating *bi-ghayr takht wa-lā maḥw bal bi-dawāt wa-qirṭās*, "without board and erasing, but with ink and paper," a technique, he adds, that nobody else in Baghdad in his time was versant with. All this shows that the *takht*, the dust board of the Arabs, was really used in practice—though for myself I have some difficulty to imagine what it looked like—and that it was basically different from the Latin *abacus*.

Let me add here that the Eastern Arabic forms of the numerals also penetrated the European East, in Byzantium. Woepcke has printed facsimiles of the Arabic numerals appearing in four manuscripts of Maximus Planudes' treatise on Hindu reckoning, *Psephophoria kat' Indous*.[21]

So far, at least for the Arabic East, matters appear to be reasonably clear. But now we have to turn to the Arabic West, that is, North Africa and Muslim Spain. Here we are confronted with two major questions, for only one of which I think an answer is possible, whereas the second cannot safely be answered for lack of documentary evidence.

Question number one concerns the notion of *ghubār*. This term, meaning "dust" (in reminiscence of the dust board), is understood by most of the modern authorities as the current designation for the Western Arabic forms of the numerals; they usually call them "*ghubār* numerals."

It is indeed true that the term *ghubār*—as far as I can see—does not appear in book titles on Hindu reckoning or applied to the Hindu-Arabic numerals in the arithmetical texts of the early period in the Arabic East. On the contrary, in the Arabic West we find book titles like *ḥisāb al-ghubār* (on Hindu reckoning) and terms like *ḥurūf al-ghubār* or *qalam al-ghubār* for the numerals used in the Hindu reckoning system. The oldest occurrence so far noticed of the term is in a commentary on the *Sefer Yeṣira* by the Jewish scholar Abū Sahl Dunas ibn Tamīm. He was active in Kairouan and wrote his works in Arabic. This commentary was written in 955/6. In it Dunas says the following: "Les Indiens ont imaginé neuf signes pour marquer les unités. J'ai parlé suffisamment de cela dans un livre que j'ai composé sur le calcul indien connu sous le nom de *ḥisāb al-ghubār*, c'est-à dire calcul du *gobar* ou calcul de poussière."[22]

The next work to be cited in this connection is the *Talqīḥ al-afkār fī ʿamal rasm al-ghubār* by the North African mathematician Ibn al-Yāsamīn (died about 1204). Two pages from this text were published in facsimile in 1973;[23] on page 8 of the manuscript (= page 232 in the publication) the author presents the nine signs (*ashkāl*) of the numerals which are called *ashkāl al-ghubār*, "dust figures"; at first they are written in their Western Arabic form, then the author goes on: *wa-qad takūnu ayḍan hākadhā* [here follow the Eastern Arabic forms] *wa-lākinna l-nās ʿindanā ʿalā l-waḍʿ al-awwal*, "they may also look like this . . . , but people in our [area] follow the first type." (It should be noted that the manuscript here reproduced—Rabat K 222—is in Eastern *naskhī* and of a later date.) Another testimony is found in Ṣāʿid al-Andalusī's *Ṭabaqāt al-umam* (written about 1068 in Spain). In praising Indian achievements in the sciences this author writes: *wa-mimmā waṣala ilaynā min ʿulūmihim fī l-ʿadad ḥisāb al-ghubār alladhī bassaṭahu Abū Jaʿfar Muḥammad ibn Mūsā al-Khwārizmī* etc.,[24] "And among what has come down to us of their sciences of numbers is the *ḥisāb al-ghubār* [dust reckoning] which . . . al-Khwārizmī has described at length. It is the shortest [form of] calculation . . . , etc." This paragraph was later reproduced by Ibn al-Qifṭī in his *Tārīkh al-ḥukamāʾ* (probably written in the 1230s), but here the most interesting words of Ṣāʿid's text were shortened; in Ibn al-Qifṭī it merely reads: *wa-mimmā waṣala ilaynā min ʿulūmihim ḥisāb al-ʿadad alladhī* . . . , "And among what has come down to us of their sciences is the *ḥisāb al-ʿadad* [calculation of numbers] which al-Khwārizmī . . . etc."[25]

From these testimonies it is clear that in the Arabic West since the middle of the tenth century the system of Hindu reckoning as such was called "dust reckoning," *ḥisāb al-ghubār*—certainly in reminiscence of what the eastern arithmetical texts mentioned about the use of the *takht*, the dust board. It will then further be clear that the terms *ḥurūf al-ghubār* or *qalam al-ghubār* (dust letters or symbols) for the nine signs of the numerals used in this system of calculation basically described the written numerals as such, without specification of their Eastern or Western Arabic forms. This is corroborated by some known texts that put the *ḥurūf al-ghubār*, written numerals, in opposition to the numbers used in other reckoning systems that had no written symbols, such as finger reckoning and mental reckoning. In favor of this interpretation may be quoted some of the texts first produced by Woepcke. One supporting element here is what Woepcke derives from the *Kashf al-asrār* [or: *al-astār*] *ʿan ʿilm* [or: *waḍʿ*] *al-ghubār* of al-Qalaṣādī (in Muslim Spain, died 1486).[26] Further, in Woepcke's translation of a treatise by Muḥammad Sibṭ al-Māridīnī (*muwaqqit* in Cairo, died 1527), where the author cites words from the *Kashf al-ḥaqā'iq fī ḥisāb al-daraj wa-l-daqā'iq* of the Cairene astronomer Shihāb al-Dīn Ibn al-Majdī (died 1447), we read (of Ibn al-Majdī), "Cependant (Chehab Eddîn), . . . , s'est étendu dans l'ouvrage cité sur l'exposition de la méthode des (mathématiciens des temps) antérieurs, en fait du *maftoûh* et du *gobâr*."[27] Here the two systems, *ḥisāb maftūḥ* (mental reckoning) and *ḥisāb al-ghubār* (Hindu reckoning, with written numerals), are clearly set apart. In another paper Woepcke gave the translation of a treatise *Introduction au calcul gobârî et hawâï* (without mentioning an author or the shelf-mark of the manuscript) where, again, the "calcul *gobârî*" (Hindu reckoning, with written numerals) is opposed to the "calcul *hawâï*" (mental reckoning, i.e., without the use of written symbols).[28]

From these testimonies it can be derived that the written numerals in the Hindu reckoning system were called *al-ḥurūf al-tis'a* (the nine letters, or *litterae*) or, in *Mafātīḥ al-'ulūm*, *al-ṣuwar al-tis'* (the nine figures) or *ashkāl al-ghubār* (dust figures, in Ibn al-Yāsamīn) and *ḥurūf* or *qalam al-ghubār* (dust letters) by other Western Arabic authors. The designation thus refers to the written numerals as such, as opposed to numbers in other reckoning systems that did not use written symbols. I should think that, therefore, it is no longer justified for us to call the Western Arabic forms of the Hindu-Arabic numerals "ghubār numerals." Rather we should speak of the Eastern and the Western Arabic forms of the nine numerals.

The second, most difficult, question in connection with the Arabic West concerns the forms of the written numerals in that area, their origin and their relationship with the "Arabic numerals" that came to be used in Latin Europe.

Here one might ask why the Arabic West developed forms of the numerals different from those in the East. It is hard to imagine a reason for this development, especially when we assume—in conformity with our understanding of the birth and growth of the sciences in the Maghrib and al-Andalus in general—that the Hindu reckoning system came to the West like so many texts and so much knowledge from the Arabic East. About the mathematician and astronomer Maslama—in Spain, died 1007/1008—for example we learn from Ṣā'id al-Andalusī[29] that he studied the *Almagest*, that he wrote an abbreviation of al-Battānī's *Zīj* and that he revised al-Khwārizmī's *Zīj* (this work has survived in a Latin translation by Adelard of Bath and has been edited); he also knew the Arabic version of Ptolemy's *Planisphaerium* and wrote notes and additions to it that survive in Arabic and in several Latin translations.[30] Thus he, or his disciples, will certainly also have known al-Khwārizmī's *Arithmetic* and, together with it, the Eastern Arabic forms of the numerals. Not quite a century later Ṣāʿid al-Andalusī knows of al-Khwārizmī's *Arithmetic* under the title *ḥisāb al-ghubār*, as we have just heard.

Certainly, in this connection one has to consider that also some more elements of basic Arabic erudition took a development in the West different from that in the Arabic East: first, the script as such—we think of the so-called Maghrebi ductus in which, beyond the general difference in style, the letters *fāʾ* and *qāf* have their points added differently; second, the sequence of the letters in the ordinary alphabet; and, third, the sequence of the letters in the *abjad* series where the West deviates from the old Semitic sequence that was retained in the East and assigns to several letters different number values.[31] As far as I can see, linguists have also not brought forward plausible arguments for these differences.

That the Eastern Arabic numerals were also known in al-Andalus is demonstrated by several Latin manuscripts that clearly show the Eastern forms, for example, MSS Dresden C 80 (2nd half 15th century), fols. 156v–157r; Berlin, fol. 307 (end of 12th century), fols. 6, 9, 10, and 28; Oxford, Bodleian Library, Selden sup. 26; Vatican, Palat. lat. 1393; and Munich, Clm 18927, fol. 1r, where the Eastern figures are called *indice figure*, whereas the Western forms are labeled *toletane figure*;[32] the zeros are here called *cifre*.

However that may be, the evidence for the Western Arabic numerals in Latin sources begins in 976; a manuscript—the "Codex Vigilanus"—written in that year and containing Isidor's *Etymologiae* has an inserted addition on the genius of the Indians and their nine numerals, which are also written down in the Arabic way, that is, proceeding from right to left, in Western Arabic forms.[33] The same was repeated in another Isidor manuscript, the "Codex Emilianus," written in 992.[34] Hereafter follow, in Latin, the "apices," the numeral notations on abacus counters, which render similar forms of the numerals.[35] Here, the Western Arabic forms are still drawn in a very rough and clumsy manner. A third impulse came in the twelfth century with the translation of al-Khwārizmī's *Arithmetic*; from now on the forms of the numerals become smoother and more elegant.[36]

Unfortunately, the documentary evidence on the side of Western Arabic numerals is extremely poor. So far, the oldest specimen of Western Arabic numerals that became known to me occurs in an anonymous treatise on automatic water-wheels and similar devices in MS Florence, Or. 152, fols. 82r and 86r (the latter number also appears on fol. 81v). Two other texts in this section of the manuscript are dated to 1265 and 1266, respectively (figures 1.3a–b).[37] Here we have the symbols for 1, 2, 3, 4, 5, 8, and 9. The figures for 2 and 3 look like the corresponding Eastern Arabic forms and are not turned by 90° as in other, more recent, Maghrebi documents. The meaning of these numerals

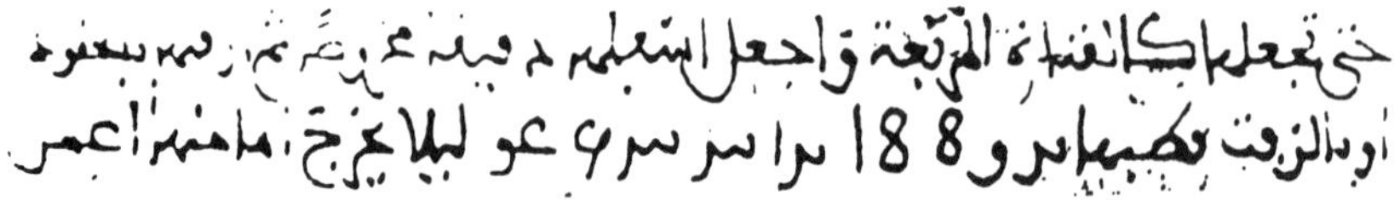

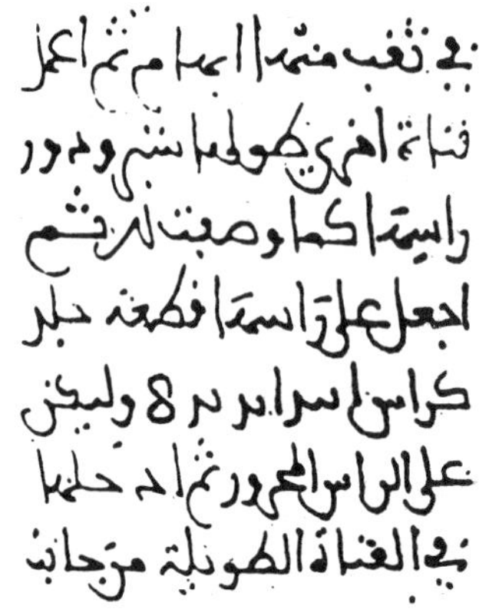

Figure 1.3a
MS Florence, Or. 152, fol. 82r
(dated 1265–1266)

Figure 1.3b
MS Florence, Or. 152, fol. 86r
(dated 1265–1266)

in the present context remains unexplained to me. The numerals in two other Maghrebi manuscripts that fell into my hands (figures 1.4–1.5)[38] resemble the forms found in the specimens reproduced in facsimile by Labarta—Barceló from Arabic documents in Aragon and Valencia from the 15th and 16th centuries.[39]

While specimens of Western Arabic numerals from the early period—the tenth to thirteenth centuries—are still not available, we know at least that Hindu reckoning (called *ḥisāb al-ghubār*) was known in the West from the tenth century onward: Dunas ibn Tamīm, 955/6; al-Dānī, before 1053; Ṣāʿid al-Andalusī, 1068; Ibn al-Yāsamīn, 2nd half of the 12th century. It must be regarded as natural that, together with the reckoning system, also the nine numerals became known in the Arabic West. It therefore seems out of place to adopt other theories for the origin of the Western Arabic numerals. From among the various deviant theories I here mention only two. One theory, also repeated by Woepcke,[40] maintains that the Arabs in the West received their numerals from the Europeans in Spain, who in turn had received them from Alexandria through the "Neopythagoreans" and Boethius; to Alexandria they had come from India. Since Folkerts's edition of and research on the Pseudo-Boethius[41] we now know that the texts running under his name and carrying Arabic numerals date from the eleventh century. Thus the assumed way of transmission from Alexandria to Spain is impossible and this theory can no longer be taken as serious. Recently, Richard Lemay had brought forward another theory.[42] He proposes that, in the series of the Western Arabic numerals, the 5,

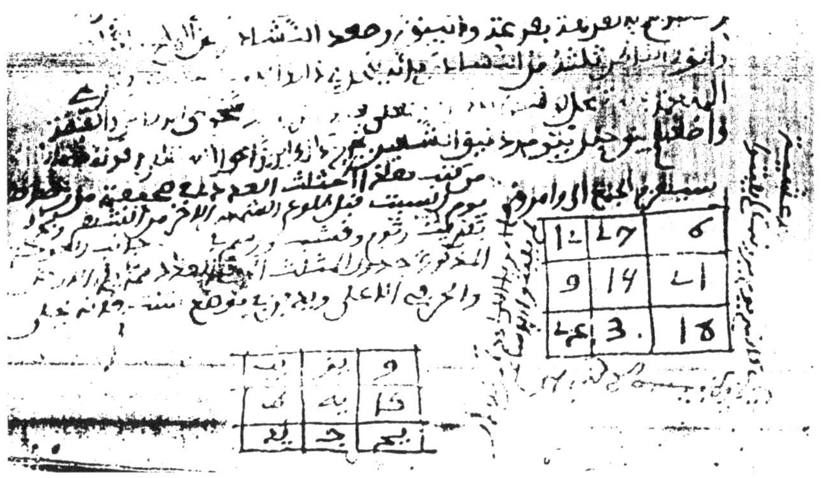

Figure 1.4
Rabat, al-Khizāna al-ʿĀmma, MS 321, p. 45 (after 1284)

6, and 8 are derived from Latin models, 5 as rendering the Visigothic form of the Roman v, 6 as a ligature of vi in the same style, and 8 as the *o* of *octo* with the final *o* placed above. This might appear acceptable for the Arabic numerals used in Latin texts. But since the Western Arabic numerals are of the same shape, that would mean that the Western Arabs broke up their series of nine numerals and replaced their 5, 6, and 8 by forms taken from European sources. This seems highly improbable. The Western Arabs received their numerals from the East as a closed, complete, system of nine signs, and it would only appear natural that they continued to use it in this complete form, not breaking the series up and replacing single elements by foreign letters.

When one compares the Eastern and the Western Arabic forms of the numerals, one finds that they are not completely different. The Western forms of 1, 2, 3, 4, 5, and 9 can be recognized as being related to, or derived from, the corresponding Eastern forms. Major difficulty arises with 6, 7, and 8. It may not be accidental that the oldest existing Latin re-working made from the translation of al-Khwārizmī's *Arithmetic* mentions just these three figures (plus 5) as being differently written.[43] As I have already said earlier, this notice can hardly stem from al-Khwārizmī himself; rather it may have been added by a Spanish-Arabic redactor of al-Khwārizmī's text. He would have been best equipped to recognize this difference. The Latin translator, or Latin adapters, would less probably have been able to notice the difference between the Eastern and Western Arabic forms of these four numerals. We cannot explain why, and how, the three Western figures were formed, especially since we have no

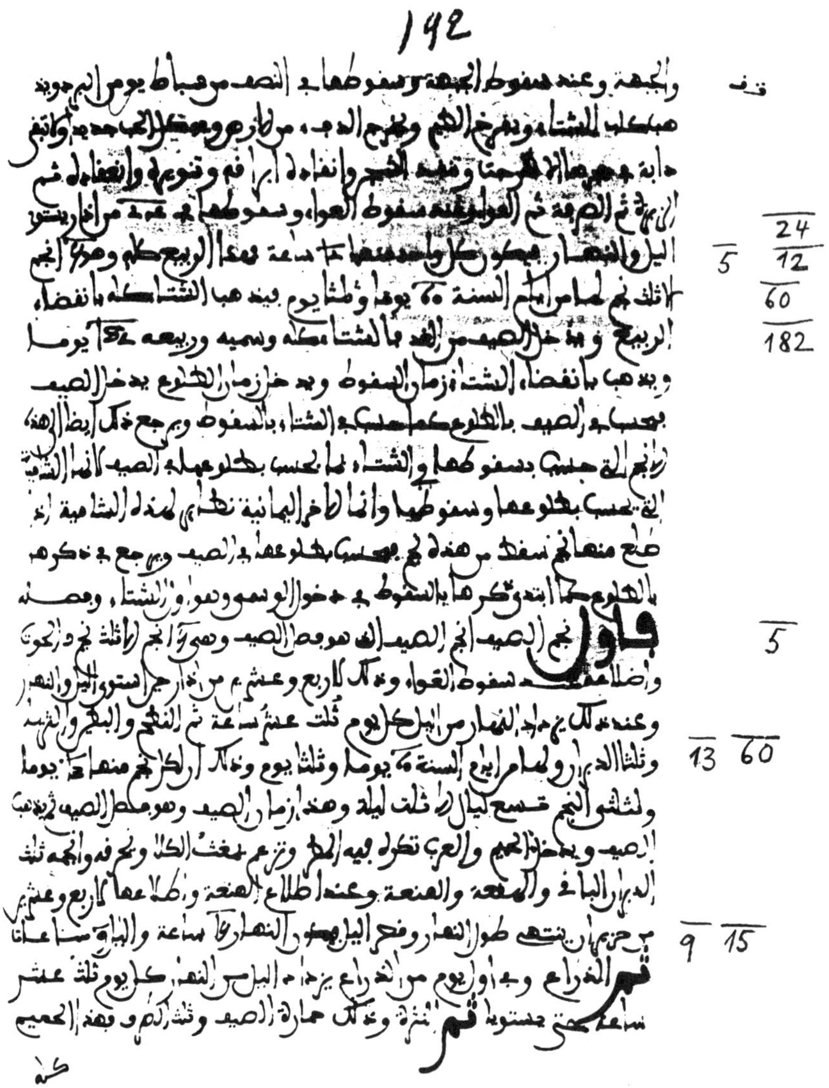

Figure 1.5
MS Ait Ayache, p. 192 (after 1344)

written specimens of Western Arabic numerals before the thirteenth century. For further research into the matter, therefore, the discovery of older, or old, documents remains a most urgent desideratum.

Lastly, I want to mention a curious piece of evidence. Somebody in the Arabic West once found out that the Western Arabic forms of the nine numerals resemble certain letters in the Maghrebi script and he organized their description in a poem of three memorial verses (in the metre *kāmil*). The poem is reported by the Spanish-Arabic mathematician al-Qalaṣādī (died 1486) in a commentary on the *Talkhīṣ fī ʿamal al-ḥisāb* of Ibn al-Bannāʾ (died 1321 or 1324) and, afterwards, by Ḥusayn ibn Muḥammad al-Maḥallī al-Shāfiʿī (died 1756, an Eastern Arabic author) in a commentary on an arithmetical work of al-Sakhāwī (died after 1592, also an Eastern author). The two *loci* are cited by Woepcke.[44] The text of the poem is as follows:

alifun wa-yāʾun thumma ḥijjun [*wa-*] *baʿdahu* ⋆
ʿawwun wa-baʿda l-ʿawwi ʿaynun tursamu
hāʾun wa-baʿda l-hāʾi shaklun ẓāhirun ⋆
yabdū ka-l-khuṭṭāfi idhā huwa yursamu
ṣifrāni thāminuhā wa-alifun baynah[*um*]*ā* ⋆
wa-l-wāwu tāsiʿuhā bi-dhālika yukhtamu

That is, 1 is compared to an *alif*, 2 to a final *yāʾ* (but to *ḥāʾ* in al-Maḥallī; both comparisons are possible), 3 to the combination *ḥāʾ-jīm*, 4 to the combination *ʿayn-wāw*, 5 to *ʿayn*, 6 to (an isolated) *hāʾ*, 7 to a *khuṭṭāf* (i.e., an iron hook), 8 to two zeros above each other and linked by a stroke, and 9 to a *wāw*. These memorial verses may be much older than al-Qalaṣādī's time. They seem to have become a topic since they are cited even by an Eastern Arabic author. Perhaps one can conclude from this standardized description that the written forms of the Western Arabic numerals were less variable than the Eastern ones.

To sum up, we can register that the history of the transmission of the Hindu numerals and Hindu reckoning to the Arabs in the East appears to be clear. For the Arabic West it is known that all the cultural and scientific achievements of the East were transferred there. In the stream of this cultural movement the knowledge of Hindu reckoning and the nine numerals must also have passed there. The oldest known testimony for the acquaintance with the Hindu system is documented for 955/6 in Kairouan. So far no written evidence of Western Arabic numerals for the tenth to the thirteenth centuries have been found; documents are only known from the thirteenth century on. But these numerals must have existed earlier since the first evidence in Latin sources—which took up these numerals from the Arabs in Spain—dates from 976. The most important task for further research would therefore be to find older Western Arabic material for the knowledge and use of the Hindu numerals in that region.

Appendix

An inspection of microfilms of the manuscripts of Leonardo of Pisa's *Liber abaci* (AD 1202) shows that a group of older manuscripts has numerals similar in shape to those in the New York MS of al-Khwārizmī's *Arithmetic* as visible in the facsimiles of its recent edition (Folkerts 1997): MSS Florence, BN, Conv. Sopp. C.1.2616 (beg. 14c.? Here the series of the nine symbols, at the beginning of the text, looks different, more "modern"; but in the text itself and in the diagrams and tables etc., they are of the Khwārizmī-MS N-type. This manuscript was used by Boncompagni for his edition, 1857–1862); Siena, Bibl. Publ. Comm., L.IV.20 (2nd half 13c.); Florence, Magliabecchi XI, 21.

On the other hand, more recent, "modern(ized)," forms of the numerals are used in MSS Florence, BN II.III.25 (16c.); Vat. Palat. 1343 (end 13c.?); Milan, I. 72 (15c.?). It thus appears evident that the numerals in the Leonardo manuscripts follow the forms current in the known Latin arithmetical texts. Contrary to what is sometimes assumed, they do not show the intrusion of new Arabic influence resulting fro Leonardo's oriental travels and his personal contacts with trade centers in the Arab world.

Postscript

For the Maghribi manuscript Ait Ayache, Ḥamzawīya 80, quoted in this article, it is now established that it was copied shortly after AD 1600; see the detailed description by Ahmad Alkuwaifi and Monica Rius, "Descripción del Ms. 80 de Al-Zāwiya al-Ḥamzawīya," *Al-Qanṭara* 19 (1998), 445–463. Therefore the manuscript can no longer serve as a testimony to early forms of Western Arabic numerals.

Notes

1. See Nau.

2. Sezgin V, 211.

3. See al-Bīrūnī, *India*, ch. 14, *apud* Woepcke 1863, 475f. (note 1), *sub* 13°, 19° and 24° (= repr. II, 407f.).

4. Al-Uqlīdisī: Saidan 1973 and 1978; Kūshyār: Levey-Petruck; al-Baghdādī: Saidan 1985.

5. About fifteen such titles up to the middle of the eleventh century are quoted by Sezgin, V.

6. Al-Yaʿqūbī I, 93; cf. Köbert 1975, 111.

7. Al-Masʿūdī I, p. 85 (§152).

8. Al-Khwārizmī, 193–195.

9. Fischer, 783–793.

10. *Fihrist*, I, 18f.; cf. Köbert 1978.

11. Fischer, 792; Köbert 1975, 111.

12. Pellat, 466b.

13. Grohmann, 453f., no. 12, and Plate LXV, 12.

14. Prof. W. Diem, Cologne, who has studied and edited such papyri for many years, informs me (in a letter dated 6 August 1996) that the understanding of the symbols as a year number is not free from doubt, because an expression like *fī sanat* ("in the year . . ."), which is usually added to such datings, is here missing. Furthermore, he confirmed that a second dating of that type in another papyrus, understood by Karabaček, 13 (no. 8), as the Hindu numerals 275 (888/9), is not formed by Hindu numerals, but rather by (cursive) Greek numeral letters. This document, therefore, must no longer be regarded as the second oldest occurrence of Hindu-Arabic numerals in an Arabic document.

15. See the quotation by Woepcke 1863, 275f. (= repr. II, 358f.).

16. Translated by Woepcke 1863, 496 (= repr. II, 428).

17. Saidan 1985, 33.

18. Cf. on this also Woepcke 1863, 482f. (= repr. II, 414f.).

19. Gandz 1927 and 1931.

20. It is in §2 of the treatise. I owe this information to Richard Lorch. Dr. Lorch is preparing an edition of al-Sijzī's text.

21. Woepcke 1859, 27, note *** (= repr. II, 191).

22. First cited by Joseph Reinaud in an Addition to his "Mémoire sur l'Inde," 565, from one of the four Hebrew translations that were made from Dunas's original Arabic text, which itself has survived only in part.

23. Ibn al-Yāsamīn, 232f.; a German translation was given by Köbert 1975, 109–111.

24. Ṣāʿid al-Andalusī, 58.

25. Ibn al-Qifṭī, 266, ult.—267,3.

26. Woepcke 1854, 359, *sub* 3° (= repr. I, 456).

27. Woepcke 1859, 67 (= repr. II, 231).

28. Woepcke 1865–66, 365 (= repr. II, 541).

29. Ṣā‘id al-Andalusī, 169.

30. Edited by Kunitzsch-Lorch.

31. Cf. *Grundriss*, 176ff., 181f., 182f.

32. For Selden and Pal. lat., cf. the table in Allard, 252; for Clm 18927, cf. Lemay 1977, figure 1a.

33. See the reproduction in van der Waerden-Folkerts, 54.

34. Reproduced also in van der Waerden-Folkerts, 55.

35. For reproductions, see, *inter alios*, van der Waerden-Folkerts, 58; Tropfke, 67; Folkerts 1970, plates 1–21.

36. See the photographs in Folkerts 1997, plate 1. etc., from the newly found and so far oldest known manuscript of a re-working of the Latin translation of al-Khwārizmī's *Arithmetic*.

37. I owe the knowledge of this manuscript to the kind help of Dr. S. Brentjes, Berlin, which is gratefully acknowledged. A detailed description of the manuscript was given by Sabra 1977.

38. Rabat, al-Khizāna al-‘Āmma, MS 321, p. 45. The preceding text, ending on p. 44, is dated in the colophon to 683/1284. P. 45 was left blank by the original writer; a later hand added in the upper part an alchemical prescription and at the bottom a magic square with directions for its use. I am grateful to Prof. R. Degen, Munich, for bringing this page to my attention, and to Prof. B. A. Alaoui, Fes, and M. A. Essaouri, Rabat, for procuring copies of the relevant pages from the manuscript.—Morocco, Ait Ayache, MS Ḥamzawīya 80. On p. 201 of the manuscript, in an excerpt from the *Zīj* of Ibn ‘Azzūz al-Qusanṭīnī, there is a calculated example for July–August 1344 (cf. Kunitzsch 1994, p. 161; 1997, p. 180).

39. It should be added that in the table of *ghubār* numerals given by Souissi, 468, the numerals in the first two lines (said to date from the 10th century and ca. 950, respectively) are not (Arabic) *ghubār* numerals, but rather Indian numerals (cf. Sánchez Pérez, the table on p. 76, lines 8–9). It should also be noted that the date given by Sánchez Pérez, 121, table 1, for the specimen in line 9 ("Año 1020") is the Hijra year (= AD 1611/12); the author there mentioned, Ibn al-Qāḍī, died in Fes 1025/1616. Similarly, the specimen in line 12, ibid., from MS Escorial 1952, must belong to the 11th century Hijra; the manuscript contains a commentary by Abu 'l-‘Abbās ibn Ṣafwān on the summary of Mālik ibn Anas' *al-Muwaṭṭa'* of Abu 'l-Qāsim al-Qurashī.

40. Woepcke 1863, 239 (= repr. II, 322).

41. Folkerts 1968, 1970.

42. Lemay 1977 and 1982.

43. Folkerts 1997, 28: MS N, lines 34–38.

44. Woepcke 1863, 60f. and 64f. (= repr. II, 297f. and 301f.).

Bibliography

Allard, A. 1992. *Al-Khwārizmī, Le calcul indien.* Paris and Namur.

Fischer, A. 1903. "Zur Berichtigung einer Etymologie K. Vollers'." *Zeitschrift der Deutschen Morgenländischen Gesellschaft* 57, 783–793.

Folkerts, M. 1968. "Das Problem der pseudo-boethischen Geometrie." *Sudhoffs Archiv* 52, 152–161.

Folkerts, M. 1970. *"Boethius" Geometrie. Ein mathematisches Lehrbuch des Mittelalters.* Wiesbaden.

Folkerts, M. 1997. *Die älteste lateinische Schrift über das indische Rechnen nach al-Ḫwārizmī.* Munich.

Gandz, S. 1927. "Did the Arabs know the abacus?" *American Mathematical Monthly* 34, 308–316.

Gandz, S. 1931. "The Origin of the Ghubār Numerals or The Arabian Abacus and the Articuli." *Isis* 16, 393–424.

Grohmann, A. 1935. "Texte zur Wirtschaftsgeschichte Ägyptens in arabischer Zeit." *Archiv Orientální* 7, 437–472.

Grundriss. 1982. *Grundriss der arabischen Philologie*, ed. W. Fischer, I: *Sprachwissenschaft.* Wiesbaden.

Ibn al-Nadīm. 1871–1872. *Kitāb al-fihrist*, ed. G. Flügel, I–II. Leipzig.

Ibn al-Qifṭī. 1903. *Taʾrīḫ al-ḥukamāʾ*, ed. A. Müller and J. Lippert. Leipzig.

Ibn al-Yāsamīn, Abū Fāris. 1973. "Dalīl jadīd ʿalā ʿurūbat al-arqām al-mustaʿmala fī l-maghrib al-ʿarabī." *Al-Lisān al-ʿArabī* 10, 231–233.

Karabaček, J. 1897. "Aegyptische Urkunden aus den königlichen Museen zu Berlin." *Wiener Zeitschrift für die Kunde des Morgenlandes* 11, 1–21.

al-Khwārizmī, Muḥammad ibn Aḥmad. 1895. *Liber Mafâtîh al-olûm*, ed. G. van Vloten. Leiden.

Köbert, R. 1975. "Zum Prinzip der *ġurāb*-Zahlen [sic pro *ġubār*] und damit unseres Zahlensystems." *Orientalia* 44, 108–112.

Köbert, R. 1978. "Ein Kuriosum in Ibn an-Nadīm's berühmtem *Fihrist*." *Orientalia* 47, 112f.

Kunitzsch, P. 1994/1997. "ʿAbd al-Malik ibn Ḥabīb's *Book on the Stars*." *Zeitschrift für Geschichte der Arabisch-Islamischen Wissenschaften* 9, 161–194; 11, 179–188.

Kunitzsch, P., and Lorch, R. 1994. *Maslama's Notes on Ptolemy's* Planisphaerium *and Related Texts.* Munich.

Labarta, A, and Barceló, C. 1988. *Números y cifras en los documentos arábigohispanos.* Córdoba.

Lemay, R. 1977. "The Hispanic Origin of Our Present Numeral Forms," *Viator* 8, 435–462 (figures 1a–11).

Lemay, R. 1982. "Arabic Numerals." *Dictionary of the Middle Ages*, ed. J. R. Strayer, vol. 1. New York, 382–398.

Levey, M., and Petruck, M. 1965. *Kūshyār ibn Labbān, Principles of Hindu Reckoning*. Madison and Milwaukee.

al-Masʿūdī. 1965ff. *Murūj al-dhahab*, ed. C. Pellat, Iff. Beirut.

Nau, F. 1910. "Notes d'astronomie syrienne." III: "La plus ancienne mentionne orientale des chiffres indiens." *Journal asiatique*, sér. 10, 16, 225–227.

Pellat, C. 1979. "Ḥisāb al-ʿAqd." *Encyclopaedia of Islam*, new ed., III. Leiden, 466–468.

Reinaud, J. 1855. "Addition au Mémoire sur l'Inde." *Mémoires de l'Institut Impérial de France, Académie des Inscriptions et Belles-Lettres* 18, 565f.

Sabra, A. I. 1977. "A Note on Codex Biblioteca Medicea-Laurenziana Or. 152." *Journal for the History of Arabic Science* 1, 276–283.

Sabra, A. I. 1979. "ʿIlm al-Ḥisāb." *Encyclopaedia of Islam*, new ed., III. Leiden, 1138–1141.

Ṣāʿid al-Andalusī. 1985. *Ṭabaqāt al-umam*, ed. Ḥ. Bū-ʿAlwān. Beirut.

Saidan, A. S. 1973. *al-Uqlīdisī, al-Fuṣūl fī l-ḥisāb al-hindī*. Amman.

Saidan, A. S. 1978. *The Arithmetic of al-Uqlīdisī*. Dordrecht and Boston.

Saidan, A. S. 1985. *ʿAbd al-Qāhir ibn Ṭāhir al-Baghdādī, al-Takmila fī l-ḥisāb*. Kuwait.

Sánchez Pérez, J. A. 1949. *La aritmetica en Roma, en India y en Arabia*. Madrid and Granada.

Sezgin, F. 1974. *Geschichte des arabischen Schrifttums*, V: *Mathematik, bis ca. 430 H.* Leiden.

Souissi, M. 1979. "Ḥisāb al-Ghubār." *Encyclopaedia of Islam*, new ed., III. Leiden, 468f.

Tropfke, J. 1980. *Geschichte der Elementarmathematik*, vol. 1, 4th ed. Berlin and New York.

van der Waerden, B. L., and Folkerts, M. 1976. *Written Numbers*. The Open University Press, Walton Hall, Milton Keynes, GB.

Woepcke, F. 1854. "Recherches sur l'histoire des sciences mathématiques chez les orientaux. . . ." *Journal asiatique*, 5ᵉ série, 4, 348–384 (= repr. I, 445–481).

Woepcke, F. 1859, *Sur l'introduction de l'arithmétique indienne en Occident. . . .* Rome (= repr. II, 166–236).

Woepcke, F. 1863. "Mémoire sur la propagation des chiffres indiens." *Journal asiatique*, 6e série, 1, 27–79; 234– 290; 442–529 (= repr. II, 264–461).

Woepcke, F. 1865–1866. "Introduction au calcul gobârî et hawâï." *Atti dell'Accademia de'Nuovi Lincei* 19, 365–383 (= repr. II, 541–559).

Woepcke, F. 1986 (repr.). *Études sur les mathématiques arabo-islamiques, Nachdruck von Schriften aus den Jahren 1842–1874*, ed. F. Sezgin, I–II, Frankfurt am Main.

al-Yaʿqūbī. 1883. *Historiae*, ed. M. Th. Houtsma, I–II. Leiden.

2

The Transmission of Arabic Astronomy via Antioch and Pisa in the Second Quarter of the Twelfth Century

Charles Burnett

This chapter considers a group of Latin astronomical texts translated from Arabic, or based on Arabic material, which share the same technical language and have the same systems of numeration, and explores the possibility that, unlike the majority of Arabic scientific works, which entered Europe from Spain, these came directly from the East and were brought to the West by scholars working in Antioch and Pisa. The texts under consideration here are a translation of Ptolemy's *Almagest*, known as the "Dresden *Almagest*," a Latin cosmology describing the Ptolemaic system, called the *Liber Mamonis*, and a version of the astronomical tables of al-Ṣūfī. All these texts appear to have been written in the second quarter of the twelfth century.

I The Dresden *Almagest* and *Liber Mamonis*

It was Charles Homer Haskins, whose intuitions usually prove remarkably accurate, who first pointed out a feature which could link the Latin version of the *Almagest* surviving uniquely in MS Dresden, Landesbibliothek, Db. 87, and the *Liber Mamonis*, a cosmology of which the only known copy is the incomplete text in the twelfth-century MS Cambrai, Bibliothèque municipale, 930.[1] The first section of this article confirms the link between the two works, and suggests that the Dresden *Almagest* represents the first attempt to translate the *Almagest* into Latin in the Middle Ages, whilst the *Liber Mamonis*, in turn, is an early attempt at replacing a cosmology based on Latin sources with the Ptolemaic system.[2]

MS Dresden, Db. 87 was written in ca. 1300;[3] it once belonged to Berthold of Moosberg, and became the property of the Dominicans of Cologne. It consists of several texts related to Ptolemy's *Almagest*: Geminus of Rhodes's *Introduction to the Phenomena* (often called "Introductio Ptolemei in Almagestum"), the *Parvum Almagestum*, and Jābir ibn Aflaḥ's *Correction of the Almagest*. The first item, however, is a unique copy of a translation of the first four books of Ptolemy's *Almagest*. It was first brought to the notice of scholars by J. L. Heiberg, who quoted the incipits and explicits of the four books.[4]

Haskins, in turn, transcribed the list of titles for the first book and the prologue. Both scholars considered that the translation was made from Greek, on the grounds of its vocabulary. However, the Arabic name that appears in the explicits to each book already makes one uneasy about a Greek origin: for example, "Here ends the first book of the mathematical treatise of Ptolemy which is called the *Megali Xintaxis* of astronomy, in the translation <and> dictation by the love of languages (?) of 'Wittomensis Ebdelmessie.'"[5]

The interpretation of the name, and the precise role of this "ʿAbd al-Masīḥ of Winchester" is unclear. But a comparison of the text with the Greek and Arabic versions of the *Almagest* reveals clearly that an Arabic text lies at the base of this version:[6]

1. The word order and terminology of the first phrase corresponds to the Isḥāq/Thābit version of the Arabic *Almagest.*[7]
2. Most of the diagrams are reversed in respect to the Greek; this may have arisen out of a mistaken notion of a translator from Arabic that, since he had to reverse the direction of the script, he also had to reverse the diagrams.
3. The terminology is based on Arabic rather than Greek.[8]
4. Several turns of phrase are reminiscent of other translations from Arabic.[9]

Heiberg and Haskins were misled by the translator's total avoidance of transcriptions from Arabic, and by his addition of a veneer of Greek or Pseudo-Greek terms. That this is a veneer is immediately obvious when one looks for these Greek terms in the Greek text of the *Almagest*; for in most cases they are simply not there. Where the Dresden *Almagest* has 'praxis' (fol.1r), the Greek text has 'πρακτικόν' (Heiberg, 4.9), where it has 'phisialoica' (fol.1r), the Greek has 'φυσικόν' (Heiberg, 5.9), where it has 'organum' (fol. 3r), the Greek has 'κατασκευάς' (Heiberg, 13.12), and so on. To the category of a Greek veneer may also be ascribed the word occurring in the explicits of each book: 'philophonia' ('love of languages'?), which is meant, no doubt, to sound stylish, but makes little sense. Nevertheless, it is possible that the translator attempted to look at a Greek manuscript, or at least consulted someone who knew Greek. Otherwise it is difficult to explain how he came to use the nonce-word 'aretius' where the Greek text has 'ἀρετῶν.'[10] Moreover, he transcribed quite accurately Greek proper names, and the names of Egyptian months,[11] as well as the Greek title of Ptolemy's work: 'megali xintaxis' (i.e., μεγάλη ξύνταξις). It is clear that the translator wished to give the impression that he had taken the whole text from the Greek.

Whether the translation is literal can only be ascertained through a close comparison with the extant Arabic versions of the *Almagest*, which (except for the star-tables of Books 7 and 8) are not yet edited. When one compares the text with the Greek *Almagest* one finds that

1. Although there are references to tables,[12] none of Ptolemy's tables has been copied into the manuscript. It is difficult to tell whether they were originally included.[13]

2. In the theorems on trigonometry and spherical astronomy in Books 1 and 2, references to the relevant theorems in Euclid's *Elements* have been added. That this is an addition in the original Arabic text is suggested by the presence of exactly the same forms of reference to the *Elements* in the text of al-Nasawī (see item 6 below). But it is worth noting that here again the translator refers to Euclid's work either with a transcription of the Greek title—e.g., 'liber estichie (= στοιχεῖα) euclitis,' fol. 7v—or a literal translation of that title: 'liber elementorum euclithis' (fol. 6v),[14] and not as 'liber geometrie'/'liber institutionis artis geometrie' *vel sim.* which is found in most of the translations of the *Elements* made from Arabic.

3. Of similar status are the cross-references to other theorems in the *Almagest*: e.g., fol. 17r (*Almagest*, II.3): 'demonstracione .xiiii. figure primi sermonis huius libri'; this cross-reference is not in Ptolemy's text, but such cross-references also occur in al-Nasawī's work.

4. Sometimes only the geometrical elements of a theorem are given, and the numerical values have been omitted; e.g., in Book II, chapters 2 and 3.

5. Sometimes a more precise value replaces Ptolemy's 'rounded' value: e.g., fol. 15r: '11; 39, 59°' where Ptolemy has 'approximately 11; 40°' (I, 14, Heiberg, 78.12).

6. The only substantial addition *vis-à-vis* the Greek text is that of several theorems on the sector-figure appended to the end of *Almagest*, I, chapter 13 (Dresden MS, fols 13v–14v). As has been pointed out by Richard Lorch, this passage corresponds to a section of chapter 2 of *al-Ishbāᶜ fī sharḥ al-shakl al-qaṭṭāᶜ* by the eleventh-century Arabic mathematician, al-Nasawī, who served the Būyid amīrs in Baghdad.[15] The verbal equivalence between the Dresden *Almagest* and this passage of al-Nasawī suggests that the translator was translating literally from an Arabic text.[16]

On the whole, aside from the tables, little appears to be missing in the first four books of the *Almagest*.[17] There is nothing, either, to indicate that the translation stopped after the fourth book, rather than that the scribe of the Dresden manuscript decided at this point not to copy any more of the translation.

There are two characteristic features which separate this translation from other Latin translations of the *Almagest* (whether from the Arabic or from the Greek): the terminology, and the notation for numerals. It was the notation for numerals that attracted the attention of Haskins, and this needs more careful analysis.

The translator[18] began by using roman numerals only. Then, on fol. 15v (in *Almagest*, I, chapter 16), he makes his first attempt to use alphanumerical notation. That alphanumerical notation is being introduced at this point is indicated by the fact that here, and here only, the numerical values of the letters of the alphabetic are spelt out:

> quinquaginta f sex partium et unius a sexagenarie et viginti e quinque secund-arium ad centum t et viginti k partes ('<the ratio of> fifty *f* (six) degrees and *a* (one) minute and twenty *e* (five) seconds to *t* (one hundred) and *k* (twenty)[19] degrees')

Thereafter, alphanumerical notation is used, sometimes on its own, sometimes in combination with roman numerals, and sometimes alternating with them. One can construct the following key for the numerical values of the letters:

1	a	10	k	100	t	1000	a mille
2	b	20	(l)	200	u	2000	b milia
3	c	30	(m)	300	x	3000	c milia
4	d	40	(n)	400	y	4000	d milia
5	e	50	(o)	500	z	etc.	
6	f	60	(p)	600	Θ		
7	g	70	(q)	700	Φ		
8	h	80	(r)	800	.n.[20]		
9	i	90	(s)	900	[Θ][21]		

On the whole, in comparison with the Greek text, the numbers in the Dresden *Almagest* are accurate, whether they are written in roman numerals or alphanumerical notation. There is, however, one fundamental flaw in this copy, namely that, for some reason or other, the scribe completely omits any digit in the 10s or 10000s unless it is followed by a '0', in which case it is invariably written as 'k' Thus:

10	k	20	k	30	k	40	k	50	k etc.
11	a	21	a	31	a	41	a	51	a etc.
12	b	22	b	32	b	42	b	52	b etc.
13	c	23	c	33	c	43	c	53	c etc.
etc.									

10000	k milia	20000	k milia etc.
11000	a milia	21000	a milia etc.

Thus, the only numbers which are correctly written in the 10s and 10000s are 10 and 10000 themselves. As examples one may take the following:

> fol. 54r: c partes et k sex. et d secunde et h tercie = 13;10,34,58°

fol. 66r: i dies et b dies et k recte hore = 122 days and 10 hours
fol. 64v: x milia et a milia et Φ et c dies = 311,783 days (see figure 1)

That the 10s (and consequently the 10000s) were originally represented is clear from the fact that the missing letters of the alphabet, 'l' to 's' (placed in round brackets in the table above), exactly fit between 10 (= k) and 100 (= t). We must also presume that, given the accuracy of the numbers when they are written in roman numerals, the original translator also used the alphanumerical notation in an accurate way. The parallel examples of Greek and Arabic alphanumerical notation illustrate how the letters, used in their Semitic order, progressively represent the units, 10s and 100s, and these are the parallels the translator of the Dresden *Almagest* would have been following. It seems, however, that, in one respect, his system differed from the normal Greek and Arabic systems: he appears to have added "k" when a 10 or 10000 was not followed by a unit, giving something like:

10	k	20	lk	30	mk	40	nk	50	ok	etc.
11	ka	21	la	31	ma	41	na	51	oa	etc
12	kb	22	lb	32	mb	42	nb	52	ob	etc.

In each case (excepting always "10") the first letter was dropped at some time in the copying process, perhaps because it was originally written (or intended to be written) in rubric, and the rubrics were not filled in.[22]

When one turns to the *Liber Mamonis* one finds a very similar alphanumerical system, but this time, the expected letters for the 10s are used:

1	a	10	k	100	t
2	b	20	l	200	u
3	c	30	m	300	x
4	d	40	n		
5	e	50	o		
6	f	60	p		
7	g	70	q		
8	h	80	r		
9	i	90	s		

In the *Liber Mamonis* the author has no occasion to use any number between 360 and 1000. Therefore, it is impossible to know what symbols he would have used for the 100s between 400 and 900. Moreover, he uses "k" only for '10' and not in combination with the letters for 20 to 90.

For the thousands and above, the author of the *Liber Mamonis* usually writes out the numbers in a way which is also found in the Dresden *Almagest*:

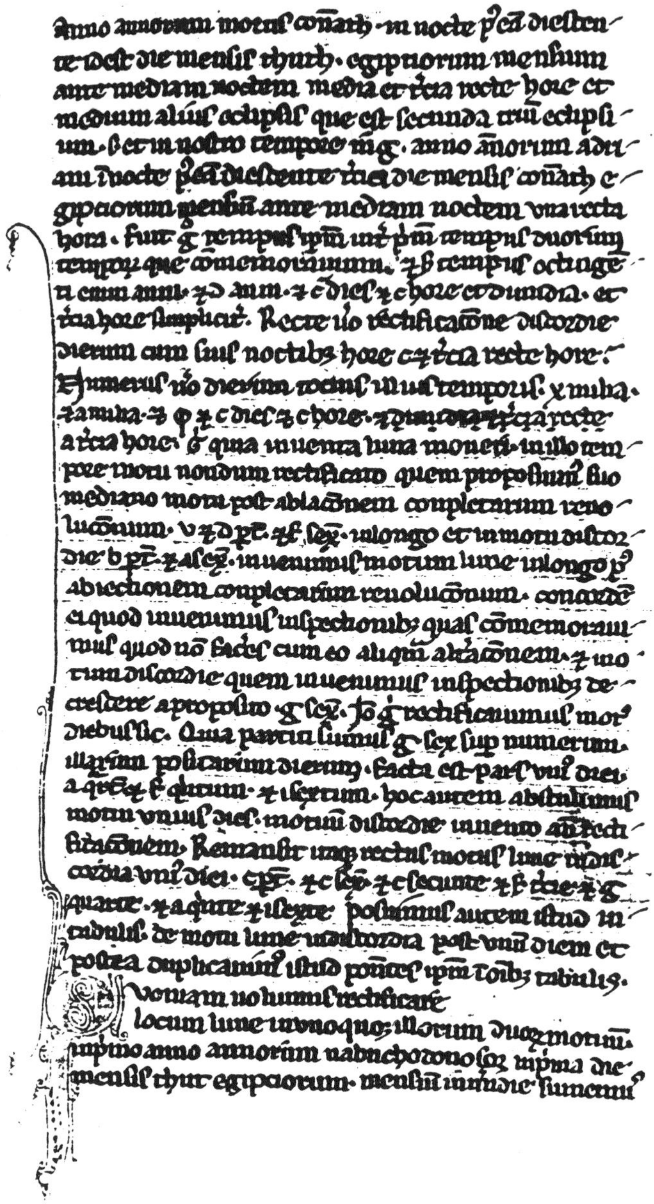

Figure 2.1
Dresden, Landesbibliothek, Db. 87, fol. 64v. A passage from the translation of Ptolemy's *Almagest*, Book IV, ch .7, showing the use of alphanumerical notation. In lines 13–14 the number 311,783 is written as 'x milia et a milia et Φ et c.'

Liber Mamonis, fol. 27v: mille septingenti .l.h. [= 1728].
Dresden *Almagest*, fol. 51r: quatuor milia et z et c [= 4573]

However, on two folios (27v–28r), he experiments with using a different system of numerals for high numbers, namely Hindu-Arabic numerals. He is writing at a time when Hindu-Arabic numerals were only just beginning to be used by Latin scholars, and it is significant that he (or perhaps, rather, the scribe) assimilates their shapes to Latin letters.[23] Moreover, he uses the Eastern forms of the numerals, as can be seen from the following table:

	Eastern forms
1 = ı *or* t	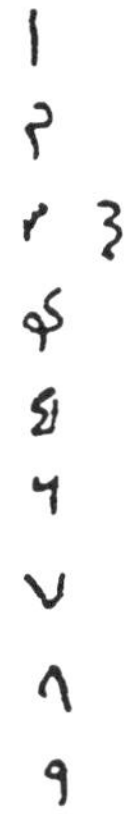
2 = p	
3 = Ψ	
4 = ꝗ· (possibly the abbreviation of 'quia')	
5 = g	
6 = 7 (the tyronian 'et')	
7 = u	
8 = a *or* ∂	
9 = q	

This active use of the Eastern forms of the Hindu-Arabic numerals is quite remarkable. The numerals in current use in Western scripts nowadays, which are usually referred to as "Arabic," were known in Castile by the late tenth century,[24] and appear to have become the standard system used by translators and scholars working in Toledo in the later twelfth century: one variant of these Western forms is described specifically as *figure toletane* ("Toledan symbols").[25] The Eastern forms of the Hindu-Arabic numerals, on the other hand, are closer to the original Sanskrit shapes, and developed into the numeral forms used nowadays in Arabic. They are called *figure indice* ('Indian symbols') in the same Latin manuscript that called the Western forms *figure toletane*.[26] The Eastern forms are found only in copies of a small number of interrelated Latin works, as will be discussed later.

It would be attractive to think that the author of the *Liber Mamonis* was influenced by the example of the Dresden *Almagest* and refined the alphanumerical notation he found there, whilst also experimenting with using Hindu-Arabic numerals for the higher numbers. If the similarity between the two texts stopped here, then this would remain only a weak hypothesis. However, of even greater significance is the fact that the majority of the astronomical terms in the *Liber Mamonis* are the same as those of the Dresden *Almagest*.[27] Much of this terminology is unique to these two works and quite remarkable; e.g., the use of the terms 'sexagenaria' for 'minute,'[28] 'synodos' for 'conjunction,' and '(circulus) rotunditatis' for epicycle. The two works must, therefore, be related.

The question arises, however, as to whether the author of the *Liber Mamonis* was using the contents as well as the terminology of the Dresden *Almagest*. He refers to Ptolemy's work or its author on several occasions:

1. "Ptolemy and the other <authorities> who are more sensible in their thinking <concerning climes> . . . ."[29]

2. "We know from the discovery of the best men in astronomy—Ptolemy and the others—that this parallax is greater in <the case of> Mercury than in Venus, and that Venus has a greater <parallax> than the Sun."[30]

3. "These two <equinoctial> points are fixed and do not move, according to what Ptolemy stated in the *Syntaxis*. However, the opinion of others who came after <Ptolemy> and who took it upon themselves to investigate this and other things more intimately, is that they move eastwards with a slow motion—namely one degree every 106 years. In this matter it is amazing that the intelligence of Ptolemy was deceived. For they *do* move. But what caused him to stumble has been proved to be a certain astronomer who preceded him, who made a false observation when seeking the position of the height of the Sun. For he had said that the apogee of the Sun was in the same place in his time as Ptolemy, who dealt with and investigated all the secrets of the stars more perspicaciously, correctly discovered it was in, in his own time. Hence Ptolemy is said to have been led into error, although he said nothing more than that <the point> was not moving. He made no mistake at all in finding the place; rather, his predecessor made a mistake. Therefore, no doubts should be cast on the accuracy of Ptolemy, but one should blame the ignorance of the man who, by committing to writing something that he did not know, made a wise man stumble. Now that we have said enough in support of <Ptolemy's> diligence, let us return to the subject."[31]

4. "The movement by which this sphere moves and carries the other <spheres> with it is a slow one from west to east, completing—as Ptolemy said in his *Megale Syntasis*—the path of one degree in the period of 100 years; or 106 years in the opinion of others."[32]

5. "The spheres of the three planets, Saturn, Jupiter and Mars, are very similar to each other in the number and divisions of their spheres, their movement and their circles, as Ptolemy is proved to have said in his *Syntaxis*."[33]

6. "The measurement of the maximum latitude of the planets and the Moon to the north and the south of the zodiac is: Moon: 4° 45′ (when the Great Circle is divided into 360°); Saturn 3°; Jupiter 2°; Mars, 4° 20′ to the north, 7° to the south; Venus, as Ptolemy says in his *Syntaxis*, 6° 2′, but, as other astronomers have said, 9°; Mercury 4°."[34]

7. "This lowest sphere of Saturn moves eastwards round two fixed poles with a slow motion which is one degree in 100 years. These are the two poles of the zodiac which the centre of the Sun circles, according to what Ptolemy says has been discovered by himself and his predecessors."[35]

The author of the *Liber Mamonis* gives the impression (or would like to give the impression) that he knows the *Almagest* directly, and there are certain passages which follow the *Almagest* closely in their argumentation and/or in the values that they give:

1. The sequence of the argument for working out the relative distances of the planets from parallax on fol. 27v (item 2 above) corresponds to that in *Almagest*, IX, chapter 1.
2. The description of finding the solar anomaly on fol. 31v corresponds to *Almagest*, III, chapters 5–6.
3. The limits of the movement in latitude quoted in item 6 above are rounded-off figures from *Almagest*, XIII, 5.

If we are to suppose, on the basis of this evidence, that the author of the *Liber Mamonis* knew the *Almagest* directly, then we have to conclude that he was familiar as much with the later books as with the first four books which alone are found in the Dresden *Almagest*.

However, the *Almagest* is not the principal model for the *Liber Mamonis*, nor, perhaps, is it the principal source for the numerical values given in the Latin work. For the *Liber Mamonis* is a cosmology (in the Arabic *hay'a* tradition),[36] not a mathematical work, and in some respects, the work is closer to Ptolemy's *Planetary Hypotheses* (e.g., in talking in terms of contingent spheres on fols. 28v–29r). Moreover, the author refers specifically (though not by name) to Ptolemy's successors and to astronomers who have criticized or corrected Ptolemy. One of these astronomers is, presumably, the unnamed Arab whom he describes, in the preface to the fourth book, as being his principal guide throughout the work:

> But since in other <books> we followed for the most part a certain Arab, in this also we will follow <him> through much, although we have found certain things concerning the number of the spheres and their epicycles, and he has touched upon the truths about the circles and the obliquities of the planets with which the number of spheres is dissonant.[37]

It is presumably from these successors of Ptolemy that he gets the values which differ from those in the *Almagest*, such as 23° 35′ for the obliquity of the ecliptic (fol. 29r; Ptolemy's value is 23° 51′20′′), and a precession of 106 years per degree.[38] Also, not in the *Almagest* is his list of values for the "completion

Figure 2.2
Cambrai, Bibliothèque municipale, 930, fol. 27v–28r, showing both the alpha-numerical notation and the Eastern forms of the Hindu-Arabic numerals.

qb; sublatis .t. u. p. a. scilicet spatio atia usq; ioue remanet a ioue usq; satrni. Sublato
de spatio ioui spatiu martis restat a marte. usq; iouem fiet. reliqs. Et si spatio
a luna usq; sole. a gdus a sole ad uenerem .d. ꝫ mcurii R .b. A mcurio ad matte
.t. s. b. a marte ad ioue. i. g. i. p. a ioue ad saturnu. qz qz. q. p. i. cui reliqt ouitas spa
tiis iunctis por sing nis. Que spatioꝝ assignatio mltis rōnib; improbaē. Si ui-
ur id dic una eadq; omniu celitas. duplo tpis sol suu pagraret ccelm q luna eas
ccuit. duoꝝ ꝟ ectōz si altiui diametru duplu sit diametᵃ altiui ꝫ celi tue sehit.
Sequitur q si ui eent assiguata spatia eadq; citatio sol biuisus tota zodiacu. uel
f. mē. l. ꝫ. mars. u. R. E. iupit. t. n. d. annis. Saturnu. y. i. a. a b anni q cui uia
sit aut falsa ꝫ spaᵗ assignatio .aut celitas ñ erit eade. Cetterarum q; ñ eē eande aps
eunde facile ꝫ inueniri. ui de celesti armonia loquitur. Posthꝭ ꝝ uocē matre ex
pcussione fieri. acutu aut ꝫ ꝯuem p m̄ ꝫ ipetu pcussionis atᵗ acuᵗiores sonos ab
exᵗimis ꝫ striᵗiorib' spis emitti q ab infimis. q si uerē ꝫ ñ eas celitas. Ma si uelocitᵃ
pari ipetu aere pcusso equalitᵗ ꝯsonantes emitterent uocet. Sed hee q̄ q̄uel ille
acutas. Non q̄ pari feruntur uelocitate. Sociat-ꝫ cursus sol uenis ꝫ mcurii ñ pa
rem ipsi inee festinantia q phat. Nā c̄ñ sit i ead spa positi s; sol spe uenis. uenis
micurua supposita sit ꝫ he l'modica gcedante hᵗe ꝗnatá duitᵃ. cui buioi ꝫ ocula
paucioribz; debet uenis eeduci mcurii. plib' uii sol uenie medru teneute. Sed eñ sit
li nec illud. ñ it ꝫ illud gcedimi. q ap quosda inuenitē. micurui paucionib ueuie.
ueuem. sole diebz; suis eemeare ecut. Toto naq; hemispio si ꝗstaret illo a sole se
ceber sic ꝫ alii planete ñ reliqs q̄pe sic. iii. statioues. mars. saturni. iupit. tardi a
uelocitate sol. s; tardiores luna citiores sole. usq; du opponerent pceder; ex inde
ꝯsequeutur. q ñ sit ñ pmissa. Req; it ueni ante aut p̄a sole plēq. n. h. gȝdib ñ
curi au plēq. l. f. posse recede. nun si pcesserit tardi ꝫ reȝȝdi si subsequentur
uelociores ad ipm reituenē. ꝫ rotundicate. motu ñ maioz spꜩ. mui ria. pa
ter q̄ clari p pmissas argumentones macbru falsa ꝫ ꝫ ab ustancia de ordine ꝫ
spatiis ꝫ uelocitate ditisse. ꝫ nē l; nē illa siē captui. y sueudo ꝫ Iro de ordine qd
sit de spatiis in seȝuirib' pauca alias pleui. de uelocitate i postᵗȝib' ꝫ furi lo
cis disputabini. Hc u q̄rōne ab ouente i occideutē ferri pbetur śr ñ ue inuis
uideni. In oriente planetaꝝ. qꝫ spa ab occedite rotari omniu ꝫ eiusq̄ q̄
plise huic acui ꝫ freqni studuere sentencia. Ad cui q̄batione macbu tale
indux arguntu. Sole poico i capite arietis cu occbit librē uideni oriri.
T lapsis au. m. dieb' in occatsu sol ñ cap librē s; scorpii oritur qꝫ ñ u sol i ariete
ꝫ s; ad tauru migure. Taur aut p̄ariete omēs illa ab ouente ꝫ Irq̄ manifest

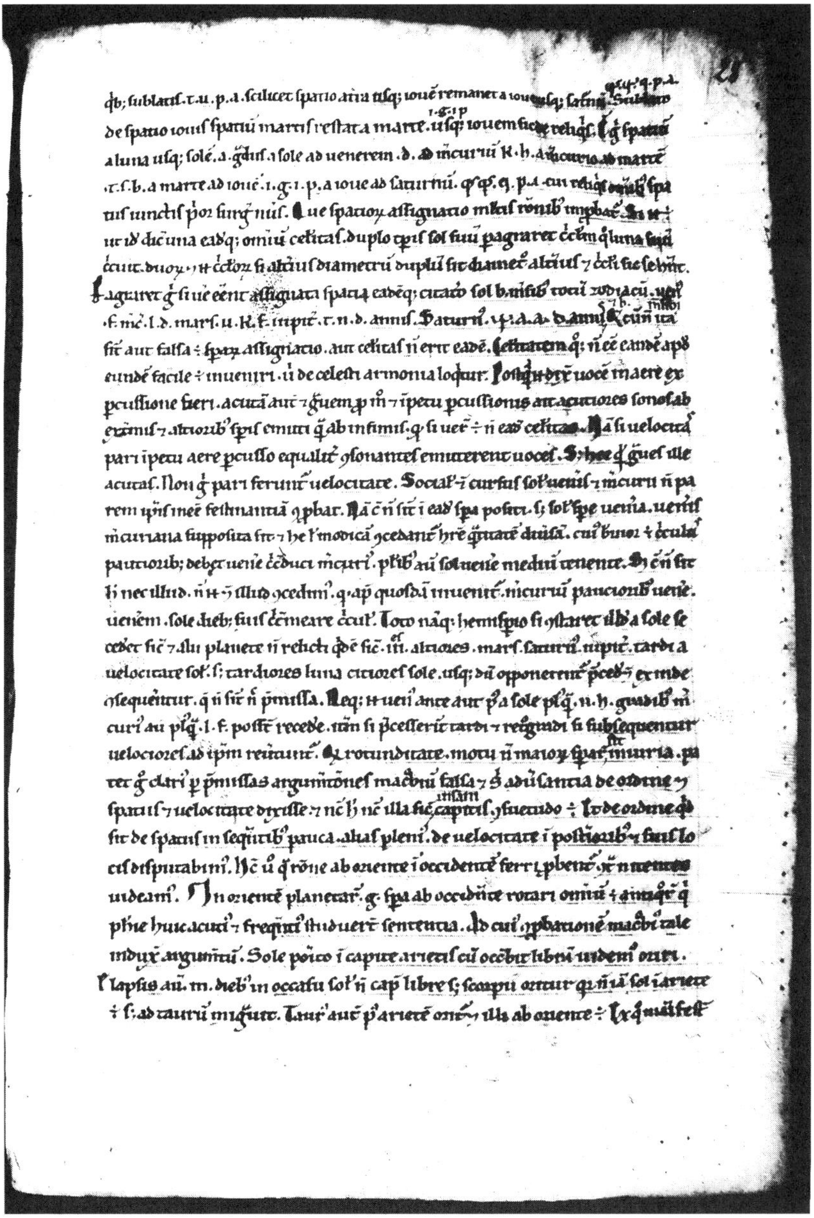

of movement in latitude" ("Completur enim Saturni lati motus . . .") of each of the planets on fol. 48v:

> Saturn: 29 years, e.d. (5.4. ?) days, 15 hours, 24 minutes.
> Jupiter: 11 years, 315 days, 14 hours, 29 minutes.
> Mars: 1 year, 322 days, 24 hours.
> Venus and Mercury: 1 year, 5 hours, 49 minutes.

There do not appear to be any direct quotations from the *Almagest*.

A clue to the direct source of the *Liber Mamonis* may be hidden in its title itself. For it is tempting to see in the "Mamon," implied by the Latin genitive "Mamonis," a reference to the caliph al-Maʾmūn (ruled 813–833 A.D.), on whose command the earliest Arabic translations of the *Almagest* were made. It is unlikely that the *Liber Mamonis* is an allusion to the *Almagest* itself. However, it is possible that "the book of al-Maʾmūn" is, rather, the correction of the *Almagest*, commissioned by the same al-Maʾmūn, and resulting in *al-zīj al-mumtaḥan* (referred to as the *tabule probationum* or *tabule probate* in Latin translations). These astronomical tables were specifically associated with the name of al-Maʾmūn, and at least one of the values that the *Liber Mamonis* gives corresponds to a value attributed to al-Maʾmūn's 'correctors,' over against Ptolemy's value: i.e., 23° 35′ for the value of the obliquity of the ecliptic. One may compare the relevant passage of the *Liber Mamonis* with the reference to al-Maʾmūn's correction in John of Seville's translation of al-Farghānī's *Rudimenta*, c. 5 (translated in 'Limia' on March 11, 1135):

> *Liber Mamonis*: "The movement of this sphere of the spherical circling (i.e., the ecliptic) is from the west to the east over two fixed poles and a fixed radius to the poles. These poles are neither the radius of the universe nor the radius (?), but each of them is distant from the pole of the universe closest to it by 23 degrees and 35 minutes."[39]
>
> Al-Farghānī: "<The obliquity of the ecliptic> is, according to what Ptolemy discovered, 23° 51′ (when the circle is 360°), but, from the most accurate proof by which Almemon proved it, and with the agreement of the majority of wise men, it is 23° 35′."[40]

If the *Liber Mamonis* gives the value of al-Maʾmūn's 'correctors' for the obliquity of the ecliptic without comment, it is possible that the other values that he gives and approves are from the same source. This remains to be checked. However, the title given in the Cambrai manuscript is not simply "Liber Mamonis," but rather "Liber Mamonis in astronomia a Stephano philosopho translatus." This would imply that a certain philosopher called "Stephen" translated a work associated with al-Maʾmūn from Arabic into Latin. Yet the *Liber Mamonis* is not a translation, but rather a treatise written in a literary style, which has as its leitmotif a criticism of the current doctrines in Latin cosmology, epitomized in the theories of Macrobius, and the need for their replacement by the Ptolemaic system. The word 'translatus' in the title, then, would mean not 'translated' (especially since no source or target language is mentioned) but rather 'transmitted', a meaning supported by classical usage, and, more important, in another work by "Stephen the Philosopher."[41] For, there is little doubt that the "Stephanus philosophus" of the *Liber Mamonis* is the "Stephanus philosophie discipulus" who translated the comprehensive medical work, *al-kitāb al-malakī* (*Regalis dispositio*) of ʿAlī ibn al-ʿAbbās al-Majūsī.

The reasons for this identification have been explained in detail elsewhere,[42] but can be summarized here. We find the same style of literary Latin, including whole phrases, in both works; we find consultation of Greek as well as Arabic sources; but, above all, we find the same system of alpha-numerical notation. Moreover, we find a place and a date, or rather, several dates, attached to different books of the translation of the *Regalis dispositio*. The place is Antioch, and the dates all fall within the year 1127. This makes it very likely that Stephen the Philosopher is "Stephanus thesaurarius Antiochie" for whom a copy of the *Rhetorica ad Herennium* (now MS Milan, Ambrosiana, Cod. E. 7 sup.) was written in 1121.[43] For, this manuscript also uses the alphanumerical notation in the same way as in the *Liber Mamonis*. Richard Hunt pointed out that there was a treasurer called Stephen at the Benedictine monastery of St Paul, one of the principal religious foundations in Antioch, who had been given a house in the city between 1126 and 1130.[44] This is probably our "Stephen." Further biographical details are given by the medical writer, Matheus Ferrarius, who states that "Stephen, a certain Pisan, went to those parts (*meaning the Orient?*), and, learning that language, translated the whole of the *Practica* (i.e., the practical portion of the *Regalis dispositio*),"[45] and from an unknown twelfth-century supporter of the medical school of Montpellier who calls Stephen the "nephew of the Patriarch of Antioch."[46]

The correspondence in language and the form of the alphanumerical notation in the *Liber Mamonis*, the *Regalis dispositio* and the Milan copy of the *Rhetorica ad Herennium* (alphanumerical notation only) corroborates that the "Stephen" mentioned in all three works is the same man, and we can be reasonably sure that he was working in Antioch in the 1120s. He had a strong interest and competence in astronomy; he translated from Arabic, but also had some knowledge of Greek; he had the help of Arabic-speaking colleagues. The Dresden *Almagest* is a little different from these three works: the name Stephen is not attached to it; all Arabic transliterations are avoided whereas, in the *Regalis dispositio* at least, Stephen deliberately transcribes the Arabic terms when he does not know the Latin equivalent; the alphanumerical notation is not as advanced, and no Hindu-Arabic numerals are used. Nevertheless, the similarities between the Dresden *Almagest* and the *Liber Mamonis* are such that the work must, at least, have arisen in the same milieu, if it was not directly used by Stephen the Philosopher.[47] That milieu could have been Pisa, which was not only an important centre for Greek-Latin translations, but also had close links with the Arabic world through its quarters in Antioch, Laodicea, Acre, and elsewhere.[48] But it was more likely Antioch itself, where Stephen was actively engaged in translating works from Arabic.

One last clue associating the Dresden *Almagest* and the *Liber Mamonis* must be taken into account. At the beginning of the *Liber Mamonis* the author states: "Since we have fulfilled our promise, having written a treatise on the rules we had proposed for the canon of astronomy, I approach this second task. . . ."[49] He has, then, already written rules for astronomical tables. These he appears to refer to in the body of the *Liber Mamonis*,[50] and he makes frequent references to the tables themselves.[51] Altogether, four tables are mentioned, which are to be used in succession for ascertaining the position in longitude of the planets; a further table is used for their latitudes. These tables and their rules have not been identified,[52] but the late thirteenth-century astronomer, Henry Bate, in listing a number of authorities that agree in measuring the movements of the planets in respect to the "ninth sphere," mentions among the tables written by the followers of the *magistri probationum*, 'tabule pisane wintonienses' ('the tables of Pisa <and> Winchester).[53] It is true that one of the sets of instructions for using the tables of Pisa mentions the longitude of Winchester.[54] But it is intriguing that this title should nicely join the names of Stephen (of Pisa and Antioch), the author of the *Liber Mamonis* and ʿAbd al-Masīḥ of Winchester, the translator of the Dresden *Almagest*.[55]

II Astronomical Tables Connected with Pisa and Lucca

There is a very strong reason to connect the Tables of Pisa and the *Liber Mamonis*: namely, that the earliest copy of the Tables and the instructions for their use—Berlin, Staatsbibliothek, lat.fol. 307 (Rose, no. 956)—are written entirely in the Eastern forms of the Hindu-Arabic numerals. The Tables of Pisa appear to be based on the lost Arabic tables of ʿAbd-al-Raḥmān ibn ʿUmar al-Ṣūfī (d. 986),[56] and the Berlin version gives 1149 completed years as their starting point (i.e., they were written in 1150 A.D.). The Eastern forms are actively used[57] only in a very restricted range of works. Most of these are Latin texts attributed to, or based on the work of the Jewish polymath, Abraham ibn Ezra (1089/92–after 1160).

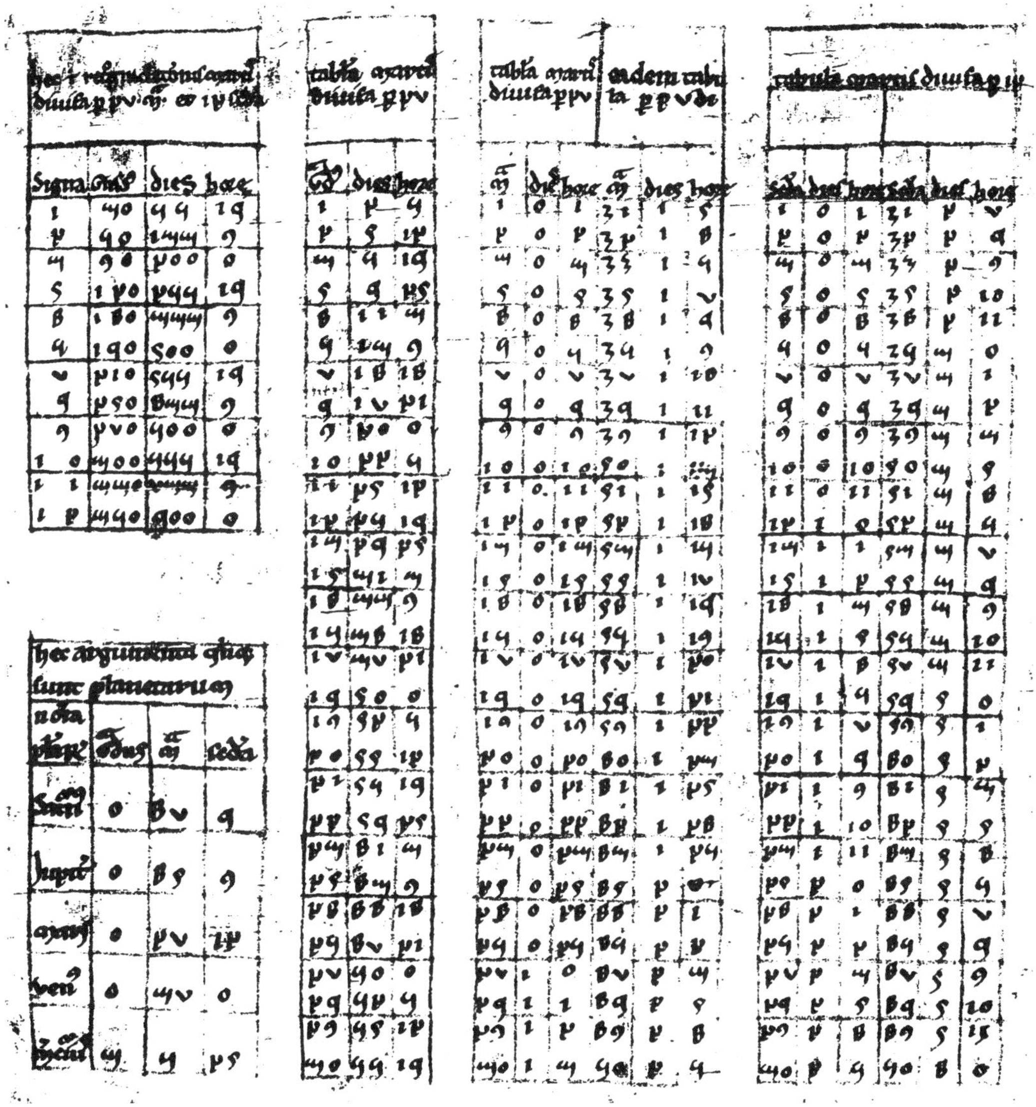

Figure 2.3
Berlin, Staatsbibliothek, lat. fol. 307, fol. 32r. A page from the Pisan Tables showing the use of the Eastern forms of Hindu-Arabic numerals.

A text very similar to the anonymous instructions for the use of the Pisan Tables in the Berlin manuscript occurs in London, British Library, Arundel 377, fols 56v–63r. This text is entitled "Tractatus magistri Habrahe de tabulis planetarum," and is immediately followed by a text on the astrolabe according to the words of "Abraham, outstanding among the philosophers of his time, and our master, on whose dictation we wrote this account of the astrolabe."[58] This copy of these two texts employs the Western forms of the Arabic numerals, which (as we have seen) became the norm, but the presence of a key (fol. 56r) for converting the Eastern forms to the Western forms implies that an earlier copy had been written using the Eastern forms, and this is what we find in another manuscript of the astrolabe text: MS British Library, Cotton Vespasian A.II, fols 37v–40v. Yet another text was written to accompany the Tables of Pisa: this text, known nowadays as the *Fundamenta tabularum*, is a scholarly discussion of the theory of astronomical tables, drawing on many Arabic sources. In certain manuscripts it is called "Abrahismus" and Millás Vallicrosa has shown that it was written by Abraham ibn Ezra in Dreux in 1154. In the earliest manuscripts of the work the Eastern forms of the numerals are used.[59]

Abraham Ibn Ezra was born in Tudela in the Muslim kingdom of Saragossa between 1089 and 1092 and spent the earlier part of his life in Northern Spain, though he also visited other parts of Spain and North Africa. In the early 1140s he started to visit Jewish communities in Christian Europe, first in Rome, followed soon after by Lucca (1142–1145), which is only some fifteen kilometers from Pisa.[60] We are fortunate in possessing a manuscript to which have been added some notes which mention the year 1160 and were presumably written then, or soon after: MS British Library, Harley 5402, fols 69r–v.[61] These notes include a handy way of calculating the position of the Moon, in which the writer uses Hindu-Arabic numerals in their Eastern form, and instructions for tables for the meridian of Lucca which are similar to those accompanying the tables of Pisa, as we read them in the Berlin manuscript.[62] What is striking is that these notes are written in a mixture between Italian and ungrammatical Latin. Examples of written Italian before the end of the twelfth century are rare; the most substantial document, in fact, is a religious poem written by a Jew in Hebrew script.[63] We may, therefore, be in the presence of a Jewish scholar writing in Lucca and using the Eastern forms of the Hindu-Arabic numerals: i.e., someone just like Abraham ibn Ezra.

But the Harley manuscript tells us more. The original scribe, who presumably wrote the text before 1160, did not use Hindu-Arabic numerals. He transcribed a copy of Pseudo-Ptolemy's *Iudicia* (an astrological work) and the Latin corpus of Sahl ibn Bishr's astrological texts, using, where necessary, roman numerals. But between these two texts he added an astronomical table in which the numerical values (aside from those of the first column which are in roman numerals) are written entirely in alphanumerical notation. The notation is that of the *Liber Mamonis*, *Regalis dispositio*, and the Milan *Rhetorica ad Herennium*, but in this case, and in this case only, a key is provided (see figure 2.4).[64]

We see, then, in the Harley manuscript the two kinds of notation for numerals which appear in the *Liber Mamonis*: the Latin alphanumerical system and the Eastern forms of the Hindu-Arabic numerals. The writer of the *Liber Mamonis* used alphanumerical notation for lower numbers, and the Eastern forms for numbers consisting of more than three digits. This mixed system occurs regularly in Islamic astronomical tables, and occasionally in Greek tables, in which alphanumerical notation is used for all numbers up to 360 (the number of degrees in the circle), but Hindu numerals are used where the numbers do, or in principle could, reach the higher hundreds or exceed 1000.[65] It is quite possible that the tables to which Stephen, the author of the *Liber Mamonis*, refers, were written in this mixed system just as the *Liber Mamonis* itself is. For some reason or other, however, neither the alphanumerical notation nor the mixed system caught on. Rather, as it seems, the Eastern forms, as used in the mixed system, were used for *all* numerals, as we see in the Berlin manuscript of the Pisan Tables.

It is unlikely that Ibn Ezra himself introduced the Eastern forms of the Hindu-Arabic numerals. At the beginning of his Hebrew work on arithmetic, *Sefer ha-Mispar*, he mentions Hindu-Arabic numerals and substitutes for them the letters of the Hebrew alphabet. Of the several manuscripts of this text that I have seen, only one gives the Hindu-Arabic numerals in their oriental form, as an alternative to the Western forms which Ibn Ezra could have been familiar with in Spain.[66] It is only in the Latin versions of Ibn Ezra's works that the Eastern forms are preferred, and this preference must have been due to Ibn Ezra's Latin collaborators or students. It could also be questioned whether Ibn Ezra was responsible for drawing up the Pisan Tables, which is inferred by modern scholars from the fact that he bases his instructions on how to use astronomical tables on the Pisan Tables. The Berlin manuscript does not mention his name, and Henry Bate differentiates between "the tables of Abraham" and the "Pisan Tables."[67] Since Henry adds "of Winchester" to the to the Tables of Pisa, it is at least worth considering whether "ʿAbd al-Masīḥ of Winchester" had anything to do with the transmission of al-Ṣūfīs tables to the West. An Antiochene origin would also solve the problem that, apparently, al-Ṣūfī's tables were not known to Andalusi astronomers.[68]

The use of the Eastern forms of the Hindu-Arabic numerals, even in the small number of works I have mentioned, implies at least a community of scholars who understood them. The anonymous Latin scholar who collaborated with Abraham ibn Ezra on writing his text on the astrolabe, and the other Latin scholars who presumably put into good Latin his instructions for the use of the Pisan Tables, belong to this group. It is possible that Hermann of Carinthia, Robert of Ketton, and Hugo of Santalla who were working "on the banks of the Ebro" and in Pamplona, Tarazona and Tudela, and were all involved in compiling compendia on astrological judgments (the *Liber trium iudicum* and the *Liber novem iudicum*) also belonged to this group.[69] Abraham ibn Ezra, as we have seen, came from Tudela, and could well have known these scholars. Subsequently both Hermann (in 1143) and Ibn Ezra (in 1148) are attested in Béziers, where Hermann's student Rudolph of Bruges wrote a text on the astrolabe (in 1144), which is sandwiched between Abraham's *Fundamenta tabularum* and his treatise on the astrolabe in MS London, Cotton, Vespasian A.II (all three texts, and these only, have been written in the same hand), though it uses only roman numerals.[70]

There is some evidence that these scholars also knew, and may have used the Eastern forms of the numerals. Three early copies of works by Hugo of Santalla preserve these forms, in two cases in a table only,[71] in the third case in the text.[72] It is clear that the scribes of these three manuscripts were not familiar with the Eastern forms, and either copied them wrongly (in the table), or abandoned them (in the text). But in one copy of the *Liber trium iudicum* the scribe uses the Eastern forms, confidently and consistently, not only for writing the *Liber trium iudicum* itself, but also for the copy of Hyginus's *Astronomicon* which accompanies that text. This is a twelfth-century manuscript incorporated into MS London, Arundel 268, which is remarkable because it is one of the earliest examples we have of a Western manuscript made of paper. The *Liber trium iudicum* is an astrological compendium addressed in this manuscript to "karissime R.," who is presumably Robert of Ketton, which would make the compiler Hermann of Carinthia; in another manuscript the same *Liber trium iudicum* is dedicated to Michael, bishop of Tarazona, who was Hugo of Santalla's patron.[73] The use of paper at this date suggests a close connection with the Islamic world. The hand-writing suggests that it was written in Italy, as does that of Hugo's text in MS Digby 50.

Hermann of Carinthia was a student of Thierry of Chartres, and some of his translations arrived in Chartres; the Berlin manuscript of the Pisan tables is said to have originated from Chartres, though the instructions for the Pisan tables mention the longitudes of Angers and Toledo, and another text in the manuscript implies that Paris is place of writing.[74] Ibn Ezra himself traveled North, to the Jewish communities in Rouen, Dreux (1153–1156), and Evreux, and eventually to London (1158 and 1160), where, according to some sources,

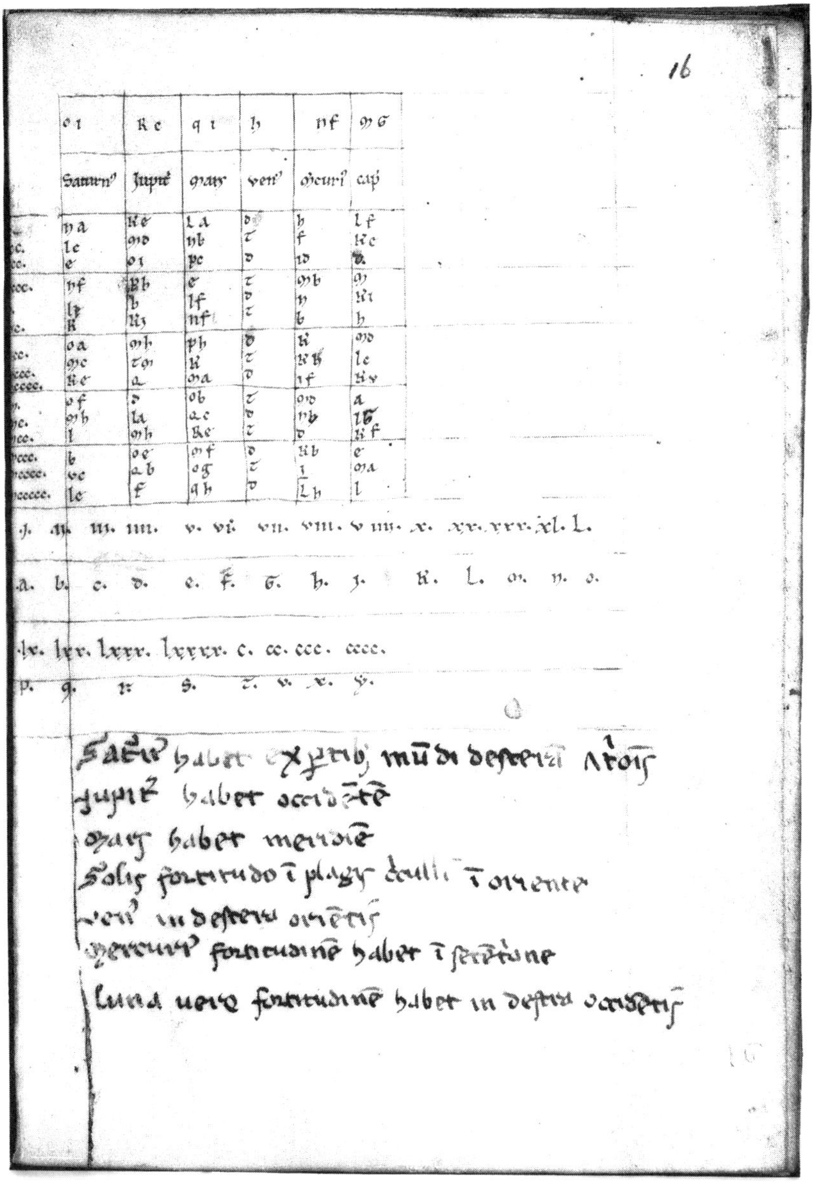

Figure 2.4
British Library, Harley 5402, fol. 16r, written in alphanumerical notation.

he died. Thus, it is not surprising that the majority of the manuscripts of Ibn Ezra's works (and hence including the Eastern forms of Hindu-Arabic numerals) are Norman and English.[75]

Much more work needs to be done on this subject. That the works of Ibn Ezra, Robert of Ketton, Hermann of Carinthia, Rudolph of Bruges, Hugo of Santalla, and Stephen the Philosopher and ʿAbd al-Masīḥ of Winchester, and the Eastern forms of Hindu-Arabic numerals associated with these works, did not remain in currency for very long may be due to the fact that, after 1150, Toledo became the centre of the translating activity from Arabic into Latin, and this 'industry' in Toledo was so successful that it replaced or drove underground much of what went before. It is becoming increasingly obvious that, presumably under the supervision of Gerard of Cremona, there was a concerted program of revising and translating afresh those works that were perceived to be most important in astronomy and astrology. The mechanics of the transmission of the Toledan translations and the reasons for their success are still not clear. At first sight it would seem that the translations of Greek astronomical works directly from Greek should have been preferred to the translation of their Arabic versions: but in the case of Ptolemy, Theodosius, Menelaus, Archimedes, and so many other authors, although Greek-Latin translations were made, and often made before the Toledan Arabic-Latin translations, nevertheless it was the latter which succeeded in becoming established in the scholarly community.

The same conclusion can be reached in regard to the works of Stephen the Philosopher and his colleagues. The Dresden *Almagest* and the *Liber Mamonis* both survive incomplete in one manuscript each, and both the alphanumerical notation and the Eastern forms of the Hindu-Arabic numerals failed to catch on. Nevertheless, some credit must be given to Stephen and his friends. If the arguments that are put forward in this chapter are convincing, then the Dresden *Almagest* is likely to predate the earliest hitherto known translation—that made from the Greek in Sicily in ca. 1160 A.D.[76]—by up to forty years, and the *Liber Mamonis* shows a remarkably advanced understanding of Ptolemaic astronomy and its Arabic developments at a correspondingly early date. Both works are indicative of a considerably higher level of astronomical learning among certain Latin scholars in the second quarter of the twelfth century than has hitherto been recognized, and they invite us to look to the Eastern Mediterranean, rather than to Spain, as a source for this learning.

Acknowledgment

I am very grateful for the help of Richard Lorch, Paul Kunitzsch, Fritz Saaby-Pedersen, Tzvi Langermann, and Tony Lévy.

Notes

1. C. H. Haskins, *Studies in the History of Mediaeval Science*, 2nd ed. (New York, 1927), p. 109, n. 155: "There is also a confusing form of numerals [in the Dresden *Almagest*]: b = β = 2, etc. Cf. supra, n. 128 [discussion of the numerals in the *Liber Mamonis*]."

2. The fullest account of the *Liber Mamonis* up to now is that of Haskins, Studies, pp. 98–103. An edition of the work has been promised by Richard Lemay.

3. I am very grateful to Menso Folkerts for lending me a microfilm of Dresden, Db. 87, and to Richard Lorch for information on the text. For a description of the manuscript see Bertoldo di Moosburg, *Expositio super elementationem theologicam Procli*, 184–211, *De animabus*, ed. L. Sturlese (Rome, 1974), pp. xlvii–xlix. I am also indebted to Gerhard Brey for sending me a copy of the on-line description of the same manuscript from the ICCMSM database.

4. J. L. Heiberg, "Noch einmal die Mittelalterliche Ptolemaios-Übersetzung," in *Hermes*, 46, 1911, pp. 207–216 (see pp. 215–216).

5. MS Dresden, Db. 87, fol. 15v: "Explicit primus sermo libri mathemathice Ptolomei, qui nominatur megali xintaxis astronomie translacione dictamine (*elsewhere* 'dictaminis') philophonia wittomensis ebdelmessie" (*with variants in the other explicits* 'wintomiensis' and 'wuttomensis'). The interpretation of these *explicits* and of the name of the translator/dictator has been explored in C. Burnett, "'ʿAbd al-Masīḥ of Winchester," in *Between Demonstration and Imagination: Essays on the History of Science and Philosophy Presented to John D. North*, ed. L. Nauta and A. Vanderjagt (Leiden, 1999), pp. 159–169. "'ʿAbd al-Masīḥ" ("Servant of the Messiah") is sometimes a generic name for a Christian in a Muslim society.

6. This fact was first pointed out to me by Richard Lorch.

7. "Preclare fecerunt qui corrigentes scienciam philosophie, o Syre, diviserunt theoricam partem philosophie a practica" = *niʿma mā faʿala fimā arā lladhīna staqṣaw ʿilm al-falsafa yā Sūrus fī ifrādihim juzʾ al-falsafa an-naẓarī ʿan al-ʿamalī* (P. Kunitzsch, *Der Almagest*, Wiesbaden, 1974, p. 133). Contrast the Greek: Πάνυ καλῶς οἱ γνησίως φιλοσοφήσαντες ὦ Σύρε, δοκοῦσί μοι κεχωρικέναι τὸ θεωρητικὸν τῆς φιλοσοφίας ἀπὸ τοῦ πρακτικοὺ (Ptolemy, *Syntaxis mathematica*, ed. J. L. Heiberg [Leipzig, 1898], p. 4, lines 7–8) = 'The true philosophers, Syrus, were I think, quite right to distinguish the theoretical part of philosophy from the practical' (trans. G. J. Toomer, Ptolemy's *Almagest* [London, 1984], p. 35).

8. 'discordia' (fol. 36r; 'anomaly') corresponds to *ikhtilāf* ('difference'), rather than to ἀνωμαλία; 'discordia visus' (fol. 50r; 'parallax') = *ikhtilāf al-manẓar* ('difference of vision'), rather than παραλλάξις; 'circulus rotunditatis' (fol. 36v; 'epicycle') = *falak al-tadwīr* ('circle of roundness'), rather than ἐπίκυκλος; 'longinqua longinquitas' (fol. 37r; 'apogee') = *al-buʿd al-abʿad* ('the most distant distance'), rather than ἀπόγειον, etc.

9. E.g., fol. 2r: "rememorat qualitatem communem," which is followed immediately by another translation of the same phrase: "vel commemoratio communis qualitatis" and

is followed soon after by another double translation: "rememorabimus vel commemorabimus ergo." The words "rememoratio" and "rememorare" are particularly common in the Arabic-Latin translations of John of Seville and Gerard of Cremona, corresponding to forms of the Arabic root *dh-k-r*.

10. Fol. 1r: "propter quod aretius morum anime possit esse in pluribus sine doctrina"; see Heiberg, p. 4, line 12: διὰ τὸ τῶν μὲν ἠθικῶν ἀρετῶν ἐνίας ὑπάρξαι δύνασθαι πολλοῖς καὶ χωρὶς μαθήσεως. The Latin word-order and sense follow the Arabic text here.

11. Hipparchus is always "ypparcus," Kalippus is "caliphi" (fol. 70r), Phanostratos, the Athenian archon, appears as "Phanastrato Athenarum custode" and "Phanastri primatis Athanarum" (both fol. 69r; cf. trans. Toomer, pp. 211–212), the Egyptian months appear as "Thuth" (Thoth), "Phamenoth" (Phamenoth), "Messurem" (Mesore), and "Mechir" (Mechir) on fols 69r–70v. The Greek months, however, either betray the influence of Arabic transcription (fol. 69r "Bussiothos" for Poseideon) or are omitted altogether. A list of month names, including the Egyptian and the Greek, preceded the *Almagest* text in Wolfenbüttel, Gud. lat. 147, which added the preface of the Sicilian Greek-Latin translation of the *Almagest* to a text of the Toledan Arabic-Latin translation (see Kunitzsch, *Almagest*, p. 95).

12. E.g., fol. 9v: 'Ponemus igitur istud in tabulis' (= *Almagest*, I.11); fol. 20v: 'in tabulis que hoc verbum secuntur' (= *Almagest*, II.6, tables for the third parallel up to the North Pole); fol. 22v: "posuimus autem tabulas orientalium .x. et decem graduum" (= *Almagest*, II.8); fol. 30v: "ponentes eorum tabulas" (= *Almagest*, II.13); fol. 55r: "et posuimus in prima tabularum tempora collectorum annorum et displicatorum (sic) et mensium et dierum et horarum et post istud motum in longo et post illum in discordia et post illum in lato et post illum in longinquitatem que est inter Solem et Lunam" (= *Almagest*, IV.4).

13. It is possible to imagine that the tables were left in the original form, in Greek or Arabic (or merely with their rubrics glossed in Latin), and that the reader was expected to be able to interpret the numeral values: this would provide a possible reason for the adoption of alphanumerical notation (the notation of the Greek and Arabic tables) in the Latin text; see below.

14. The combination of both forms occurs in "liber estichie elementorum" (fol. 7r) and the translator also uses the abbreviations "l.e.e." and "l.he.e." It is unlikely that a scholar who knew Ptolemy's *Almagest* did not have access to Euclid's *Elements*, and it is worth considering whether our translator also translated the *Elements* and if any extant version shows the characteristics of his style.

15. MS Istanbul, Ahmet III, 3464, fols 216r–217r. The correspondence between the two works begins with the theorem which precedes the addition, which is the last theorem in *Almagest*, I, chapter 13. For an edition of the relevant section of the Dresden manuscript see R. Lorch, *Thābit ibn Qurra On the Sector-Figure and Related Texts*, Frankfurt 2001, pp. 362–375.

16. The relationship between the Dresden *Almagest* and the work of al-Nasawī needs further investigation. Both texts refer to the same earlier theorems of Ptolemy (e.g.,

Dresden *Almagest*, fol. 13v: "quod est ostensum in .xii. figura istius sermonis"), and the style of writing in the Dresden *Almagest* gives the impression that the "addition" is integral to the work: the same system of cross-references and appeals to the relevant theorems of Euclid's *Elements* are found throughout the section of the *Almagest* on trigonometry and spherical astronomy.

17. I have noticed only that figure 4.5 (the rejected eccentric model for the Moon's motion) is missing.

18. "Scribe" and "translator" need not be distinguished here in that the scribe of this unique manuscript appears to have tried to copy what was before him without alteration. They need only be distinguished when (as sometimes happens) it is clear that the scribe misunderstood what was in front of him.

19. The use of ".k." for twenty here is due to the flaw in the system mentioned below, and not to confusion with the Greek and Semitic alphanumerical notations in which "k" *is* 20.

20. The scribe uses the abbreviation for "enim," but in two forms: the capital H with the middle bar extended in both directions (often described as an English practice) on fol. 54r, and .n. on fol. 66v. The first form, which is a distinctive symbol, may have been the original one. It is also possible that the Greek capital eta (H) was originally intended, and that, in copying the text on fol. 66v, the scribe mistook this sign for the abbreviation for "enim."

21. The isolated use of Θ where the Greek text has "900" on fol. 63v is apparently a mistake, since there are several examples of Θ being used for "600." In other cases where the Greek text has "900" the number is written out in the Latin translation as "nongenti." Whether the scribe's letters are closer to Greek majuscules or miniscules can only be determined by comparison with Greek letters occurring both in Greek and Latin manuscripts of the period. Latin authors tended to use the Greek majuscule, and we have several examples of the knowledge among them of the numerical values of the Greek letters; see W. Berschin, *Greek Letters and the Latin Middle Ages*, revised and expanded edition, trans. J. C. Frakes, Washington, D.C., 1988, pp. 29–30 and 289, n. 40.

22. The alphanumerical notation in *Liber Mamonis* (see below) is generally rubricated.

23. On Hindu-Arabic numerals in Arabic manuscripts see Kunitzsch's chapter in this volume. The Eastern forms in this manuscript are more closely assimilated to letters of the Latin alphabet than in other manuscripts. The fact that different letters are used for the same numerals in the case of "1" and "9" in the Cambrai manuscript of the *Liber Mamonis* might suggest that the assimilation was made by the scribe, who *expected* to find letter-forms, rather than by the author, from whom one would expect a one-to-one equivalence of symbols.

24. Richard Lemay, in "The Hispanic Origin of Our Present Numeral Forms" (*Viator*, 8, 1977, p. 435–62) suggests that this "Western form" arose out of a mixture of the original Sanskrit numerals and Visigothic forms, but the question of origin is not yet settled.

25. MS Munich, Bayerische Staatsbibliothek, clm 18927, f. 1r; see Figure 1a in R. Lemay, "The Hispanic Origin . . ."

26. The two forms are also distinguished in Arabic in the twelfth-century Maghrib by Ibn al-Yāsamīn: see Abū Fāris, "Dalīl jadīd ʿalā ʿurūbat al-arqām al-mustaʿmala fī al-maghrib al-ʿarabī," *Al-lisān al-ʿarabī*, 10 (1392/1973), pp. 232–234; the full text of Ibn al-Yāsamīn has been edited in a dissertation by a student of Ahmed Djebbar, see p. 345.

27. Of the terms already mentioned (n. 8 above), the *Liber Mamonis* (M), gives "visus discordia" (fol. 27v) for "parallax", "rotunditas" or "rotundus circulus" (fol. 39r *et passim*) for "epicycle" (cf. D[resden *Almagest*], "circulus rotunditatis"), and "longinqua longinquitas" and "propinqua longitudo" (M, fol. 29v) for "apogee" and "perigee." Other parallels in terminology are "speralis" (rather than "spericus") for "spherical" (M and D *passim*), "austrum" (rather than "meridies") for "south" (M, fol. 49v, D, fol. 10v), "circulus recti diei" for "equator" (M, fol. 8v; cf. D, fol. 6r, "circulus rectitudinis diei"), "punctum capitis" for "zenith" (M, fol. 18v, D, fol. 16r), "synodos" for "conjunction" (of planets) and "new moon" (M, fol. 39r, D, fols 55r, 65r and 68r), "pansilini" for "full moon" (M, fol. 34r, D, fols 55r and 65r), "sexagenaria" for "minute" (M and D *passim*), "latum" for "latitude" (M, fol. 49v, D, fol. 51r), "circulus signorum" for "zodiac" (M, fol. 49v, D, fol. 10r), and, finally, the word for Ptolemy's work itself: "megali sintasi" (ablative; M, fol. 39v), "megali xintaxis" (nominative; D, fol. 15v etc.). Occasionally, the author of the *Liber Mamonis* uses different terminology, perhaps because of his greater familiarity with the Latin astronomical and geometrical tradition: e.g., "liber geometrie" (M, fol. 7v), rather than "liber elementorum" for Euclid's work; "circulus extra centrum" (M, fol. 39r), rather than "circulus forinseci centri" (D, fol. 36v) for "eccentric"; "caput draconis" and "cauda draconis" rather than "ligans capitis" and "ligans caude" (D, fol. 67r) for the ascending and descending nodes.

28. To call a "minute" a "sixtieth part (of a degree)" would seem a natural thing to do, but such terminology does not appear to have been used either in the Greek or in the Arabic versions of the *Almagest*: see Kunitzsch, *Almagest*, pp. 156–160.

29. Fol. 22v: "Tholomeus autem et ceteri quibus sanior est intellectus. . . ."

30. Fol. 27v: "Deprehensum autem a probatissimis in astronomia viris Tptolomeo (sic) et reliquis cognovimus maiorem hanc esse visus discordiam Mercurio quam in Venere, Veneremque maiorem habere Sole."

31. Fol. 29v–30r: "Sunt autem hii duo puncti fixi nec moventur sicut Ptolomeus posuit in *Sintaxi*. Aliorum autem qui secuti sunt deinceps astronomicorum quibus fuit etiam et hec et alia secretius rimari, sententia est in orientem moveri tardo motu, in .t. scilicet et .f. annis uno gradu. Qua in re mirum est Ptolomei deceptam fuisse sollertiam. Moventur enim. Set illi casus causa extitisse probatus <est> quidam sui temporis precessor astronomus cuius inquirendo altitudinis Solis loco falsa fuit inspectio. Illo enim in loco longuinquam Solis longinquitatem suo tempore dixerat, quo Ptholomeus perspicatius omnia astrorum secreta discutiens et investigans suo tempore illam esse veracissime comperuit. Hinc ergo deceptus Ptolomeus fuisse dicitur, tametsi non ipse nisi quod non moveri dixit. In loco enim inveniendo nichil peccavit, sed qui precesserat ipsum deceptus fuit. Non ergo Ptholomei sagacitas cecidisse arguenda est, sed illius depra-

vanda ignorantia qui quod nescivit scripture tradens sapienti viro cadendi causa fuit. Nunc satis quod pro illius defendenda industria diximus, ad nostra redeamus."

32. Fol. 39v: "Motus autem is quo hec spera movetur et alias secum movet tardus est ab occidente in orientem, complens, sicut dixit Ptolomeus in *Megali sintasi* sua in .t. annorum curriculo unius cursum gradus. Aliorum sententia est in .t.f. annis."

33. Fol. 44v: "Trium autem planetarum, Saturni scilicet Iovis et Martis, spere simillime sunt adinvicem in numero et divisionibus sperarum, motu quoque et circulis fere, sicut dixisse Ptolomeus in sua probatur *Sintaxi*."

34. Fol. 49v: "Mensura autem maioris lati planetarum et Lune in septentrionem et austrum a circulo signorum est Lune quidem .d. graduum .n.e. sex., per mensuram qua dividitur magnus circulus in .x.p. partes, Saturni trium, Iovis duorum, Martis in septentrione .d. graduum .l. sex., in austrum .g. graduum, Veneris, sicut dicit Ptolomeus in sua *Sintaxi*, .f. graduum .b. sex., sicut autem alii dixerunt astronomi, novem, Mercurii .d.".

35. Fol. 49v: "Ima vero spera Saturni hec movetur in orientem tardo motu qui est in .t. annis uno gradu super duos fixos polos. Hii sunt duo poli circuli signorum quem Solis centrum circinat, sicut dicit Ptolomeus sua et precedentium ratione inventum."

36. For an account of the *hay'a* tradition see Y. T. Langermann's introduction to his edition and translation of Ibn al-Haytham, *On the Configuration of the World*, New York and London, 1990, especially pp. 25–34, and A. I. Sabra, "Configuring the Universe: Aporetic, Problem Solving, and Kinematic Modeling as Themes of Arabic Astronomy," *Perspectives on Science*, 1998, pp. 288–330.

37. Fol. 38r: "Verum cum in aliis Arabem quendam plurimum secuti sumus, in hoc quoque per multum sequemur, licet quedam de sperarum numero et rotunditatum invenerimus et de circulis quidem et inclinationibus planetarum vera perstrinxit a quibus sperarum numerus dissonat."

38. See items 3 and 4 above. This is possibly a mistake for 66 years, al-Battānī's value, unless the author has confused the precession rate with the small fraction of a day (1/106) by which a year is less than 365 1/4 days long, according to al-Ṣūfī (*see* J.-M. Millás-Vallicrosa, "El magisterio astronómica," n. 59 below, p. 314).

39. Fol. 29r: "Motus autem huius spere speralis circuitus est ad orientem ab occidente super duos fixos polos et fixum polorum radium. Non sunt autem hii poli aut radius poli mundi aut radius, set horum uterque a sibi proximo mundi polo .l.c. gradibus .m.e. sex. distat."

40. John of Seville's translation, printed Paris, 1546, p. 16 and in MS Florence, Con. soppr. J.II.10, fol. 154r: "Et est secundum quod invenit Ptholomeus .xxiii. graduum et .li. minutorum, cum fuerit circulus .ccclx. graduum, probatione autem certissima qua probavit Almenon . . . et convenerunt in ea plures sapientes esse .xxiii. graduum et .xxx. minutorum." The errors in this text can be corrected from the translation by Gerard of Cremona (ed. R. Campani, *Il "libro dell'aggregazione delle stelle,"* Città di Castello, 1910, p. 229): " . . . secundum considerationem vero consideratam et expertam quam Johannes filius Almansoris (*This is Yaḥyā ibn Abī Manṣūr, who was one of al-Ma'mūn's*

correctors) consideravit in diebus Maimonis et convenit in ea numerus sapientum est 23 gradus et 35 minuta," and by the Latin translation accompanying J. Golius's edition (*Muhammedis fil. Ketiri . . . qui vulgo Alfragani dicitur, Elementa astronomica*, Amsterdam, 1669, p. 18): "et verò juxta dimensionem quam probatum (*al-mumtaḥan*) vocant, quamque piae memoriae Almámon institui jussit, adhibitis et eam rem viris doctis compluribus, ea declinatio continet gradus 23, et minuta 35."

41. See the preface to the second part of the *Regalis dispositio*, in which Stephen apologizes for leaving certain terms in Arabic: "Malui igitur paulo infirmus videri quam scientiam non transferre" ("I preferred, therefore, to seem a little infirm than not to transmit knowledge"). The prefaces to the *Regalis dispositio* are edited in the article mentioned in the following note.

42. The identity of the two Stephens is explored in detail in my "Antioch as a Link between Arabic and Latin Culture in the Twelfth and Thirteenth Centuries," in *Occident et Proche-Orient: contacts scientifiques au temps des Croisades*, ed. I. Draelants, A. Tihon, and B. van den Abeele, Louvain-la-Neuve, 2000, pp. 1–78.

43. That both this date and the dates of the translation of the *Regalis dispositio* are given in terms of "a passione domini," constitutes another link between the two works. It does not imply that the 33 years of Christ's life have to be added to the dates; the arguments for this are given in my "Antioch as a link . . .," Appendix III.

44. R. W. Hunt, "Stephen of Antioch," in *Medieval and Renaissance Studies*, 6, 1950, p. 172–173.

45. "Stephanon autem quidam Pisanus ad illas partes ivit et linguam illam addiscens eam [Practica] ex toto transtulit" (MS Erfurt, Wissenschaftliche Bibliothek, Amplon. O 62, f. 50r).

46. MS British Library, Sloane 2426, fol. 8r (within a *Theorica* attributed by a later scribe to Cophon): "Stephanus nepos patriarche Antiochensis." This preface is edited in C. Singer, "A Legend of Salerno: How Constantine the African Brought the Art of Medicine to the Christians," *The Johns Hopkins Hospital Bulletin*, 28, 1917, pp. 64–69 (see p. 67a).

47. A comparison between the translating techniques and the terminology used in the Dresden *Almagest* and the *Regalis dispositio* might reveal further links. Suffice to say here that both works give elaborate details concerning the author and translator, and use the word "sermo," in preference to "liber," for the constituent books (= Arabic *maqāla*). In the explicit to the fourth book of the Dresden *Almagest* the word "liber" has been expunged, and "sermo" has been written in its place.

48. See Burnett, 'Antioch as a link . . . ', *passim*.

49. Fol. 2r: "Quoniam in canonem astronomię quas proposueramus regularum exsequto tractatu promissum exsolvimus, secundum hoc opus . . . aggredior."

50. Fol. 44v: "Hence it is that, in finding their latitude in *the rules for the canon*, we have realized that a certain latitude is taken . . ." ("Inde est quod in eorum inveniendo lato in canonis regulis quoddam latum sumi precepimus"). Italics mine.

51. E.g., fol. 46r: 'In quarte tabula postquam rectificata fuerit . . . '; fol. 46v: 'Quare autem quarte tabule nec (?) tercia auferatur sepe quinta iungitur, evidens hoc modo fiet ratio. Secunde tabule numerus ab obliqua longinquitate circuli extra centrum usque .s. [= 90] fere decrescit, exinde usque .t.r. [= 180] recrescens.'

52. They are unlikely to be the *Regulae in Canonem Astronomiae* in Florence, Bibl. Naz., Con. Soppr. J.II.10, f. 235ra–239va (227r–231v), as maintained by R. Lemay in *De la Scolastique à l'histoire par le truchement de la philologie: itinéraire d'un médiéviste entre Europe et Islam*, in *La diffusione delle scienze islamiche nel medio evo europeo*, Convegno internazionale dell'Accademia Nazionale dei Lincei, Rome, 1987, pp. 399–535 (see p. 472), not only because these rules are not accompanied by tables, but also because the astronomical terminology is completely different. It is true that both texts criticize the opinion among the Latins that retrogradation is caused by the attraction of the rays of the Sun (Florence MS, fol. 238/230ra–va, Liber Mamonis, fol. 39r, preface to the fourth book), but the language of the criticisms is different and the criticised opinion was widespread.

53. Henry Bate, *Descriptio instrumenti pro equatione planetarum*, which follows his *Magistralis compositio astrolabii* (1274), printed with *Liber Abraham iudei de nativitatibus*, Venice, 1485, sig. d4 recto–verso: 'Ptholomeus vero: et Geber: Albategni: Abrahamque Iudeus: et Açophius ceteri quoque magistri probationum et maxime orientales astronomi motus planetarum secundum nonam speram considerantes radices suas super hoc fundauerunt: et hoc patet in tabulis Ptholomei: Albategni et Abrahe: in tabulis Pisanis vuintoniensibus et aliis.' The punctuation of the Renaissance printing has been carefully observed, and corresponds to that of MS Oxford, Digby 48, fols 152v–155v.

54. British Library, Arundel 377, fol. 56v; see Burnett, *The Introduction of Arabic Learning into England*, London, 1997, pp. 48–49. The tables are not given in this copy.

55. The implications of this coincidence are discussed in my "'Abd al-Masīḥ of Winchester.'"

56. For the arguments for believing that the Latin tables preserve those of al-Ṣūfī see R. Mercier, 'The Lost Zīj of al-Ṣūfī in the Twelfth Century Tables for London and Pisa,' in *Lectures from the Conference on al-Ṣūfī and Ibn al-Nafīs*, Beirut and Damascus, 1991, pp. 38–72. Al-Ṣūfī served several members of the Būyid dynasty and is attested in Dinawar and Isfahan.

57. I discount (1) the 'fossilized' forms in which a Latin translator or scribe seems to be copying Eastern forms from Arabic manuscripts without understanding their significance (though these are valuable as giving evidence for which forms *were* used in Arabic manuscripts), and (2) the instances in which a Latin scribe gives the series of Eastern forms as an alternative to the Western forms, the latter of which he uses exclusively. For a fuller account of the use of Eastern forms in Latin manuscripts and their Arabic exemplars, see C. Burnett, "Indian Numerals in the Mediterranean Basin in the Twelfth Century," in *From China to Paris: 2000 Years' Transmission of Mathematical Ideas*, ed. Y. Dold-Samplonius *et al.* Stuttgart, 2002, pp.237–288.

58. British Library, Cotton Vespasian A.II, fol. 40v: 'Ut ait philosophorum sibi contemporaneorum Habraham magister noster egregius quo dictante et hanc dispositionem astrolabii conscripsimus.'

59. See J.-M. Millás-Vallicrosa, 'El magisterio astronómica de Abraham ibn 'Ezra en la Europa latina,' in id., *Estudios sobre historia de la ciencia española*, Barcelona, 1949, pp. 289–347 and the same author's edition of the *Fundamenta tabularum: El libro de los fundamentos de las Tablas astronómicas de R. Abraham Ibn Ezra*, Madrid and Barcelona, 1947. This publication includes illustrations of the numeral forms in the manuscripts.

60. Shlomo Sela, in a valuable article on Ibn Ezra's versions of astronomical tables and instructions for their use, adduces arguments for Abraham being in Pisa itself: see his 'Contactos científicos entre judíos y cristianos en el siglo XII: el caso del *Libro de las tablas astronómicas* de Abraham ibn Ezra en su versión latina y hebrea,' *Miscelánea de estudios árabes y hebraicos, sección Hebreo*, 45, 1996, pp. 185–222 (see pp. 212–222). Further information is given in id., 'Algunos puntos de contacto entre el *Libro de las tablas astronómicas* en su versión latina y las obras literarias hebreas de Abraham ibn Ezra,' *Miscelánea de estudios árabes y hebraicos, sección Hebreo*, 46, 1997, pp. 37–56 and *Astrology and Biblical Exegesis in Abraham ibn Ezra's Thought* (in Hebrew), Ramat-Gan, 1999.

61. For a study of this manuscript see C. Burnett, 'Latin Alphanumerical Notation and Annotation in Italian in the Twelfth Century: MS London, British Library, Harley 5402', in *Sic Itur ad Astra*, Festschrift Paul Kunitzsch, ed. M. Folkerts and R. Lorch, Wiesbaden, 2000, pp. 79–90; this includes editions of the mixed Italian-Latin texts and the astrological table mentioned in n. 64 below.

62. One may compare the opening of these instructions—'Sciatis quod tabule iste facte sunt super ciuitas luce. Et da angle da'occidente usque ad illam abet (i.e., habet) G <radus> .xxxiv;—with the statement by José Bonfils that Ibn Ezra 'in the tables which he composed in Lucca in the land of Lombardy, affirmed that the longitude of Lucca, that is, the distance from the West, is 34 degrees': see Sela, 'Contactos científicos,' p. 206.

63. B. Migliorini, in *Storia della lingua italiana*, Florence, 1985, p. 111. I am grateful to Francesca Ziino and Laura Lepschi for advice on early examples of Italian.

64. This table occurs elsewhere only in a later manuscript, Pommersfelden 66, fol. 84r.

65. See 'Antioch as a Link . . .', Appendix II.

66. Two sets of numeral forms are given in the margin of f. 1v of Paris, BNF, hébreu 1052 (fifteenth century); in the text of this manuscript, however, and in the copies of *Sefer ha-Mispar* in MSS Paris, BNF, hébreu 1049, 1050 and 1051, only the Western forms are used. I am grateful to Tzvi Langermann for alerting my attention to these manuscripts.

67. See quotation in n. 53 above. On the other hand, MS Cambridge, University Library, Kk.I.1, fol. 145v mentions the 'tabule mediorum cursuum Solis ad meridiem Winton. ab Abrahamo condite' (I owe this reference to Fritz Saaby Pedersen).

68. The tables are not mentioned, as far as I know, in any Arabic text, though there remains the problems of how, in the *Fundamenta tabularum*, they are described as

'secundum Azofi compositum, et Azarchelis probatio fuit in anno 482 ab helligera' (*El libro de los fundamentos*, ed. J. M. Millas Vallicrosa, p. 87). With this should be compared a passage in the *Sefer ha-Olam* in which Abraham says that al-Zarqālluh's value for the amount by which the year falls short of 365.25 days is the same as that of al-Ṣūfī, without explicitly saying that the later astronomer was indebted to the earlier; I owe this information, once again, to Tzvi Langermann.

69. See C. Burnett, 'A Group of Arabic-Latin Translators Working in Northern Spain in the mid-twelfth Century,' *Journal of the Royal Asiatic Society*, year 1977, pp. 62–108.

70. See R. Lorch, "The Treatise on the Astrolabe by Rudolf of Bruges," in *Between Demonstration and Imagination: Essays in the History of Science and Philosophy Presented to John D. North*, ed. L. Nauta and A. Vanderjagt (Leiden, 1999), pp. 55–100. (This article includes an edition of the text.)

71. That is, in two copies of Hugo's translation of Ibn al-Muthannā's commentary on al-Khwārizmī's astronomical tables: Cambridge, Gonville and Caius, 456/394, fol. 73r and Oxford, Bodleian Library, Selden, Arch. B, 34, fols 32v–33r; the third copy, ibid., Savile, 15, does not include the tables.

72. Hugo's translation of a work on geomancy, in Oxford, Bodleian, Digby 50, fols 94r–101r.

73. See Burnett, 'A Group of Arabic-Latin Translators.'

74. For a thorough description of the manuscripts, see V. Rose, *Verzeichniss der lateinischen Handschriften . . . der Königlichen Bibliothek zu Berlin*, II, 3, Berlin, 1905, pp. 1177–1185. The Chartres provenance is suggested in a modern note on the cover of the manuscript itself (Rose gives "France" as the provenance). For the inhabitants of Pisa and Angers, see fol. 27r: "Et scias quod iste tabule conposite sunt secundum meridiem Pissanorum quorum longitudo occidentalis 33 graduum est, Andegavensium vero longitudo occidentalis 24 graduum"; on fol. 30r there is a "Tabula diversitatis aspectus lune ad longitudinem Toleti 39 graduum 54 minutorum et eius hore 14 minuta horarum 51"; Paris is mentioned on fol. 1r: "nos sumus Parisius existentes." I am grateful to Raymond Mercier for some readings of this manuscript.

75. For the diffusion of these manuscripts in Normandy and England see C. Burnett, *The Introduction of Arabic Learning into England*, pp. 46–60.

76. Haskins, *Studies*, pp. 157–163, J. E. Murdoch, "Euclides Graeco-Latin," *Harvard Studies in Classical Philology*, 71, 1966, pp. 249–302 (see pp. 263–270; the date 1165 is proposed on p. 269).

II

Transformations of Greek Optics

3

The Many Aspects of "Appearances": Arabic Optics to 950 AD

Elaheh Kheirandish

The section on "the science of optics" (*ʿilm al-manāẓir*) in al-Fārābī's (d. 339/950) *Catalogue of the Sciences* (*Iḥṣāʾ al-ʿulūm*)—a work representing the state of many fields up to about the mid-4th/10th century—covers five distinct subjects within the full text of its Arabic edition (hereafter I):[1] the need for such an autonomous science and its distinctions from geometry (I.1), its function and methods (I.2), its applications and instruments (I.3), its assumptions and mechanisms (I.4), and its domains of inquiry (I.5).[2] The respective passages, on the other hand, cover their intended subjects all with reference to some aspect of "appearances" (their veracity in I.1, justifiability in I.2, fallibility in I.3, and elements and effects in I.4–I.5), the appearances involved being in most cases those *other than commonly experienced*, either in terms of perceived shapes or sizes (I.1 and I.2) or in the form of appearance through mediums other than air (I.4 and I.5).

From the standpoint of the history of optics before 339/950, the long passage is an indispensable source for providing a most fitting outline and timeframe for discussion: on the one hand, it contains evidence on the state of early Arabic optics: its sources and problems, traditions and orientations, and boundaries and methods,[3] along with the terminological aspects of the "appearances" that it covers; on the other hand, its treatment of optics from the perspectives *originally* intended (its classification as a science and a *mathematical* science) deserves close attention, if only for the partial, and potentially misleading, nature of "evidence" provided by it, from a *historical* perspective.

Take the case of the opening passage (I.1), which addresses the problem of the veracity of appearances with reference to the age-old example of the circular appearance of far rectangular objects (see below). It is true that what that passage contains provides valuable historical evidence for the identification of one of the most central problems of early optics (*the* most central, if one goes by the prominent place it occupies in that passage alone). But it is also true that what the passage *does not* contain (in this case, reference to other central problems of optics at this early period), is, at best, incomplete in the light of what is available through other early sources. Whether al-Fārābī is not fully aware of the optical problems of his time (including problems as central as clarity of vision, on how *clear* something is seen, rather than how *real* the nature of its appearance is), or it is the case that he is not referring to these in the context of his intended discussion (in this case, distinguishing between the two mathematical sciences of optics and geometry, with the same objects but different functions), one may argue that the five passages, marked here I.1–I.5, are all notable for *both* what they contain and what they do not. Indeed, what is contained in the texts of the middle and closing passages too is quite reflective of many aspects of "the science of aspects"(*ʿilm al-manāẓir*): in the case of the second and third passages (I.2 and I.3), its methodological and practical aspects, and in the case of the fourth and fifth passages (I.4 and I.5), its theoretical and physical aspects respectively. It is the purpose of the present essay to include a discussion of what is not contained in all the respective passages (I.1–I.5) as well, with a historical analysis of their full texts in the light of early optical sources previously treated by the present author from other perspectives.[4]

What follows is an individual treatment of the passages forming the optics chapter of al-Fārābī's *Catalogue* in historical context, as the basis for the argument that its coverage of the "science of aspects" (*ʿilm al-manāẓir*) in general and of "appearances" (*mā yaẓhar*) in particular, while extensive and informative, is far from exhaustive of the full range of problems and concepts present within the *manāẓir* (optics), *marāyā* (catoptrics), and *misāḥa* (surveying) core of the early Arabic optical tradition.

The "manāẓir" tradition, which in its earliest stages was primarily Arabic,[5] is a rich tradition where the Euclidean concept of "appearance" (from φαίνεσθαι),[6] takes on many aspects of its own, from what "appears to sight" (*yaẓhar fī al-baṣar*), "seen" or "seen as" (*yurā*), or else "viewed" (*yunẓar:* from the same verbal root as *manāẓir*)[7] in al-Fārābī (I.1–I.2), to what appears (*ẓahara*), "seen" (*yubṣar*), "thought to be seen" (*yuẓannu annahu yurā*), "imagined to be seen" (*qad yutawahhamu an yubṣar*), "more accurately seen" (*aṣdaqu ruʾyatan*)—and outside of the Euclidean text, also what sight or the eye (*baṣar*) "perceives" (*yudrik*).[8] The *manāẓir* tradition is also a versatile and complex tradition, one that even if the particular "agenda" of al-Fārābī's *Catalogue* allowed for the full coverage of its versatility (I.3–I.5), it could not possibly be expected to reflect the many shades of its complexities.

The *misāḥa* (surveying) and *marāyā* (catoptrics) components of the *manāẓir* tradition are similarly illuminated by al-Fārābī's discussions, but understandably limited in their scopes of representation: The case of *misāḥa* is, despite the involvement of a wide range of methods (determination of heights, widths, and lengths) and objects (from trees, walls, valleys, and rivers to mountains and the heavenly bodies) in the corresponding al-Fārābī passage (I.3), still short of a range of problems posed, not only by the one condition mentioned in passing—that of "sight (*baṣar*) falling on (*yaqaʿu ʿalā*)" objects—but especially by the most common (and as we shall see, problematic) instrument for surveying it fails to mention—a plane mirror. As for *marāyā*, treated by al-Fārābī at once as an integral part and separate branch of optics (I.5), the rather unusual categorization of "indirect" (*ghayr mustaqīm*) rays, first as "deflected" (*munʿaṭif*), "reversed" (*munʿakis*) and "bent" (*munkasir*) next to "direct" (*mustaqīm*) rays, and then, to the exclusion of *refracted* rays,[9] is both wanting and confusing, this time to be examined in the light of the largely unsettled, and particularly problematic state of Arabic optics during the earliest phases of its development.

I Veracity and Accuracy of Vision

> I.1: "The science of optics (*ʿilm al-manāẓir*) investigates the same things as does the science of geometry, such as figures, magnitudes, order, position, equality and inequality, but not in so far as these exist in abstract lines, surfaces and solids, whereas geometry investigates them insofar as they exist in abstract lines, surfaces and solids. Thus geometrical investigation is more general. *But there was a need for a separate science of optics*, although [its objects] are included among the objects of geometry, *because many of the things which are proved to be of a certain shape or position or order or the like, acquire opposite properties when they become objects of vision: thus objects which are really square are seen (r-ʾ-y) as circular when seen (n-ẓ-r) from a distance* and equal objects appear to be unequal and unequal ones appear equal, and many objects which are placed in the same plane appear to be some lower and some higher, and many foreground objects appear to be farther back. And such things are many. . . ."

The opening passage (I.1) poses the question of the need for a science of optics (*ʿilm al-manāẓir*) in terms of the problem of the truth correspondence (*al-ḥaqīqa*) of "what appears" (*mā yaẓhar*) a certain way. In discussing the problem of the veracity of appearances, the present passage addresses a historically important aspect of appearances, while leaving out another pro-minent aspect: that of the accuracy of vision (*ṣidq al-ruʾya*).

In the selection above, the discussion of the appearance of visible properties other than what they *really* are, begins with the case of shapes, and the circular appearance of far rectangular objects, a problem that is, by itself, notable for the extensive and diverse explanations it had received beginning with the Greek sources. The author of the passage in the corpus *Problemata Physica* attributed to Aristotle, treats the case of a "square (τετράγωνον) [literally, a figure with four angles] that appears (φαίνεται) to have sundry angles," but from a distance, "looks like a circle,"[10] an effect that is explained in terms of the "cut-off" shape of the angles owing to the uneven strength and distribution of rays within the base of the visual cone. There is also the proposition in Euclid's *Optics* that orthogonal magnitudes (ὀρθογώνια μεγέθη) appear (φαίνεται) circular from afar,[11] with a "geometrical" proof based on the disappearance of the object's angles beyond a certain distance (i.e., where visual rays lose contact with objects).

In the Arabic tradition, various treatments of the problem before al-Fārābī include different versions of the same Euclidean proposition, starting with its supposed translation that "sight (*baṣar*) by moving (*yantaqil*) from one point on the object's outline to another, skips some points in between (*mā bayn*)."[12] Among other curious formulations, some emphasizing the temporal aspects of appearance and disappearance by pointing to angles as the *first* part of a figure to disappear,[13] there is an apparently early treatment by Aḥmad ibn ʿĪsā[14] in his book on optics (Ḥ in Appendix), that is of interest on many levels. On one level, it is a "critique" of the Euclidean proposition on both physical and epistemological grounds, the assumption of the "leap" (*ṭafra*) of sight and the inability to perceive the figure's real shape (*al-shakl al-ḥaqīqī*) being called "the most amazing of amazements" (*min aʿjab al-ʿajab*);[15] on another level, geometrical demonstrations are offered for a variety of "angular" figures appearing as circles from a distance, demonstrations in which the "circle" itself is no longer conceived as a figure without angles, but rather, as a figure equidistant from its own center.[16]

The treatment is also of interest beyond what was or was not available to an apparently early optical author such as Aḥmad ibn ʿĪsā.[17] In a book, bearing the striking title of *Kitāb al-Tarbīʿ wa al-tadwīr (Book of Rectangularity and Circularity)*, al-Jāḥiẓ (d. 255/868–869), the famous ninth-century literary figure and mutakallim, makes a number of relevant and revealing remarks, including the statement that: "if we say a quadrilateral structure is seen from a distance as circular, then perhaps the sun is polygon-shaped (*muḍallaʿa*) [*muṣallaba* (cross-shaped) in the published edition)] and the stars are quadrilateral (*murabbaʿa*)."[18] The addressee, a certain Aḥmad ibn ʿAbd al-Wahhāb, whom al-Jāḥiẓ describes as "quadrilateral (*murabbaʿ*) in form, but estimated as round (*mudawwar*) due to his figure!," may be compared to Aḥmad ibn ʿĪsā beyond an overlap of names and apparent intellectual orientations.[19] If it is no incident that Aḥmad Ibn ʿĪsā devotes exceptionally long and thorough discussions in his *Optics* to the Euclidean proposition on the circular appearance of far rectangular objects, then, the problem of "the circle and the square" would take us much beyond questions of identity and onto the exciting milieu of the 3rd/9th century itself.[20]

But for early Arabic optics, the problem of the appearance of shapes other than what they actually are, was not limited to the case of the circle and the square, just as the problem of "appearance" itself was not limited to the determination of its truth correspondence. Al-Kindī (d. ca. 257/870), himself the subject of attack by none other than al-Jāḥiẓ, focused, for example, on the case of the rectilinear appearance of circular shapes as part of his arguments against intromission in the *De aspectibus* (A in Appendix),[21] and spoke, not so much of *veracity* of appearances, but rather, of *clarity* and *accuracy* of vision, using two distinct expressions: "clearer" (*abyan*) and "more accurate" (*aṣdaq*) vision (*ru'ya*), in his later *Taqwīm* (*Rectification*), extant in Arabic[22] (Q in Appendix).

"Clear" and "accurate" vision were, in fact, subjects treated extensively and variously by other authors on optics: in the period before al-Fārābī, by Ibn ʿĪsā (before 250/846?), Ḥunayn ibn Is-ḥāq (d. 264/877) and Qusṭā ibn Lūqā (d. 300/912) in addition to al-Kindī (d. ca. 257/870,), and in the centuries following the composition of *Iḥṣā' al-ʿulūm*, by Ibn al-Haytham (d. ca. 432/1040), Naṣīr al-Dīn al-Ṭūsī (d. 672/1274) and Kamāl al-Dīn al-Fārisī (d. ca. 718/1318). And yet, of the two aspects of appearances that were prominent in the early optical tradition, namely veracity of appearances (as presented through the passages of Ibn ʿĪsā and al-Jāḥiẓ), and clarity and accuracy of vision (as developed through the transmission of another Euclidean proposition), it is the latter that is left out of al-Fārābī's account altogether, while the former occupies a prominent place, not just in al-Fārābī's opening passage, but also in the one immediately following it.

II Justifiability and Variety of Demonstrations

> I.2: "*By means of this science discrimination (m-y-z) is made between what is seen (ẓ-h-r) as different from what it truly is and what is seen (ẓ-h-r) as it truly is; and the reasons (s-b-b) why all this should be so are established by certain (y-q-n) demonstrations.* And with regard to all that can be subject to visual error (*gh-l-ṭ*) this science explains various devices (*ḥ-y-l*) for avoiding error and apprehending what the seen thing truly is in respect to size, shape, position, order, and all that can be mistaken by sight (*b-ṣ-r*) . . ."

While the opening passage (I.1) is about the justification for having a science of optics as a separate branch from geometry (with which it shares its subjects: figures, magnitudes, order, position, equality and inequality), the second passage (I.2) is about the justifications offered by the science of optics itself: not only is this science one that discriminates (*yumayyiz*) between what literally "appears to sight" (*yaẓhar fī al-baṣar*) according to its real (*ḥaqīqī*) properties and what does not; it is the science by which one establishes the reasons (*asbāb*) behind appearances (whether these are consistent with or at odds with real properties), and why they are as such (*lima hiya*) by means of "certain demonstrations" (*barāhīn yaqīniyya*). The explanation of various "devices" (*ḥiyal*), this time for avoiding "errors (*ghalaṭ*) of sight (*baṣar*), with respect to size, shape, position, order, and all that may be subject to error" is, once again, expressed in terms of "coincidence with the truth" (*yuṣādif al-ḥaqīqa*).

In specifying demonstrations as distinct forms of explanation with reference to appearances, the present passage well reflects aspects of the field that were central to the early optical tradition. Demonstrations were indeed among the key features of the science of optics; but these were expressed in terms that were much more specific than al-Fārābī's "certain" (*yaqīnī*) demonstrations:[23] Demonstrations (*barāhīn*)—often meaning mathematical proofs to an early figure such as al-Fārābī—could be geometrical demonstrations (*barāhīn handasiyya = demonstrationes geometricae*), as referred to by al-Kindī in distinction from "philosophical demonstrations" (*barāhīn falsafiyya*); there were also such variations as "demonstrations by lines" (*barāhīn khuṭūṭiyya*), in the words of Qusṭā ibn Lūqā with specific reference to "the science of rays" (*ʿilm al-shuʿāʿāt*), as there was the genre, "illustration" (*mithāl*), itself used by someone like Ibn ʿĪsā, in the form of both "geometrical illustration" (*mithāl handasī*) and "sensible illustration" (*mithāl ḥissī*). The latter pair, corresponding to Ibn Lūqā's "illustration by lines" (*mithāl khuṭūṭī*) and "sensible explanation" (*bayān ḥissī*) respectively, were to act as justification by means of textual proofs (in the first case), and experience-based setups (in the second case). But what best reveals the orientation of a wide range of explanations in early Arabic optics, is neither in the Galenic language of "demonstration by lines" (as found in book X of *De usu partium*), nor in the Euclidean language of geometrical illustrations (as a formal division of a Euclidean proof), but rather, in the Aristotelian language of scientific reasoning (as set out in Aristotle's *Posterior Analytics*).[24]

With the Aristotelian distinction between knowledge of fact (τὸ ὅτι) and of the reason why (τὸ διότι) in a well-known passage in the *Posterior Analytics* (I.13: 78a34–35), and especially, the reservation of the privileged knowledge of the cause (or reasoned fact) for the mathematician, and the optician (with respect to inferior sciences), a method of providing "demonstrations" had found its way quickly into optical texts. The terminology of *ʿilla* after the Aristotelian expression for "knowledge by reasoning = τὸ διότι" (*ʿilm bi al-ʿilla*),[25] was accordingly more common in optics than the "demonstration" (*burhān*) of the Arabic title of *Posterior Analytics* (*Kitāb al-Burhān*), the text on which many authors on optics wrote commentaries, including al-Fārābī.[26] So central was the process of *reasoning* to optics from early on, that the term *ʿilla* was even included in the full Arabic titles of a few early optical texts.[27]

Discussions on demonstrative sciences in general, and optics in particular, were in fact transmitted through two Aristotelian works: The *Posterior Analytics* and *Physics* were both widely circulated in Islamic lands, through figures such as al-Kindī, Ibn Lūqā, and among al-Fārābī's (d. 339/950) own contemporaries, Abu 'l-Ḥasan al-ʿĀmirī (d. 382/992), who all wrote on optics and on methodology in different forms and capacities: [28] Al-Kindī, characterized optics as a "science in which geometrical demonstrations proceed in accordance with the requirements of physical things," and supplemented geometrical demonstrations throughout his text with others from the world of experience.[29] Qusṭā ibn Lūqā, on the other hand, spoke explicitly about the "cooperation (*ishtirāk*) between natural philosophy, from which we acquire sense perception (*idrāk ḥissī*), and geometry and its demonstrations through lines (*barāhīn khuṭūṭiyya*)," and about the incomparable "excellence of this coming together in the science of rays" (*ʿilm al-shuʿāʿāt*). Ibn ʿĪsā similarly used, in addition to Euclidean illustration (*mithāl handasī*), "sensible illustration" (*mithāl ḥissī*), with reference to devices like tubes (*unbūb*).[30] In this way, aspects of appearances that became subject to demonstrations often involved repeated demonstrations, typically supporting mathematically-based arguments with experience-based set ups. The subjects of demonstrations themselves were focused more frequently on problems such as clarity or accuracy of vision, than the reality or veracity of appearances.

The dominance of the subject of visual accuracy in the Arabic optical tradition may be measured by the abundance of demonstrations that involve quantified treatments of visual clarity in terms of the *amount* of radiation involved. Al-Kindī, who treats the subject in more than one place, demonstrates geometrically that what determines the effectiveness of the central region of the visual field is the falling of the greatest amount of radiation on the central region.[31] Other cases include demonstrations by Ibn Lūqā, where the amount of radiation falling on a region determines *how many* objects are seen, rather than how *clear* they are seen.[32] In yet another variation by Ibn ʿĪsā, alternative demonstrations for the privileged position of the visual axis for "more accurate vision" (*aṣdaqu ruʾyatan*) is based on there being "more of the ray" falling (*mā waqaʿa min al-shuʿāʿ akthar*).[33] All these demonstrations (including those involving repetitions) are somehow related to the problematic transmission of Euclidean optics,[34] which in this case amounts to a missing causal premise in the proposition demonstrating that "that on which more of the ray falls is seen more accurately" (*mā waqaʿa ʿalayhi al-shuʿāʿu akthara fa ruʾyatuhu aṣdaqu*)."

III Versatility and Fallibility of Applications

> I.3: *"By means of this art (ṣināʾa) too, one can determine the size of distant and inaccessible bodies, the magnitudes of their distances from us and their distances from one another. Examples are: the heights of tall trees and walls and the widths of valleys and rivers; even the heights of mountains and the depths of valleys, provided that sight (b-ṣ-r) can reach (w-q-ʿ) their limits; the distances of clouds and other objects from our location and above any place on the earth* . . . In general, every visible magnitude of which the size or distance from something else we seek to know, [can be determined] sometimes by means of instruments which are made for guiding the passage of sight (*b-ṣ-r*) so that it may not err, and sometimes without such instruments."

The passage on the *applications* of optics as an art (*ṣināʿa*) rather than its *explanations* as a science (*ʿilm*) (I.3) points to an important, and often overlooked, aspect of the discipline: The *determination* of the size and distance of objects was, just as the *explanation* of their appearance, an integral part of optics. In including among the applications of the *manāẓir* tradition (i.e., optics), the *misāḥa* problems, based on a set of four propositions in Euclid's *Optics* (i.e., the "surveying" problems), al-Fārābī acknowledges an aspect of the discipline that was far from marginal to the practices of early Arabic optics. But the absence of the one Euclidean proposition in that set where the determination of a magnitude involves the use of a plane mirror, gives al-Fārābī no occasion to move from the discussion of instruments and errors to the more problematic aspects of each case, the mirror and the principle of reflection respectively.

The omission is all the more surprising in the light of the thoroughness of al-Fārābī's account regarding the many applications of "the art" (*ṣināʿa*). The inclusion of applications such as the determination of unknown heights, depths and lengths from known values through measurements and calculations is indeed close to the common practices of the discipline, both before and after al-Fārābī, as is the extension of the objects and distances involved to include *natural* objects and *far* distances. A good example of an early work that combined the methods of magnitude determination with principles in optics is a short treatise by Sinān ibn al-Fatḥ (ca. 4th/10th cent.) with the self-explanatory title of *al-Misāḥāt al-manāẓiriyya* (Ẓ in Appendix).[35] Another text that extended its methods of magnitude determination to the case of natural objects and far distances, is a treatise by al-Kindī, entitled *On Clarification of Finding Distances between an Observer and the Centers of Mountain Heights* (*Risāla fī Īḍāḥ wijdān abʿād mā bayn al-nāẓir wa marākiz aʿmadat al-jibāl* (W in Appendix).[36] Comparable works of a period slightly later than al-Fārābī include two short tracts bearing the name of Ibn al-Haytham (d. ca. 432/1040), an author much better known for his influential *Optics* (*Kitāb al-Manāẓir*):[37] one is *On the Determination of the Height of Upright Bodies, Mountains and Clouds* (*Maqāla fī Maʿrifat irtifāʿ al-ashkhāṣ al-qāʾima wa aʿmidat al-jibāl wa irtifāʿ al-ghuyūm*), and the other, *On the Extraction of the Elevation of Mountains* (*Qawl fī Istikhrāj aʿmidat al-jibāl*).[38]

But the *miṣāha* tradition had close associations, not just with the *manāẓir* tradition, most directly through the visual and solar rays involved in magnitude determinations; it also had direct links with the *marāyā* tradition, as it involved indirect forms of radiation, for example, through the Euclidean proposition involving a plane mirror for height determination. The close association is clear from a title such as *Mirʾātiyya* (Related to Mirrors), an alternative title for a treatise by Badr al-Dīn al-Ṭabarī (ca. 824/1421) called *Height* (*Irtifāʿ*),[39] in this case involving the determination of the height of lower objects, once by means of a "rod" (*ʿamūd*), and once by a plane mirror. The late author of the *Mirʾātiyya* reveals a useful piece of information about the transmission of the methods involved, when he says that "regarding the determination of the height of tall objects by means of a mirror, what the people of the art (*ahl-i-ṣināʿat*) may have offered on the subject, has not reached him or has not been seen by him anywhere." The curious omission of this aspect of the *misāḥa* tradition from al-Fārābī's extensive account of the "art," may be related to the reportedly poor transmission of surveying techniques using mirrors; but this is something that must also be viewed in the light of other problematic cases in transmission.

The transmission of the Euclidean propositions on height determination by means of reflecting visual rays is a case that had already taken a misdirected course as a result of the Arabic terminology used for the principle of reflection from a plane mirror. In the Arabic version of that proposition, the term reflection was *not* translated into Arabic in the later standard form *inʿikās*, but rather, as *inʿiṭāf*, a form also employed by al-Fārābī with reference to the deflection of rays at a polished surface such as mirrors, just as in both the early and late Arabic versions of Euclid's *Optics* (3rd/9th and 7th/13th centuries respectively).[40] The terminological twist is particularly surprising, because the more standard term for reflection (*inʿikās*) was used as early as in Ḥunayn ibn Is-ḥāq's *Ten Treatises on the Eye*, in the Arabic versions of the Pseudo-Euclidean *De speculis* and Aetius' *Placita philosophorum*,[41] as well as in late works by Ṭūsī and his commentator, Ṭabarī.[42] The exact source of al-Fārābī's combined formulation where "deflected" (*munʿaṭifa*) (i.e., reflected) rays become "reversed" (*munʿakisa*) (i.e., on their return paths from a mirror), is not quite clear. Neither does the extensive literature on the subject contain frequent references to "bent" (*munkasira*) rays or to "seeing behind oneself," as we find in al-Fārābī's next passage.

IV Elements and Mechanisms of Vision

I.4: *"Now all that can be looked at (n-ẓ-r) and seen (r-ʾ-y) is seen (r-ʾ-y) by means of a ray (sh-ʿ-ʿ) that penetrates (n-f-dh) the air or any transparent body in contact (m-s-s) with our eyes (b-ṣ-r) until it reaches (w-q-ʿ) the object seen (n-ẓ-r). And rays that pass through transparent bodies to a visible object are either straight (mustaqīma) or deflected (munʿaṭifa) or reversed (munʿakiṣa) or bent (munkasira).* Straight rays are those that, having issued (*kh-r-j*) from the eye, extend rectilinearly on the line (*s-m-t*) of sight (*b-ṣ-r*) until they weaken and come to an end. . . . Deflected rays are those that, having passed out of the eye, meet on their way, and before they weaken, a mirror that precludes them from passing through in a straight line, thereby causing them to be deflected (*ʿ-ṭ-f*) and turned (*ḥ-r-f*) to one side of the mirror. They then extend in the direction into which they have turned towards the beholder (*n-ẓ-r*). Reversed rays are those that return from the mirror on the path they traversed at first, until they fall (*y-q-ʿ*) on the body of the beholder (*n-ẓ-r*) from whose eyes they have issued (*kh-r-j*), and it is by means of this [kind of] ray that the beholder (*n-ẓ-r*) sees (*r-ʾ-y*) himself. Bent rays are those that return from the mirror towards the beholder (*n-ẓ-r*) from whose eyes (*b-ṣ-r*) they have issued (*kh-r-j*), but extend obliquely beside him until they fall (*y-q-ʿ*) on something else behind the beholder or on his right or his left or above him, and it is thus that we see (*r-ʾ-y*) what lies behind or beside us."

Al-Fārābī's detailed account of the elements and mechanisms of vision points to complex problems in transmission: the elements of vision, being in this case, the four cases of visual radiation, and the mechanisms, their respective features and functions. By containing explanations that clearly represent a *combination* of the various formulations of visual radiation and their distinct features and functions, the account well reflects the problematic nature of the theoretical aspects of the discipline. But by excluding some common formulations, while including uncommon ones, the detailed account is still a faint reflection of the range and complexity of the problems characterizing the early Arabic optical tradition.

Al-Fārābī's account may be considered primarily Euclidean, though it is neither identical with any of the two most common formulations of the Euclidean visual-theory, nor comparable to any one particular variation to reveal the exact source of its own "mixed" formulations. Transmitted through a textual tradition, noted for having been not just problematic, but physically defective in both Greek and Arabic,[43] the Euclidean tradition itself has a complex history, beginning with the statement of the Euclidean visual-ray hypothesis. Of the two most distinct formulations of that statement passing under the name of Euclid,[44] none are fully represented in al-Fārābī's version: "a ray (*shuʿāʿ*) passing through (*yanfudh*) a transparent body in contact (*yumāss*) with our eyes (*baṣāʾir*) until it falls (*waqaʿa*) on the viewed object (*manẓūr ilayh*)," does *not* correspond exactly, with the elements and mechanisms of vision as described in the opening lines of the Arabic versions of Euclid's *Optics* (*Kitāb al-Manāẓir li-Uqlīdis*, M in Appendix), and largely followed in the late Arabic versions of the same text (i.e., *Ṭūsī's Taḥrīr al-Manāẓir* and Ibn Abī Jarāda's *Tajrīd al-Manāẓir*); neither does it match the version found in the pseudo-Euclidean *De speculis* (*Kitāb al-Manāẓir li-Uqlīdis*, S in Appendix), the latter corresponding almost word by word, and only in this part, to the formulations of al-Kindī's *Taqwīm*, and Ibn ʿĪsā's *al-Manāẓir wa al-marāyā al-muḥriqa* (Ḥ and Q in Appendix, respectively).[45]

A critical part for the discussion of appearances, including al-Fārābī's distinct formulations, is the many forms of the third Euclidean assumption in the *Optics* that only those things upon which rays fall are seen (ὁρᾶται). It is not insignificant that the corresponding Arabic verb for "seeing" is, in the *Optics* (M), *yubṣar*, in the *De speculis* (S) *yudrak* and in al-Fārābī's account, *yunẓar* and *yurā*: "All that can be looked at (*yunẓar ilayh*) and seen (*yurā*)," states al-Fārābī, "is seen (*yurā*) by means of a ray (*shuʿāʿ*) that . . . falls (*waqaʿa*) on the viewed object (*manẓūr ilayh*)." It is clearly the case that al-Fārābī's formulation is free from the occurrence of the terminology of perception (*idrāk*), a conception that appears in similar terms in the statements of predecessors like al-Kindī, Ibn ʿĪsā, and Ibn Lūqā (as part of their reformulation of the third Euclidean definition), as well as the subsequent elaborations of Avicenna and Ibn al-Haytham.[46] In the case of direct vision, therefore, al-Fārābī's account of "appearances" remains partial by virtue of being strictly Euclidean. This is a tradition in which various expressions occur for both the passive and active modes that are involved in vision [what is viewed (*yunẓar*), looked at/seen (*yubṣar*), or appears (*yaẓhar*); and what is seen (*yurā*), thought to be seen (*yuẓannu annahu yurā*), or imagined to be seen (*qad yutawahhamu an yubṣar*)], with the similarly striking exception of the term for perception (*idrāk*).

The Euclidean character of al-Fārābī's account of indirect vision represents restrictions in other theoretical directions. This time, what is striking is the absence of any reference to vision through refraction.[47] In the part of the passage on appearances through mediums (i.e., indirect vision) rather than those through air (i.e., direct vision), everything—including the term "*inʿiṭāf*" (later, standard for refraction)—is a reference to reflection: visual radiation passes out of the eye (*nāfidh min al-baṣar*) . . . meets a mirror (*al-mirʾāh*), is deflected (*tanʿaṭif*) and turned (*inḥarafat*) to one side of the mirror; and then, the reversed (*munʿakisa*) rays return (*tarjiʿ*) from the mirror on the path they first traversed. Clearly, al-Fārābī's account, where *al-shuʿāʿāt al-munʿaṭifa* stand for deflected rays (as in the Arabic Euclidean tradition), next to non-Euclidean cases such as reversed (*munʿakisa*) or bent (*munkasira*) rays, lacks mention of refracted rays in its most standard form. And when we read in the closing passage that *ʿilm al-marāyā* is the division within *ʿilm al-manāẓir* that investigates what is visible through indirect (*ghayr al-mustaqīma*) rays, this does not include refracted rays as the reference to transparent mediums such as water or glass would have us believe.

V The Modes and Mediums of Operation

> I.5: "The medium that lies between the eye (*b-ṣ-r*) and what is looked at (*n-ẓ-r*) is, in general, a transparent body, whether air, or water, or celestial body or an earthly composite body such as glass and the like. And mirrors, which send back the rays and prevent them from rectilinearly passing through, are either those made by us of iron or the like, or they consist of a thick moist vapour, or water, or some other body similar to these. *The science of optics, then, inquires into all that is looked at (n-ẓ-r) and seen (r-ʾ-y) by means of these four rays* {straight (*mustaqīma*), deflected (*munʿaṭifa*), reversed (*munʿakisa*), and bent (*munkasira*)} *and into every kind of mirror and all that pertains to the object of vision (n-ẓ-r). It is divided into two parts, the first of which investigates what is visible (n-ẓ-r) through rectilinear rays, and the second is visible through non-rectilinear rays, and this [latter] is specially called the science of mirrors (ʿilm al-marāyā).*"

The closing passage (I.5) concludes with optics' domains of inquiry (*faḥṣ*), as distinct from the subjects of investigation of the opening lines (I.1) that were meant to make optics itself distinct from the "more general" field of geometry. In extending the discussion of indirect or mediated appearances (introduced in the previous passage), to cases involving reflecting surfaces other than mirrors (thick moist, vapor, or water) on the one hand, and mediation of transparent bodies other than air ("water, celestial sphere and earthly composite bodies like glass and the like") on the other, the final passage provides a faithful account of mediated appearances, insofar as it distinguishes between mediation through opaque bodies (such as mirrors) and transparent bodies (water or glass). But insofar as the passage includes one type of medited appearance (through reflection) to the exclusion of the other (through refraction), the account does not fully represent the stage reached by optical writings of the period before or during al-Fārābī's compositions.

The two-fold and hierarchical division of the discipline, the first (*ʿilm al-manāẓir*) investigating what is viewed (*yunẓar ilayh*) through direct (*mustaqīma*) rays, and the second, through indirect (*ghayr mustaqīm*) rays, must itself be understood in terms of the mediums—and not just modes—of propagation. On the one hand, *mustaqīma* and *ghayr mustaqīma* represent *direct* (= unmediated) and *indirect* (= mediated), rather than *rectilinear* and *non-rectilinear* (i.e., in terms of a *medium* of propagation other than air, rather than the rays being in the *mode* of rectilinearity), simply because the rays involved in appearances through a mirror, for example, are *both* rectilinear (= not bent or curved) and indirect (= changing course), this being true of both deflected (*munʿaṭifa*) and reversed (*munʿakisa*) rays. On the other hand, the part, *marāyā* in *ʿilm al-marāyā*, itself commonly translated as "the science of mirrors," can neither be reduced to mirrors, nor understood to exclude surfaces now commonly considered refractive mediums, especially since al-Fārābī's *marāyā* stands for "deflecting" surfaces such as vapor or water thick enough to produce such an effect.

The limited knowledge of optical refraction in general and of the treatment of enlarged objects in water in particular has long been noted by A. I. Sabra with reference to specific works by al-Kindī, Aḥmad ibn ʿĪsā, even early Ibn al-Haytham (d. ca. 432/1040), for their lack of understanding of the phenomenon of refraction as found already in Ptolemy's *Optics*.[48] Ironically enough, the so-called "pre-Ptolemaic" stage represented by the astronomer's explanations of the enlarged appearance of bodies through mediums in terms of visual angles, is, in some sense, more advanced than the stage represented by the optical tradition itself, a tradition where the appearance of the principle of reflection in a single proposition of Euclidean optics came and circulated with the vocabulary of *inʿiṭāf* to "confuse" the phenomena of reflection and refraction in the works of Naṣīr al-Dīn al-Ṭūsī (d. 672/1274) and Quṭb al-Dīn al-Shīrāzī (d. 711/1311), and all the way up to Kamāl al-Dīn al-Fārisī (d. 718/1318) who noted the puzzles involved.[49]

The nonstandard terminology of "*inʿiṭāf*" as "reflection" (or reflection) has already been mentioned with reference to the part of al-Fārābī's previous passage (I.4), where deflected (*munʿaṭifa*) rays appear alongside the standard form *munʿakisa*, itself used to mean *reversed*, rather than reflected rays (i.e., the *returned* rays on the same path of incidence). That the applications of the two principles remained inconsistent is immediately clear from the relevant writings of a late author such as Ṭūsī: in his short treatise, *Inʿikās al-shuʿāʿāt wa inʿiṭāfuhā*, where the two terms are used jointly (and in some manuscript transcriptions, also interchangeably), as well as in the Persian work, *Shuʿāʿ*, the word "*inʿikās*" is used for reflection, while in Ṭūsī's *Taḥrīr al-Manāẓir* (Recension of Euclid's *Optics*), the form *inʿiṭāf* (by then, standard for refraction) is still the term used for reflection from a mirror; this is all the more curious, because *inʿiṭāf* appears in the first of these works as a form of refraction a cone of rays undergoes at the surface of transparent bodies like still water, so that "*zāwiyat al-inʿiṭāf*" is no longer Ṭūsī's "angle of deflection" at the surface of polished bodies such as mirrors (as in the *Taḥrīr*), but rather, the angle that the refracted end of the visual cone makes with the refracting surface, and in such a way that it is still equal with the angle of incidence (*zāwiyat al-shuʿāʿ*).[50]

With the treatment of refraction, in particular, the confusion between the principles of reflection and of refraction, and especially their nonstandard, inconsistent, and orthographically comparable terminology, are factors to be considered, *in addition* to the apparently poor transmission and circulation of Ptolemy's relevant treatments. The important statement of A. I. Sabra about all the historical evidence pointing to the limited use of Ptolemy's *Optics* in both Antiquity and the Islamic Middle Ages,[51] must then be combined with no less qualified statements that would also take into account, not just what was transmitted, but also how whatever did get transmitted *was* transmitted. The difficulty is that historical evidence may successfully reveal the first (the *what* of transmission), but not all historical evidence would reveal the second (its *how*). In the case of early Arabic optics, we are fortunate to have a good number of texts, including al-Fārābī's passage, that may still act as *historical* documents, to determine what sources or concepts, were transmitted up to about the year AD 950. But it takes a close examination of the available sources from the perspective of the transmitted terms and expressions, *in addition* to the sources and concepts, and these through extant manuscripts in addition to published editions, to determine the exact nature and manner of the effect that all of these have had on the state of Arabic optics during a critically important stage in its development.

To conclude with remarks that take into account the entire passage and overall plan of the optics section in al-Fārābī's *Catalogue*, it should be remembered how optics is presented throughout that text: as an established scientific discipline (*ʿilm*) within the mathematical sciences (*taʿālīm*), supplied not just with "reasons" (*asbāb*) but "certain" (*yaqīnī*) demonstrations, and not just with "explanations" (*maʿrifa*) but also "devices" (*ḥiyal*), a discipline in search of "conformity with reality" (*yuṣādif al-ḥaqīqa*), one that at once demystifies and justifies "what appears to sight" (*mā yaẓhar fī al-baṣar*) both within and beyond the ordinary realms of vision. All this leaves little doubt about how al-Fārābī conceived of, or at least presented, the "program" of the early optical tradition.

But how close is such a "program" to the character of the early optical tradition itself? There is no question that al-Fārābī's account is extremely valuable as a historical document, especially in the light of the rarity of such general accounts on the early history of any scientific discipline; but with the focus of the text on disciplinary and pedagogic concerns, rather than historical, or even scientific ones, the coverage of the "science of aspects," remains inexhaustive, in regards to the many aspects of the field, in terms of both the orientations and associations of the discipline, and the forms and expressions of its concepts. Appearances are treated, in the opening passage, in terms of problems of veracity to the exclusion of the slightly different, and more common, themes of clarity and accuracy; in the second passage, in terms of demonstrations to the exclusion of their sense-perceptible dimensions; in the third passage, in terms of applications to the exclusion of their more problematic extensions; in the fourth passage, in terms of the elements and mechanisms of vision to the exclusion of their multiple variations; and finally in the closing passage, in terms of modes of investigation to the exclusion of all the modes and mediums of operation. The terminological aspects of appearances are also at once reflective and restrictive, as the concept of "appearance" itself emerges from the active involvement of an observer (*nāẓir*) viewing (*n-ẓ-r*) or seeing (*r-ʾ-y*), to the passive presence of an object appearing (*ẓ-h-r*)—all to the exclusion of other forms, including the form *al-manāẓir*, meaning appearances (and not just visual rays), as in the "science of aspects" (*ʿilm al-manāẓir*) itself.

Finally, the few extant early Arabic texts examined in the present study, themselves act as important historical documents in demonstrating that it is not only the case that al-Fārābī's coverage of "aspects" and "appearances," is not fully representative of the concepts and problems of the early Arabic texts often covering the very same items; it is also the case that such a coverage, leaves out, in effect obscures, the extremely complex character of an early tradition, to whose "unsettled" aspects, as well as rich dimensions, al-Fārābī's account itself is sufficient testimony.

Appendix

Primary Sources within the Early Arabic Optical Tradition

M = كتاب أقليدس فى اختلاف المناظر

Kitāb Uqlīdis fī Ikhtilāf al-manāẓir = Euclid's *Optics* (Arabic version) [Krause (1974); Sezgin (1974) = GAS V; Kheirandish (1991, 1996). ed. Kheirandish (1999): vol. 1: pp. 1–225; Rashed (1997)].

S = كتاب المراءى لأوقليدس

Kitāb al-Mir'āh li-Uqlīdis = [Pseudo-] Euclidean *De speculis* (Arabic version) [Sabra (1979); Kheirandish (1991, 1999); Rashed (1997); Latin text: Björnbo and Vogl (1900), pp. 97–173; Theisen (1972)].

Ḥ = كتاب المناظر والمرايا المحرقه
(تأليف أحمد بن عيسى على مذهب أقليدس فى علل البصر)

Aḥmad ibn ʿĪsā, *Kitāb al-Manāẓir wa al-marāyā al-muḥriqa ta'līf Aḥmad ibn ʿĪsā ʿalā madhhab Uqlīdis fī ʿilal al-baṣar* (Arabic text) [Krause (1936); Kheirandish (1991, 1996, 1999; Sabra (1989); Rashed (1997), includes edition of section on Burning Mirror, Sabra and Kheirandish, edition in preparation].

Q = كتاب أبى يوسف يعقوب بن إسحاق الكندى إلى بعض إخوانه
فى تقويم الخطأ والمشكلات التى لأوقليدس فى كتابه الموسوم بالمناظر

Al-Kindī, *Kitāb Abī Yūsuf Yaʿqūb ibn Isḥāq al-Kindī ilā baʿḍ ikhwānihi fī Taqwīm al-khaṭa' wa al-mushkilāt allatī li-Uqlīdis fī Kitābihi al-mawsūm bi al-Nāẓir* [*al-Manāẓir*] (Arabic text) [Marʿashī (v. 19); Rashed (1997), ed. pp. 162–335].

A = كتاب فى علل اختلاف المناظر والبراهين الهندسية عليها ؟

Al-Kindī, *De aspectibus* (= *Ikhtilāf al-manāẓir?*) (Arabic text extant in Latin) *Kitāb fī ʿIlal ikhtilāf al-manāẓir wa al-barāhīn al-handasiyya ʿalayhā? = De causis diversitatum aspectus et dandis demonstrationibus geometricis super eas*). [Björnbo-Vogl (1912); ed. Hugonnard-Roche, tr. Jolivet, Sinaceur: Rashed (1997)].

L = كتاب فى علل ما يعرض فى المرايا من اختلاف المناظر

Qusṭā ibn Lūqā, *Kitāb fī ʿIlal mā yaʿriḍu fī al-marāyā min ikhtilāf al-manāẓir allafahu . . . Qusṭā ibn Lūqā al-Yūnānī* (Arabic text) [Gulchīn Maʿānī (1350=1971), vol. 8; Toomer (1976); Kheirandish (1991, 1996, 1999); Rashed (1997), ed. pp. 572–646].

W = رسالة يعقوب بن اسحق الكندى فى ايضاح وجدان أبعاد ما بين الناظر ومراكز أعمدة الجبال وعلو أعمدتها وعلم عمق الآبار وعروض الأنهار وغيرذلك

Al-Kindī, *Fī Īḍāḥ wijdān abʿād mā bayn al-nāẓir wa marākiz aʿmidat al-jibāl wa ʿuluww aʿmidatihā . . . wa huwa yusammā Mūrīsṭus*. (Arabic text) [Ritter and Plessner (1932); Krause (1936); Brockelmann (1937) = GAL S I; Kheirandish (1991, 1999)].

Ẓ = المساحات المناظريه

Sinān ibn al-Fatḥ, *al-Misāḥāt al-manāẓiriyya* (Arabic text) [King (1986a) = *Cairo Catalogue*, v. 2, p. 1030; King (1986b) = *Cairo Survey*, p. 39; Sezgin, (1979), GAS VII; Kheirandish (1991, 1999)].

NOTES

1. Al-Fārābī, Abū Naṣr, *Iḥṣāʾ al-ʿulūm*, ed. ʿUthmān Amīn [= Osman Amine], Cairo: Librairie Anglo-Égyptienne, 1968 [earlier editions, Cairo, 1931, 1949; Arabic edition from Escorial manuscript by A. González Palencia, *Alfarabi Catálogo de las ciencias* (ACLS), includes two medieval Latin and a modern Spanish translation (Madrid, 1932)].

2. The English translation of the full passage is quoted from A. I. Sabra's *The Optics of Ibn al-Haytham: Books I–III On Direct Vision, Translated with Introduction and Commentary*, 2 vols., Warburg Institute, 1989, vol. 2, pp. lvi–lvii. The translation is described by Sabra as "made from a composite text constructed from two editions of the Arabic text and the Latin version . . . in the absence of a single satisfactory edition" (emphasis and verbal root indications are added to passages quoted in this chapter by the present author). An earlier English translation of a large part of the same passage based on the Arabic edition in ACLS (and using the Cairo edition) is published as "The Science of Aspects" in Marshall Clagett, "Some General Aspects of Physics in the Middle Ages," *Isis*, 1948, 39: 29–44, pp. 32–35.

3. The section "Aim and Scope of The *Optics*" in Sabra's *The Optics of Ibn al-Haytham*, vol. 2, pp. liii–lxiii, contains discussion of this and other key passages.

4. Kheirandish, Elaheh, *The Arabic Version of Euclid's Optics: Kitāb Uqlīdis fī Ikhtilāf al-manāẓir*, Edited and Translated with Historical Introduction and Commentary, 2 volumes, Springer-Verlag: *Sources in the History of Mathematics and Physical Sciences*, no. 16, 1999; also, "The Arabic 'Version' of Euclidean Optics: Transformations as Linguistic Problems in Transmission," *Tradition, Transmission, Transformation: Proceedings of Two Conferences on Pre-modern Science Held at the University of Oklahoma*, ed. F. Jamil Ragep and Sally P. Ragep with Steven Livesey, Leiden: Brill, 1996: 227–243.

5. Kheirandish, Elaheh, *The Arabic Version of Euclid's Optics*, 2 volumes, Springer-Verlag, 1999 (see above); also, "The 'Manāẓir' Tradition through Persian Sources," *Les*

sciences dans le monde iranien, ed. Ž. Vesel, H. Beikbaghban et B. Thierry de Crussol des Epesse, Tehran: Institut Français de Recherche en Iran (IFRI), 1998: pp. 125–145.

6. Some aspects of appearances are discussed with reference to the Greek and Latin traditions: appearance versus visual perception, by C. D. Brownson in "Euclid's *Optics* and Its Compatibility with Linear Perspective;" vision versus reality, by Vasco Ronchi in "Classical Optics is a Mathematical Science;" image reception versus perception, by Richard Tobin in "Ancient Perspective and Euclid's Optic;" objective and subjective elements in vision, by Gérard Simon in "The Notion of the Visual-Ray," by Wilfred R. Theisen in *The Mediaeval Tradition of Euclid's Optics*, and by Kim Veltman in *Optics and Perspective: A Study in the Problems of Size and Distance*, see Bibliography.

7. On the general and specific senses of the plural *manāẓir* as well as the singular forms *manẓar* and *manẓara* with examples from relevant literature; see Sabra, "Manāẓir, or ʿIlm al-manāẓir," *EI*[2] 6, p. 376 and Sabra, "Ibn al-Haytham," *DSB*, 4, p. 203, n. 9; see also, Kheirandish, *The Arabic Version of Euclid's Optics*, vol. 2, Index of Arabic Terms.

8. See ed. Amīn, pp. 79–83, tr., Sabra, *The Optics of Ibn al-Haytham*, v. 2, pp. lvi–lvii and Kheirandish, *The Arabic Version of Euclid's Optic*: Index of Arabic Terms.

9. The exclusion is noted by Sabra, *The Optics of Ibn al-Haytham*, vol. 2, p. lviii.

10. The full passage in the facing translation of the Loeb edition is as follows: "Similar to this is the phenomenon that a square appears to have sundry angles, but if we stand farther off it looks like a circle. For as the fall of the rays is in the form of a cone, when the figure is removed to a distance, those rays that are at the angles are cut off and do not see anything because they are weak and few, when the distance grows greater, but those that fall on the center persist because they are collected together and strong. When the figure is near they can see also the parts at the angles, but when the distance becomes greater they cannot," *Problems*, bk. XV, 911b19–21, Loeb edn., p. 335. The Arabic version is now available in the edition of L. S. Filius: *The Problemata Physica Attributed to Aristotle, The Arabic Version of Ḥunain ibn Isḥāq and the Hebrew Version of Moses ibn Tibbon*, Aristoteles Semitico-Latinus, Leiden: Brill, 1999, pp. 658–659.

11. Heiberg, J. L., *Euclidis Optica*, *Euclidis Opera Omnia*, ediderunt J. L. Heiberg et H. Menge, vol. VII, Leipzig: Teubner, 1895, pp. 16 and 166.

12. Kheirandish, *The Arabic Version of Euclid's Optics*, vol. 1, pp. 30–34, vol. 2, pp. 44–48.

13. There is a statement in the margin of one variant of the early Arabic version, where a circle is also defined as "a figure for which there are no angles," as well as in Ṭūsī's recension of that proposition, that act as alternative causal premises that this is "because (*li-anna*) the smallest parts of an object (namely angles) are the first to disappear (*yaghīb*) from sight (*ʿan al-baṣar*) at a far distance. Ibn al-Haytham's explanation of "why a polygonal (*muḍallaʿ*) figure is perceived to be circular (*mustadīr*)," treats the problem with the terminology of the concealment (*khafāʾ*) of the angles owing to their relative smallness (*ṣighar*) at a distance (*buʿd*): *Kitāb al-Manāẓir*, bk. III, sec. 9, ed. Sabra, p. 416; tr. Sabra, vol. 1, p. 281. The Arabic version of Ptolemy's *Optics*, which contains such a problem, has not reached us for specific or linguistic comparisons; for

the Latin version based on Arabic, see Lejeune, *L'Optique de Claude Ptolémée*, 1956, 1989.

14. Max Krause, who first reported the two Arabic manuscripts of Ibn ʿĪṣā's *Optics and Burning Mirrors* (*Kitāb al-Manāẓir wa al-marāyā al-muḥriqa*) [Ḥ in Appendix], dates it as "before 250H" (= 864 A.D.) without further specification: "Stambuler Handschriften islamischer Mathematiker," pp. 513–514; A. I. Sabra, elaborates on Krause's "conjecture" by noting the "peculiar vocabulary" of the text and its lack of any mention of Arabic authors alongside Greek authors: *The Optics of Ibn al-Haytham*, vol. 2, p. xxxvii, and n. 39; Roshdi Rashed, who includes part of the text in a recent publication (*Œuvres Philosophiques et Scientifiques d'Al-Kindī*, vol. I: *L'Optique et la Catoptrique in Islamic Philosophy Theology and Science*, Texts and Studies edited by H. Daiber and D. Pingree, vol. xxix), insists that the text is a "relatively late" compilation preserving some works by al-Kindī, and that the "discretion" with regard to naming Arabic authors is "deliberate," *Œuvres Philosophiques et Scientifiques d'Al-Kindī*, vol. 1, pp. 57–60, see also note 17 below.

15. The full passage is as follows: "Euclid said in his book *Ikhtilāf al-manāẓir* that figures having angles, like a quadrilateral (*murabbaʿ*), are seen from a certain distance as circular (*mustadīr*), so if they become distant from the eye they are seen (*yurā*) as round (*mudawwar*)". . . this is "the most amazing of amazements" (*aʿjab al-ʿajab*) because if it is in the nature (*ṭabʿ*) of sight (*baṣar*) to make a leap (*ṭafra*) and see a quadrilateral object from a distance as round, then as a result of that leap the object's real (*ḥaqīqī*) shape is not seen . . . : Ḥ [see Appendix]. The reasoning (*ʿilla*) offered follows in a much clearer text (note that the term *ṭafra* is not in the Arabic proposition of Euclid).

16. The geometrical demonstration shows that if from the center of the quadrilateral figure a line is drawn perpendicularly such that there is a point from which the excess (*faḍl*) of lines connecting the figure's center to its far corner, and to the middle of its sides is *not* a sensible magnitude (*qadr maḥsūs*), then from that point the figure is seen as circular. In contrast to other explanations of this visual effect based on the characterization of a circle as a figure with no angles, this proof is based on the conception of a circle as a figure having all its points equidistant from a center. The definition of circle in Euclid's *Elements* is "a plane figure contained by one line such that all the straight lines falling upon it from one point among those lying within the figure are equal to one another": see Heath, *The Thirteen Books of Euclid's Elements*, book 1, vol. 1, p. 153.

17. At the present state of research, it is difficult to determine the exact dates of Aḥmad Ibn ʿĪsā, a name with no few occurrences in historical records (12 in the list of al-Ṣafadī's alone, see *Kitāb al-Wāfī bi al-wafayāt*, v. 7, ed. Iḥsān ʿAbbās, Wiesbaden, 1389 = 1969, pp. 271–275). There is no concrete evidence for considering Ibn ʿĪsā's *Optics* as "pre 250/864" with Krause, though the content certainly points to an early date of composition; nor is there conclusive evidence for considering the text as a post al-Kindī "compilation" with Rashed, since with corresponding passages in particular, the chronological arrow may go either way; see note 14 above.

18. *Muḍallaʿa* (مضلّعه) makes more sense than *muṣallaba* (مصلّبه = cross-shaped) in Pellat's edition (treated as such also in Adad's French translation), see *Kitāb al-Tarbīʿ wa al-tadwīr*, ed. Pellat, p. 91 (tr. Adad, p. 308).

19. On a character whose name, dates and intellectual orientations are comparable to both the author of the optical text and the target of al-Jāḥiẓ's text, see the article under Aḥmad ibn ʿĪsā (d. 247/861) by Madelung in the *Encyclopaedia of Islam: EI², Suppl.*, pp. 48–49. The article is about a *shīʿī* scholar and leader associated with the early Abbasid court.

20. For a flavor of the early periods of intellectual activity and the rivalries involved, see Gerhard Endress, "The Circle of Al-Kindī: Early Arabic Translations from the Greek and the Rise of Islamic Philosophy," *The Ancient Tradition in Christian and Islamic Hellenism* (Gerhard Endress and Remke Kruk, eds.), Leiden: Research School CNWS, 1997.

21. *Alkindi, Tideus und Pseudo-Euklid: Drei optische Werke*, herausgegeben und erklärt von A. Björnbo und Sebastian Vogl, *Abhandlungen zur Geschichte der mathematischen Wissenschaften*, Leipzig/Berlin, 1912, 26, 3: 1–41; In Lindberg, *Theories of Vision*, reference is made to al-Kindī's use of this proposition for his refutation of the intromission theory: "If sight occurred through intromission of the forms of visible things, he [i.e., al-Kindī] argues, a circle situated edgewise before the eye would impress its form in the eye and consequently would be perceived in its full circularity;" see p. 23; for the context of discussion and the relevant references, see pp. 22–24, pp. 223–224, n. 23–27.

22. *Taqwīm al-khaṭaʾ wa al-mushkilāt allatī li-Uqlīdis fī Kitābihi al-mawsūm bi al-Nāẓir* [*al-Manāẓir*], Qum MS.: Marʾashi-yi Najafī 7580, 69b-102b, 960H (unique?), ed. Rashed, *Œuvres Philosophiques et Scientifique d'Al-Kindī*, vol. 1, pp. 162–335.

23. On the methodological aspects of optics, see Kheirandish, "The Mixed Mathematical Sciences of the Islamic Middle Ages," *The Cambridge History of Science*, 8 vols. ed. David C. Lindberg and Ronald Numbers; vol. 2: *The Middle Ages*, forthcoming.

24. See respectively: Galen, *De usu Partium, English translation: On the Usefulness of the Parts of the Body: Translated from the Greek with an Introduction and Commentary*, by Margaret T.May, 2 vols., Ithaca: Cornell University Press, 1968–1969, Arabic translation: *Kitāb al-Manāfiʿ al-aʿḍāʾ, Bibliothèque Nationale*: MS ar. 2583; Euclid's *Elements of Geometry*, English translation: *The Thirteen Books of Euclid's Elements*, second edition (revised with additions) Thomas, L., Heath, 3 vols, Cambridge: Cambridge University Press, 1926 (Dover, 1956), especially, "The Formal Divisions of a Proposition," vol. 1, pp.117–131; Aristotle, *Posterior Analytics*, edited by G. P. Goold, with an English translation by Hugh Tredennick (Cambridge: Harvard University Press, London: William Heinemann Ltd., 1976); Arabic version, *Kitāb al-Burhān min Manṭiq Arisṭū*, ed. A. Badawi (Cairo: Dār al-Kutub al-Miṣriyya, 1949), Islamica VII, part 2, 309–465.

25. Aristotle, *Kitāb al-Burhān*, ed. A. Badawi, *Organon Aristotelis in version Arabica Antiqua*, Part 2, pp. 349–353; on the distinction, see also Ragep, *Naṣīr al-Dīn al-Ṭūsī's Memoir on Astronomy*, p. 386, and Crombie, *Robert Grosseteste and the Origins of Experimental Science*, pp. 25–26, and pp. 53–54.

26. Al-Fārābī, *Kitāb al-Burhān*, ed. M. Fakhry, Beirut: Dār al-Mashriq, 1986. For the commentaries, see Peters, F. E., *Aristoteles Arabus: The Oriental Translations and Commentaries on the Aristotelian Corpus* (Leiden: E. J. Brill, 1968), pp. 17–19.

27. *Kitāb fīhi al-Manāẓir wa al-marāyā al-muḥriqa ta'līf Aḥmad ibn ʿĪsā ʿala madhhab Uqlīdis fī ʿilal al-baṣar; Kitāb fī ʿIlal mā yaʿriḍu fī al-marāyā min ikhtilāf al-manāẓir 'alifahu* (Qusṭā ibn Lūqā al-Yūnānī), Arabic texts and French translations in Roshdi Rashed, *Œuvres Philosophiques et Scientifiques d'Al-Kindī*, vol. 1, p. 649 (includes only the section on Burning Mirrors), and pp. 572–646 respectively (Ḥ and L in Appendix); and *Kitāb fī ʿIlal ikhtilāf al-manāẓir maʿa al-barāhīn al-handasiyya lahā* (a likely form for the original title of al-Kindī *De causis diuersitatum aspectus et dandis demonstrationibus geometricis super eas*, better known as *De aspectibus*), ed. Björnbo, in A. Björnbo and S. Vogl, "Alkindi, Tideus und Pseudo-Euklid: Drei optische Werke," *Abhandlungen zur Geschichte der mathematischen Wissenschaften*, Leipzig/Berlin, 26 (1912), 3, pp. 1–41, tr. J. Jolivet, H. Sinaceur, H. Hugonnard-Roche, in Rashed, *Œuvres Philosophiques et Scientifiques d'Al-Kindī*, vol. I, p. 437.

28. On the commentaries, see Peters, *Aristoteles Arabus*, pp. 17–19 and 30–31 respectively. On al-ʿĀmirī's *al-Qawl fī al-Ibṣār wa al-mubṣar* (*Discourse on Vision and Visual Objects*), see Khalīfāt (ed.), *Rasā'il-i Abu al-Ḥasan-i ʿĀmirī bā muqaddamih va taṣḥīḥ-i Saḥbān Khalīfāt, tarjumih-i muqqadamih, Mihdī Tadayyun*, Tehran: Markaz-i Nashr-i Dānishgāhī (University Publications), 1375=1996 [earlier edition, *Rasā'il Abī al-Ḥasan al-ʿĀmirī wa-shadharātuhu al-falsafīyah: dirāsah wa-nuṣūs,* Amman: al-Jāmiʿah al-Urdunīyah, 1988]; for references relevant to optics, see Kheirandish, *The Arabic Version of Euclid's Optics*, vol. 1, pp. xlvi, lviii, vol. 2, p. 13, p. 17.

29. Al-Kindī, *De aspectibus*, ed. Björnbo, p. 1.

30. Kheirandish, *The Arabic Version of Euclid's Optics* and "The Mixed Mathematical Sciences of the Islamic Middle Ages;" Sabra, *The Optics of Ibn al-Haytham*, vol. 2, pp. 25–26.

31. *De aspectibus*, sec. 12, 14, 22, ed. Björnbo, pp. 17–19, p. 24; sec. 22, and pp. 37–39. The clarity of a close object is discussed in terms of the strength of its illumination, which is enhanced by proximity to the visual axis, in the wording of *De aspectibus*: "illumination in *many* parts (*plures partes*) and from all sides;" and in *Taqwīm*, ed. Rashed, p. 171, as "whatever is under more intense illumination (*nūr al-shadīd*) is seen more clearly (*turā abyan*) and so more accurately (*aṣdaq*).

32. Qusṭā ibn Lūqā, *Fī ʿIlal* (L), p. 6: "If one ray falls (*yaqaʿu*) upon an object, it is seen as one, if two rays fall, it [the object] is seen as two (*ra'ā ithnayn*), and if more than two rays (*aktharu min shuʿāʿayn*) fall, it is seen as more than two (*ra'ā akthara min ithnayn*)."

33. For the relevant passage in Ibn ʿĪsā's Ḥ [see Appendix], see Sabra's related discussions in *The Optics of Ibn al-Haytham*, v. 2, pp. 25–26. Another passage in Ḥ contains the interesting combination that "whatever is seen by a large angle is seen as larger (*aʿẓam*) and its vision is more accurate (*aṣdaqa ru'yatan*)."

34. In the second proposition of the *Optics*, for example, the Arabic translation of the Euclidean vocabulary of visual clarity (ἀκριβέστερον) as visual accuracy (*ṣidq al-ru'ya*), along with the problematic form *kathra* used for the concept of clarity in the seventh Euclidean definition, gives rise to a range of treatments, see Heiberg, *Euclidis Optica*, p. 4 and p. 156; Kheirandish, *The Arabic Version of Euclid's Optics*, vol. 1, p. and p. 156, and vol. 2, pp. 30–34: see also def. 1–4, and Prop. 1, 3, 9, 23.

35. The apparently unique manuscript copy is in Dār al-Kutub; see King, *Cairo Catalogue*, v. 2, p. 1030; *Cairo Survey*, p. 39. On the author, see *Kitāb al-Fihrist*, ed. Flügel, p. 281, tr. Dodge, p. 665; Sezgin *GAS*: V, p. 301; VI, p. 207, and VII, p. 406.

36. One of the two extant manuscripts has an additional title with reference to Murisṭus: *wa ʿuluww aʿmidatihā wa ʿilm ʿumq al-ābār wa ʿurūḍ al-anhār wa ghayr dhālika wa huwa yusammā Mūrīsṭus (and the Elevation of its Height, and the Science of the Depth of Wells and the Width of Rivers and other Things, and he? is called Mūrīsṭus* [there is an entry under this Greek author in *EI*²]. The short Arabic treatise is not among the list of al-Kindī's works reported by Ibn al-Nadīm; see *Kitāb al-Fihrist*, ed. Flügel, pp. 257–261; tr. Dodge, pp. 618–620. Atiyeh, *Al-Kindi: The Philosopher of the Arabs*, 1966, p. 200, lists it as no. 230 (citing Brockelmann, *GALS I*, p. 374); it is also cited in Ritter and Plessner, "Schriften Jaʿqūb ibn Isḥāq al-Kindī's in Stambuler Bibliotheken," p. 370.

37. See Sabra, *The Optics of Ibn al-Haytham*, Books I–III: On Direct Vision, 2 vols., 1989 [Includes an English translation and commentary based on an earlier critical edition; see Bibliography. To be followed by a similar study of books IV–VII].

38. The former (attributed to Shaykh Abū ʿAlī Ibn al-Haytham), seems to be more directly in the Arabic Euclidean tradition, than the latter (bearing the more commonly encountered name, al-Ḥasan ibn al-Ḥasan ibn al-Haytham) for including references to the eye (*baṣar*) and its rays (*shuʿāʿ*) in the course of discussions on height determination.

39. The author of *Height* is reported as a commentator on two works, the Persian *Sī Faṣl* (*Thirty Chapters*) attributed to Ṭūsī and Euclid's *Elements*; see Munzavī, *Persian Manuscripts*, vol. 1, p. 132.

40. On the corresponding Greek and Arabic terms for the bending of rays, see Kheirandish, *The Arabic Version of Euclid's Optics*, vol. 2, p. xliv. References to sources containing the more common Arabic form appear on pp. 57–58, n. 205–206.

41. On the first of these, see Max Meyerhof's edition, p. 109, lines 7–8 [reference from Sabra, *The Optics of Ibn al-Haytham*, v. 2, p. lviii, n. 80], where *inʿikās* is used together with *inkisār*, in the sense of the "turning back" (*rujūʿ*) of the visual rays (*manāẓir*); in the Arabic manuscript of *De speculis*, the verb *yanʿakis* is used [fol. 104b, the Latin has *convertitur*]; see *Tractatus* [pseudo-] *Euclidis De speculis*, Björnbo and Vogl, *Alkindi, Tideus und Pseudo-Euklid*, p. 100 cited by Theisen, *The Mediaeval Tradition*, p. 294, n. 54. For *inʿikās* in Ibn Lūqā's Arabic translation of Aetius' *Placita philosophorum*, see Daiber, *Aetius Arabus, Die Vorsokratiker in arabischer Überlieferung*, p. 204.

42. In Ṭūsī's *Shuʿāʿ* and *Inʿikās al-shuʿāʿāt wa inʿiṭāfuhā*, the expression is used for the deflection of the cones of ray at equal angles when encountered by polished surfaces. In the case of Ṭūsī's commentators, there is Badr al-Dīn al-Ṭabarī's short Persian treatise, *Irtifāʿ*, with two chapters devoted to the problem of height determination by means of a plane mirror. The author, "recalling" another method by which the height of tall objects can be made known (*maʿlūm*) by means of a mirror placed at different locations on the ground, states that what the people of the art (*ahl-i ṣināʿat*) may have offered on this subject "has not reached him, nor has it been seen by him anywhere." The commentator of *Sī faṣl* (*Thirty Chapters*) attributed to Ṭūsī, and of Euclid's *Elements* expresses the

equality of visual angle (*zāwiyat al-shuʿāʿī*) and the angle of reflection (*zāwiyat al-inʿikāsī*) in the more common form and explicitly associates the principle with the science of optics (*ʿilm al-manāẓir*), see Kheirandish, *The Arabic Version of Euclid's Optics*, vol. 1, p. l.

43. On the evidence from the Arabic tradition, see Kheirandish, *The Arabic Version of Euclid's Optics*, vol. 1, pp. xxvi–xxvii, and p. xxix, and vol. 2, p. 6; on the case of the Greek tradition, see Jones, "Peripatetic and Euclidean Theories of the Visual Ray," p. 52, and Knorr, "Pseudo-Euclidean Reflections in Ancient Optics," p. 29, n. 48–49.

44. Kheirandish, Elaheh, "What Euclid Said to his Arabic Readers: The Case of the Optics," Proceedings of the XXth International Congress of History of Science, Liège, 1997, Published in *Optics and Astronomy* (Simon, G. and Débarbat, S., eds.), 2001, pp. 17–28.

45. In the first formulation, namely Euclid *Kitāb al-Manāẓir* (*Optics*), the indicators are the ray (*al-shuʿāʿ*), issuing (*yakhruj*), the eye (*al-ʿayn*), paths (*sumūt*), infinte (*lā nihāya*) multitude (*kathra*), cone (*makhrūṭ*), apex (*ra'*s), and object (*mubṣar*), while in the second formulation, that is, Euclid's *Kitāb al-Mir'āh* (*De speculis*) they are, luminous power (*quwwa nūriyya*), spreading (*yanbathth*), pupil (*nāẓir*), *ṣanawbarī* (pine-shaped), *zujj*, *mustaḥadd* (pointed). Note that the latter term (*zujj*), a non-standard form for the cone's apex, seems to be intended in the Arabic *De speculis* (S), Aḥmad ibn ʿĪsā's *Manāẓir wa al-marāya al-muḥriqa* (Ḥ), and in al-Kindī's *Taqwīm* (Q), rather than the similarly transcribed terms "*raḥb*" and "*wa bihi*" in Rashed's edition of Q and S respectively: see *Œuvres Philosophiques et Scientifiques d' Al-Kindī*, vol. 1, p. 163 and p. 338.

46. Some aspects treated by Ibn al-Haytham are perception by glancing (*idrāk bi al-badīha*), by contemplation (*idrāk bi al-ta'ammul*) by recognition (*idrāk bi al-maʿrifa*), and ascertained (*muḥaqqaq*) perception, see Sabra, *The Optics of Ibn al-Haytham*, vol. 2, p. 241; for a general discussion, see H. Wolfson, "The Internal Senses in Latin, Arabic and Hebrew Philosophical Texts," *Harvard Theological Review*, 1935, 28: 69–133.

47. Sabra, A. I., *The Optics of Ibn al-Haytham*, vol. 2, pp. lviii–lix.

48. Sabra, *The Optics of Ibn al-Haytham*, vol. 2, pp. lviii–lix (on the limited circulation of Ptolemy's *Optics*, with specific reference to relevant works including the exceptional cases of Ibn Sahl and Ibn al-Haytham); see also Sabra, "Psychology vs. Mathematics: Ptolemy and Alhazen on the Moon Illusion," *Mathematics and its Applications to Science and Natural Philosophy in the Middle Ages: Essays in the Honor of Marshall Clagett*, ed. Edward Grant and John E. Murdoch, Cambridge: Cambridge University Press, 1987: 217–247, pp. 219–221.

49. Fārisī's reference to "the shortcomings of Euclid's book [i.e., *Optics*]" and his reasons for undertaking his own optical researches appear in the introduction of his important commentary on Ibn al-Haytham's *Kitāb al-Manāẓir*, entitled *Tanqīḥ al-Manāẓir*, ed. Hyderabad, p. 16: "I saw in the statements of some leading philosophers, and in more than one of them, that light shines from a luminous object in straight lines, and when it encounters a surface such as the surface of water, it is reflected from it at angles equal to their opposite [side] and penetrates (*yanfūdh*) into it on the extension of

illumination, and is refracted (*inʿaṭafat*), in the extension of reflection (*inʿikās*), and from this four equal angles are produced, angles of direct radiation, reflection, penetration, and refraction (*zawāyā al-istiqāmah, al-inʿikās, al-nufūdh, wa al-inʿiṭāf*). So I became puzzled (*taḥayyartu*) by these rules (*aḥkām*)." Ṭūsī's treatment in "*Risāla fī inʿikās al-shuʿāʿāt wa inʿiṭafihā*" [see note below] is cited by A. I. Sabra in this connection: *The Optics of Ibn al-Haytham*, vol. 2, pp. lxix–lxxi, n. 112.

50. See English translation, Winter and ʿArafat, "A Statement on Optical Reflection and 'Refraction' Attributed to Naṣīr ud-dīn aṭ-Ṭūsī," p. 141; partial German translation, Wiedemann, "Über die Reflection und Umbiegung des Lichtes von Naṣīr al Dīn al-Ṭūsī."

51. Sabra, *The Optics of Ibn al-Haytham*, vol. 2, p. lix: "all the historical evidence we have points to the fact that Ptolemy's book [i.e., *Optics*] was little used both in Antiquity and in the Islamic Middle Ages almost up to I.H.'s time."

Bibliography

Aristotle, *Problems*, v. 1: Books I–XXI, edited by T. E. Page, E. Capps, and W. H. D. Rouse, with an English translation by W. S. Hett, Loeb Classical Library, Cambridge: Harvard University Press, and London: William Heinemann Ltd., 1936.

Aristotle, *Posterior Analytics*, edited by G. P. Goold, with an English translation by Hugh Tredennick (Cambridge: Harvard University Press, London: William Heinemann Ltd., 1976; Arabic version, *Kitāb al-Burhān min Manṭiq Arisṭū*, ed. A. Badawi (Cairo: Dār al-Kutub al-Miṣriyya, 1949), Islamica VII, part 2, 309–465.

Atiyeh, G. N., *Al-Kindī: The Philosopher of the Arabs,* Rawalpindi: Islamic Research Institute, Publication No. 6, 1966.

Björnbo and Vogl, Alkindi, *Tideus und Pseudo-Euklid = Alkindi, Tideus und Pseudo-Euklid: Drei optische Werke*, herausgegeben und erklärt von A. Björnbo und Sebastian Vogl, *Abhandlungen zur Geschichte der mathematischen Wissenschaften*, Leipzig/Berlin, 1912, 26, 3: 1–176.

Brockelmann, *GAL* = Brockelmann, Carl, *Geschichte der arabischen Litteratur*, 2nd edition, Leiden: E. J. Brill, 2 vols. and 3 suppl., 1937–1949 (*GAL I* = vol. I, 1943; *GAL II* = vol. II, 1949; *GALS I* = Suppl. I, 1937; *GALS II* = Suppl. II, 1938; *GALS III* = Suppl. III, 1943).

Brownson, C. D., "Euclid's *Optics* and Its Compatibility with Linear Perspective," *Archive for History of Exact Sciences*, 1981, 24, 3: 165–194.

Burton, H. E., "The *Optics* of Euclid," *Journal of the Optical Society of America*, 1945, 35: 357–372.

Clagett, Marshall, "Some General Aspects of Physics in the Middle Ages," *Isis*, 1948, 39: 29–44.

Crombie, Alistair C., *Robert Grosseteste and the Origins of Experimental Science: 1100–1700*, Oxford: At the Clarendon Press, 1953.

Daiber, Hans, *Aetius Arabus: Die Vorsokratiker in arabischer Überlieferung*, Wiesbaden, 1980.

DSB = *Dictionary of Scientific Biography*, ed. C. C. Gillispie, 16 vols. (including Supplement and Index), New York: Charles Scribner's Sons, 1970–1980.

EI^2 = *Encyclopaedia of Islam*, 2nd edition, Leiden: E. J. Brill, 1960–.

Euclid, *Elements*, see Heath, *The Thirteen Books of Euclid's Elements.*

Euclid, *Optics*: see Heiberg, *Euclidis Optica*; Burton, "The Optics of Euclid;" Ver Eecke, *Euclide: l'Optique et la Catoptrique.*

Endress, Gerhard, "The Circle of Al-Kindī: Early Arabic Translations from the Greek and the Rise of Islamic Philosophy," in *The Ancient Tradition in Christian and Islamic Hellenism* (G. Endress and R. Kruk, eds.), Leiden: Research School CNWS, 1997.

al-Fārābī, Abū Naṣr, *Iḥṣā' al-ʿulūm*, ed. ʿUthmān Amīn [= Osman Amine], Cairo: Librairie Anglo-Égyptienne, 1968 [1st and 2nd edn., Cairo, 1931, 1949]; A. Gonzàlez Palencia, *Alfarabi Catàlogo de las ciencias*: two medieval Latin translation, modern edition and Spanish translation, Madrid, 1932.

al-Fārābī, *Kitāb al-Burhān*, ed. M. Fakhry, Beirut: Dār al-Mashriq, 1986.

al-Fārisī, Kamāl al-Dīn, *Tanqīḥ al-Manāẓir li-dhawī al-abṣār wa al-baṣā'ir*, 2 vols., Hyderabad, 1347–1348 (=1928–1930).

Filius, L. S., *The Problemata Physica Attributed to Aristotle, The Arabic Version of Ḥunain ibn Isḥāq and the Hebrew Version of Moses ibn Tibbon*, Aristoteles Semitico-Latinus, Leiden: Brill, 1999.

Galen, *De usu Partium*, Arabic translation: *Kitāb Manāfiʿ al-aʿḍā', Bibliothèque Nationale*: MS ar. 2583; English translation, see Margaret May.

Gulchīn Maʿānī, *Āstān-i Quds = A Catalogue of Manuscripts in the Astan Quds Razavi Library* (*Fihrist-i Kutub-i khaṭṭī-yi Kiṭabkhānih-i Āstān-i Quds-i Raḍavī*) by Ahmad Golchin Maani, v. 8, Mashhad: A publication of the Cultural and Library Affairs, no. 6, 1350=1971.

Heath, Thomas, L. *The Thirteen Books of Euclid's Elements*, second edition (revised with additions), 3 vols, Cambridge: Cambridge University Press, 1926 (Dover, 1956)

Heiberg, J. L., *Euclidis Optica* = Heiberg, J. L., *Euclidis Optica Opticorum recensio Theonis, Catoptrica, cum scholiis antiquis*, edidit Heiberg, *Euclidis Opera Omnia*, ediderunt J. L. Heiberg et H. Menge, vol. VII, Leipzig: Teubner, 1895: 1–121.

Heiberg, *Opticorum recensio Theonis* = Heiberg, J. L., *Euclidis Optica, Opticorum recensio Theonis, Catoptrica, cum scholiis antiquis*, edidit Heiberg, *Euclidis Opera Omnia*, ediderunt J. L. Heiberg et H. Menge, vol. VII, Leipzig: Teubner, 1895: 143–247.

Ḥunayn Ibn Isḥāq, *Kitāb al-ʿAshar maqālāt fī al-ʿayn*, see Meyerhof, *The Book of the Ten Treatises of the Eye Ascribed to Hunain ibn Is-ḥâq* (809–877 AD).

Ḥusaynī and Marʿashī, *Marʿashī* = Ḥusaynī, S. A., and Marʿashī, S. M., *Fihrist-i Nuskhih'hī-yi khaṭṭī-yi Kitābkhānih-i Haḍrat-i Āyatullāh al-ʿUẓmā Najāfī-yi Mar'ashī . . .*, v. 11, 1364 (=1985); v. 12, 1365 (=1986).

Ibn al-Haytham, *Kitāb al-Manāẓir*, ed. Sabra, A. I., *Al-Ḥasan Ibn al-Haytham, Kitāb al-Manāẓir: Books I–II–III <On Direct Vision>: Edited with Introduction, Arabic-Latin Glossaries and Concordance Tables by Abdelhamid I. Sabra*, Kuwait: The National Council for Culture, Arts and Letters, 1983. See also, Sabra, A. I. *The Optics of Ibn al-Haytham: Books I–III On Direct Vision*, London, 1989.

Ibn al-Nadīm, *Kitāb al-Fihrist*, ed. Flügel = Ibn al-Nadīm, Muḥammad ibn Isḥāq, *Kitāb al-Fihrist*, edited by Gustav Flügel, 2 vols., Leipzig: Verlag von F. C. W. Vogel, 1871–1872; reprinted, Beirut: Khayyāṭ, 1964.

Ibn al-Nadīm, *Kitāb al-Fihrist: The Fihrist of al-Nadīm: A Tenth-Century Survey of Muslim Culture*, Bayard Dodge (editor-translator), 2 vols. (Records of Civilization: Sources and Studies no. 83), New York and London: Columbia University, 1970.

al-Jāḥiẓ, *Kitāb al-Tarbīʿ wa al-tadwīr*: edition, Pellat, *Le Kitāb al-tarbīʿ wa-t-tadwīr de Ǵāḥiẓ*.

Jones, Alexander, "Peripatetic and Euclidean Theories of the Visual Ray," *Physis*, 1994, 31, 1: 47–76.

Khalīfāt (ed.) *ʿĀmirī, Rasā'il = Rasā'il-i Abu al-Ḥasan-i ʿĀmirī bā muqaddamih va taṣḥīḥ-i Saḥbān Khalīfāt, tarjumih-i muqqadamih, Mihdī Tadayyun*, Tehran: Markaz-i Nashr-i Dānishgāhī (University Publications), 1375=1996 [earlier edition, *Rasā'il Abī al-Ḥasan al-ʿĀmirī wa-shadharātuhu al-falsafīyah: dirāsah wa-nuṣūs,* Amman: al-Jāmiʿah al-Urdunīyah, 1988].

Kheirandish, Elaheh, "The Arabic 'Version' of Euclidean Optics: Transformations as Linguistic Problems in Transmission," *Tradition, Transmission, Transformation*: 227–247, Leiden: Brill, 1996.

Kheirandish, Elaheh, "The 'Manāẓir' Tradition through Persian Sources," *La sciences dans la monde iranien*, ed. Ž. Vesel, H. Beikbaghban et B. Thierry de Crussol des Epesse, Tehran: Institut Français de Recherche en Iran (IFRI), 1998: pp. 125–145.

Kheirandish, Elaheh, *The Arabic Version of Euclid's Optics: Kitāb Uqlīdis fī Ikhtilāf al-manāẓir*. Edited and translated with historical introduction and commentary (revised dissertation, Harvard University, 1991), 2 vols., Springer-Verlag: *Sources in the History of Mathematics and Physical Sciences*, no. 16, 1999.

Kheirandish, Elaheh, "What Euclid Said to His Arabic Readers: The Case of the Optics," Proceedings of the XXth International Congress of History of Science, Liège, 1997, Published in *Optics and Astronomy* (Simon, G. and Débarbat, S., eds.), 2001, pp. 17–28.

Kheirandish, Elaheh, "The Mixed Mathematical Sciences of the Islamic Middle Ages," *The Cambridge History of Science*, 8 vols. ed. David C. Lindberg and Ronald Numbers; vol. 2: The Middle Ages, Cambridge University Press, forthcoming.

al-Kindī, *De aspectibus*: see Björnbo and Vogl, *Alkindi, Tideus und Pseudo-Euklid: Drei optische Werke*. See also Rashed, *Œuvres Philosophiques et Scientifique d'Al-Kindī*, 1997.

King, David A., *A Catalogue of the Scientific Manuscripts in the Egyptian National Library*, Cairo: General Egyptian Book Organization in collaboration with The American Research Center in Egypt, and the Smithsonan Institution, Part II, 1986 [Part I, 1981].

King, David A., *A Survey of the Scientific Manuscripts in the Egyptian National Library*, published for The American Research Center in Egypt, Catalogs, vol. 5, Winona Lake, Indiana: Eisenbrauns, 1986.

Knorr, Wilbur R., "Pseudo-Euclidean Reflections in Ancient Optics: A Re-Examination of Textual Issues Pertaining to the Euclidean *Optica and Catoptrica*," *Physis*, 1994, 31, 1: 1–45.

Krause, Max, "Stambuler Handschriften islamischer Mathematiker," *Quellen und Studien zur Geschichte der Mathematik, Astronomie und Physik*, Berlin, 1936, B, Band 3, Heft 4: 437–532.

Lejeune, Albert, *L'Optique de Claude Ptolémée dans la version latine d'après l'arabe de l'émir Eugène de Sicile: Édition critique et exégétique augmentée d'une traduction française et de complément*, Leiden: E. J. Brill, *Collection de Travaux de l'Académie Internationale d'Histoire des Sciences*, no. 31, 1989. [earlier edn.: Louvain: Université de Louvain, *Recueil de travaux d'histoire et de philologie*, 4. sér., fasc. 8, 1956].

Lindberg, David C., *Theories of Vision from Al-Kindi to Kepler*, Chicago and London: The University of Chicago Press, 1976.

Madelung, Wilferd, "Aḥmad ibn ʿĪsā", *EI*[2], *Supplement*, pp. 48–49.

May, Margaret T., Galen *On the Usefulness of the Parts of the Body: Translated from the Greek with an Introduction and Commentary*, 2 vols., Ithaca: Cornell University Press, 1968–1969.

Meyerhof, Max, *The Book of the Ten Treatises on the Eye Ascribed to Hunain ibn Is-ḥâq* (809–877 A.D.), Cairo: Government Press, 1928.

Munzavī, *A Catalogue of Persian Manuscripts (Fihrist-i Nuskhih'hā-yi khaṭṭī-yi fārsī)* by Ahmad Monzavi, Tehran: Regional Cultural Institute, publication, v. 1, no. 14, 1348 (=1969).

Pellat, Charles, *Le Kitāb al-tarbīʿ wa-t-tadwīr de Ǵāḥiẓ*, Damascus: Institut Français de Damas, 1955.

Pellat, Charles, *The Life and Works of Jāḥiẓ: Translations of selected texts*, Berkeley and Los Angeles: University of California Press, 1969.

Peters, F. E., *Aristoteles Arabus: The Oriental Translations and Commentaries on the Aristotelian Corpus*, Leiden: E. J. Brill, 1968.

Ptolemy, *Optica*: see Lejeune, *L'Optique de Claude Ptolémée*, and Smith, "Ptolemy's

Theory of Visual Perception: An English Translation of the *Optics* with Introduction and Commentary."

Ragep, F. J., *Naṣīr al-Dīn al-Ṭūsī's Memoir on Astronomy: al-Tadhkira fī ʿilm al-hayʾa, Edition, Translation and Commentary*, 2 vols., New York: Springer-Verlag, *Sources in the History of Mathematics and Physical Sciences*, no. 12 (ed. G. J. Toomer), 1993.

Rashed, R., *Œuvres Philosophiques et Scientifique d'Al-Kindī*. Vol. I: *L'Optique et la Catoptrique in Islamic Philosophy Theology and Science*, Texts and Studies edited by H. Daiber and D. Pingree, vol. xxix, *Œuvres Philosophiques et Scientifiques d'Al-Kindī* editées par Jean Jolivet et Roshdi Rashed, Leiden: E. J. Brill, 1997.

Ritter, H. and Plessner, M., "Schriften Jaʿqūb ibn Isḥāq al-Kindī's in Stambuler Bibliotheken," *Archiv Orientalni*, 1932, 4: 363–372.

Ronchi, Vasco, "Classical Optics is a Mathematical Science," *Archive for History of Exact Sciences*, 1961, 1, 2: 160–171.

Sabra, A. I., "Ibn al-Haytham," DSB, 1972, 6: 189–210 [reprinted in Sabra, *Optics, Astronomy and Logic*, 1994, II: 189–210].

Sabra, A. I., "A Note on Codex Biblioteca Medicea Laurenziana Or. 152," *Journal for the History of Arabic Science*, 1977, 1, 2: 276–283.

Sabra, A. I., "Manāẓir, or ʿIlm al-Manāẓir," *EI*[2], 1987, 6, fasc. 103–104: 376–377.

Sabra, A. I. "Psychology vs. Mathematics: Ptolemy and Alhazen on the Moon Illusion," *Mathematics and its Applications to Science and Natural Philosophy in the Middle Ages: Essays in the Honor of Marshall Clagett*, ed. Edward Grant and John E. Murdoch, Cambridge: Cambridge University Press, 1987: 217–247.

Sabra, A. I., *The Optics of Ibn al-Haytham: Books I–III On Direct Vision, Translated with Introduction and Commentary by A. I. Sabra*, 2 vols. (Studies of the Warburg Institute edited by J. B. Trapp, vols. 40 i–ii), London: The Warburg Institute, University of London, 1989.

Sabra, A. I., and Heinen, Anton, "On Seeing the Stars: Edition and Translation of Ibn al-Haytham's *Risāla fī Ruʾyat al-kawākib," Zeitschrift für Geschichte der arabisch-islamischen Wissenschaften*, 1991/92, 7: 31–72.

Sabra, A. I., "On Seeing the Stars II: Ibn al-Haytham's 'Answer' to the 'Doubts' Raised by Ibn Maʿdān." *Zeitschrift für Geschichte der arabisch-islamischen Wissenschaften*, 1995/96, 10: 1–59.

al-Ṣafadī, *Kitāb al-Wāfī bi al-wafayāt*, v. 7, ed. Iḥsān ʿAbbās, Wiesbaden, 1389=1969.

Sezgin, Fuat, *Geschichte des arabischen Schrifttums*, Leiden: E. J. Brill, 1970–1979 (*GAS V*: Mathematik, 1974; *GAS VI*: Astronomie, 1978; *GAS VII*: Astrologie, Meteorologie, und Verwandtes, 1979).

Simon, Gérard, "The Notion of the Visual-Ray," presented in a conference on optics at the Dibner Institute, May, 1990, "La notion de rayon visuel et ses conséquences sur l'optique géométrique grecque," published in *Physis, Rivista Internazionale di Storia della Scienza* (Olschki, L. S., ed.), 94, pp. 77–112.

Theisen, *The Mediaeval Tradition* = Theisen, Wilfred R., *The Mediaeval Tradition of Euclid's Optics*, Ph.D. Thesis, University of Wisconsin, 1972, facsimile, Ann Arbor: University Microfilms International, 1984.

Tobin, Richard, "Ancient Perspective and Euclid's Optics," *Journal of the Warburg and Courtauld Institutes*, 1990, 53: 14–42.

Toomer, G. J., *Diocles on Burning Mirrors: The Arabic Translation of the Lost Greek Original*, Edited with English Translation and Commentary by G. J. Toomer, New York: Springer-Verlag, *Sources in the History of Mathematics and Physical Sciences*, no. 1 (ed. M. J. Klein and G. J. Toomer), 1976.

al-Ṭūsī, Naṣīr al-Dīn, *Taḥrīr al-Manāẓir, Majmūʿ al-rasāʾil*, Hyderabad: ʿUthmāniyya, part 1, 1358 (=1939), 5: 2–24.

Veltman, Kim H., *Optics and Perspective: A Study in the Problems of Size and Distance*, Ph.D. Thesis, London: The Warburg Institute, 1975.

Ver Eecke, Euclide*: l'Optique et la Catoptrique = Euclide: l'Optique et la Catoptrique*, Œuvres traduites pour la première fois du grec en français, avec une Introduction et des Notes par Paul Ver Eecke, Nouveau Tirage, Paris: Librairie Scientifique et Technique Albert Blanchard, 1959.

Wiedemann, Eilhard, "Über die Reflexion und Umbiegung des Lichtes von Naṣīr al Dīn al Ṭūsī," *Jahrbuch für Photographie und Reproduktionstechnik*, Halle, 1907, 21: 38–44.

Winter, H. J. J. and ʿArafat, W., "A Statement on Optical Reflection and 'Refraction' Attributed to Naṣīr ud-Dīn aṭ-Ṭūsī," *Isis*, 1951, 42: 138–142.

Wolfson, H. A. "The Internal Senses in Latin, Arabic and Hebrew Philosophical Texts," *Harvard Theological Review*, 1935, 28: 69–133.

al-Yaʿqūbī, Taʾrīkh: *Ibn-Wādhih qui dicitur Al-Jaʿqūbī, Historiae, pars prior historiam ante-Islamicam continens*, edidit indicesque adjecit M. Th. Houtsma, 2 vols., Leiden: E. J. Brill, 1883.

4

Ibn al-Haytham's Revolutionary Project in Optics: The Achievement and the Obstacle

A. I. Sabra

For Sir Ernst Gombrich, 1909–2001

[T]he nature of our subject being confused, in addition to the continued disagreement through the ages among investigators who have undertaken to examine it, and because the manner of vision is not ascertained, we have thought it appropriate that we direct our attention to this subject as much as we can, and seriously apply ourselves to it, and examine it, and diligently inquire into its nature.—We should, that is, recommence the inquiry into its principles and premises, beginning our investigation with an inspection of the things that exist and a survey of the conditions of visible objects. We should distinguish the properties of particulars, and gather by induction what pertains to the eye when vision occurs and what is found in the manner of [visual] sensation to be uniform, unchanging, manifest and not subject to doubt.—After which, we should ascend in our inquiry and reasonings, gradually and orderly, criticizing premises and exercising caution in regard to conclusions—our aim in all that we make subject to inspection and review being to employ justice, not to follow prejudice, and to take care in all that we judge and criticize that we seek the truth and not to be swayed by opinion.

Ibn al-Haytham, *Optics*, Preface [6]

[A]ll that sight perceives it perceives by refraction.

Ibn al-Haytham, *Optics*, Bk VII, Ch. 6.

I Introduction: A Paradox of Arabic Optics

The history of Arabic science is full of puzzles, one of which is a paradox revealed by comparing the history of optics with that of astronomy. Arabic astronomy was launched in the eighth century AD with a series of translations that included Ptolemy's *Almagest*, Euclid's *Elements*, and other Greek mathematical works deemed necessary for pursuing a serious study of the subject. This clearly concerted effort was part of a spectacular cultural movement actively supported by the ʿAbbāsid caliphate in Baghdad. The ensuing scientific endeavor, which continued with renewed surges of energy under various dynastic rules in various parts of the Islamic world for more than seven

Royal patronage, not institutional endeavours?

hundred years, always gave pride of place to astronomy as the discipline at the very top of the Greek mathematical sciences and, sometimes, even at the top of the entire hierarchy comprising the sum total of Greek theoretical knowledge, as defined in Greek antiquity and somewhat modified by later traditions represented in particular by Ptolemy in the second century AD. In most or perhaps all of the patronized scientific activities from the eighth to the fifteenth century—whether in Baghdad, Cairo, Muslim Spain, Ghazna, Marāgha, or Samarkand—astronomy tended to be favored as the pursuit most worthy of the attention of both the patron rulers and the patronized mathematicians who were keen on mastering and exploiting the Greek legacy. This must have been due in large part to the practical benefits widely expected from astronomy, in particular those promised by astrology, usually conceived as applied astronomy. But along with this practical motivation, astronomy came to be viewed, almost from the start, as a supreme form of scientific knowledge that possessed exact methods of observation and calculation and proof, and as a compelling or at least highly persuasive way of manifesting God's wisdom and the perfection of His handiwork. Many Muslim astronomers (probably most of them at first) embraced a pre-Islamic, Hellenistic view of the world and of man's place in it, but almost all of them were also willing to put their skills in observation and computation at the service of Muslim religious practice. For example, they developed exact methods for determining prayer times and the direction of Muslim prayer toward Mecca at all localities within the Islamic world. As a result of all these perceptions, expectations, and assorted claims, which were in part practical, in part theoretical or spiritual or religious, the amount of resources and intellectual energy that were spent on promoting and perfecting the science of astronomy far outweighed what was available for any of the "ancient sciences," *al-ʿulūm al-qadīma* or *ʿulūm al-awāʾil*, with the possible exception of medicine, which itself frequently received more than cursory attention from scholars deeply engaged in the mathematical/astronomical sciences.[1]

And yet, despite the many refinements, corrections, and innovations, often motivated by keen intellectual interest and real investigative and critical attitudes, Arabic astronomy never managed to break out of the Ptolemaic paradigm. The planets and the orbs that carried them around continued to move in circles, with definite preference given to traditional uniform speeds; Ibn al-Shāṭir in fourteenth-century Damascus argued for discarding Ptolemy's eccenters, but the epicycles remained, still serving the cause of circular uniform motion; and, despite repeated ventures into theoretical aporetic (raising doubts/*shukūk*/*aporiai* and attempting solutions), the earth stayed firmly put and unrotating at the center of a universe largely structured by basic tenets of Aristotelian natural philosophy. The twelfth-century "Andalusian Revolt" (itself inspired by strictly Aristotelian cosmology) against the Ptolemaic eccenters *and* epicycles, proved a non-starter that failed to induce later developments in the Islamic world (Sabra, 1984; Samsó, 1994). Arabic astronomy, whose impressive accomplishments have yet to be sufficiently appreciated by the general historian of science, never achieved what can meaningfully be called "revolution" (Sabra, 1998).

Now compare the above account with the story of Arabic optics. As in the case of astronomy, Arabic "optics"/*ʿilm al-manāẓir*," by which term I shall always refer to *the mathematical study of vision* (the meaning this term had in the Greek and Arabic traditions), also started with the translation of ancient works into Arabic. And while the early translations in astronomy included Indian and Persian, as well as Greek, materials, Arabic research in optics can definitely be said to have been originally based entirely on Greek sources. Of the two surviving Greek works bearing the title *OPTIKA*, namely those of Euclid and Ptolemy, the first is known to have circulated in Baghdad in Arabic version(s) already in the ninth century (Kheirandish, 1999). Apart from the text or texts purporting to be translations of Euclid's treatise, the Arabic optical writings that have reached us from that period (they include those of Ḥunayn ibn Isḥāq, Aḥmad ibn ʿĪsā, al-Kindī, and Qusṭā ibn Lūqā),[2] exhibit a mixture, or rather mixtures of doctrines that can be traced back to Galen and, through him, to Plato and to the Stoics, as well as, independently of these authorities, to Euclid's *Optics* itself. What *all* of these ninth-century writers on optics had in common was their adherence to one version or another of what has been called the extramission hypothesis, according to which vision of an external object was mediated by a visual emanation from the eye that extended in the shape of a cone (*makhrūṭ, ṣanawbara/konos*) all the way to the object seen. That emanation was either the sensitive material pneuma itself that had first descended from the brain into the eyes through the optic nerves or, in most cases, and almost certainly under the influence of Galen's arguments (especially in his *De placitis Hippocratis et Platonis*, see below), a sensitive power (*quwwa/dynamis*) conferred by the pneuma upon the surrounding air when it struck the air as it emerged from the pupil.[3] Besides the pneuma's sensitive capacity indicated by calling it "visual"/*baṣarī* or, literally, "seeing spirit/breath" (*al-rūḥ al-bāṣir*), the pneuma was given a further property indicated by also calling it "luminous breath" (*al-rūḥ al-nūrī or al-nayyir*) or, simply, "the light" (*al-nūr*).[4] Since everybody knew that vision does not take place in the dark, the further assertion was made that illumination of the air by a shining body (the sun, the stars, or a torch) was normally a necessary condition for the instrument of sight, that is, the pneuma or the qualitatively altered air, to perform its action. Vision happened as a result of the "contact" or "coalescence" or "cooperation" of the visual light occupying the visual cone with the external light, and it occurred along the radial lines (*khuṭūṭ al-shuʿāʿ*, also called "visual rays," *manāẓir/opseis*) constituting the cone which spread outwardly from its vertex at the eye to its base at the object.[5]

Apart from a passage in al-Kindī's *De aspectibus* (Proposition 11) that S. Vogl pointed out in 1912,[6] and a brief, ambiguous reference in al-Kindī's *Taqwīm al-khaṭaʾ wa al-mushkilāt allatī li-Uqlīdis fī Kitābihi al-mawsūm bi-al-Manāẓir*/"Rectification of Euclid's *Optics*" to Ptolemy and Theon of Alexandria,[7] Ptolemy's *Optics* is conspicuously absent from the Arabic discussions of the ninth century. In these discussions it is always Euclid's *Optics* that is cited as an authority, or sometimes, as in the case of al-Kindī, taken to task for some failing or another—but never in conjunction with an explicit mention of Ptolemy's treatise.—It would thus seem that the continuous radiation of *visual light*, which is the view definitely favored (but not necessarily invented) by Ptolemy (d. ca. 170), had reached the ninth-century Arabic writers, not directly through acquaintance with Ptolemy's treatise, but through Galen (d. ca. 214), or through sources not all of which are presently known to us.

One is particularly struck by two obseravations that strongly argue against direct acquaintance on the part of those writers with Ptolemy's *Optics*, and against any correct understanding or appreciation of his distinctive contributions. The first observation is the absence from their known compositions of the certainly non-Euclidean and importantly (though not exclusively) Ptolemaic emphasis on the primacy of color as a precondition for gaining perception of all other visible properties as qualifications of color, and of coloration as the primary "effect" (*passio/infiʿāl or taʾthīr/pathos*) produced in the organ of sight (*visus/baṣar/opsis*).[8] (See below, Section II.)

The second observation is that when some of the same ninth-century writers attempted a reasoned account of optical refraction in terms of lines and angles (we have two such accounts—one by Aḥmad ibn ʿĪsā and the other by al-Kindī), they only produced disastrously wrong arguments that could not possibly have been conceived by someone who had read and understood Book V of Ptolemy's *Optics*.[9] It is also significant that the somewhat extended exposition by al-Fārābī (d. AD 950) of the "mathematical" science of optics counts three modes of optical reflection (all of which cases of bending of the visual ray back in the direction of the viewer, and to which he assigns three different names), but fails to mention refraction (Ibn al-Haytham, 1989, II, Introduction, pp. lvi–lviii). Add to this the fact that the same erroneous account, found in Ibn ʿĪsā and in the fragment ascribed to al-Kindī, reappears, without correction, in an early astronomical work by Ibn al-Haytham ("Commentary on *Almagest*," MS Saray, Ahmet III 3329),[10] a work which may have helped to spread misunderstanding of the refraction phenomenon among mathematicians as late as the thirteenth century, such as Naṣīr al-Dīn al-Ṭūsī (d. 1274) and Quṭb al-Dīn al-Shīrāzī (d. 1311).[11]—That the Arabic tradition of Ptolemy's *Optics* differed greatly from that of his *Almagest* should be sufficiently clear from the additional fact that the optical part in the widely-used collection of "the middle books" (al-Mutawassiṭāt), prepared by al-Ṭūsī for the use of students of astronomy, represented not the advanced state of knowledge found in Ptolemy's treatise, but the stage reached four-and-a-half centuries earlier in Euclid's *Optics*.[12]

The first clear evidence we have of a correct understanding of Ptolemy's theory of refraction does not appear in the Arabic sources available to us until the second half of the tenth century, when the Persian mathematician al-ʿAlāʾ ibn Sahl was able to put Ptolemy's ideas to use in formulating entirely original geometrical arguments for the construction of burning instruments by means of refraction.[13] That was a landmark achievement, and one of the fruits of Buwayhid patronage in Iraq, at a time remarkable for its intensive scientific, especially mathematical activity. But none of the extant works of Ibn Sahl is concerned, either wholly or in part, with problems of vision, a subject that in fact is never mentioned in them: his interests, it appears, lay elsewhere, probably with an eye to practical application.

It is, therefore, still true to say today that, for the first substantial treatment of vision that was directly inspired by Ptolemy's contribution, we have to turn to the works of Ibn al-Haytham in the first half of the eleventh century, that is just short of nine hundred years after Ptolemy and some two hundred and fifty years after the ʿAbbāsid rulers lent their support to the Greco-Arabic translation enterprise.

Ibn al-Haytham was one of the most prolific and most competent geometricians in the Arabic tradition. In his younger years he wrote commentaries and/or summaries of Euclid's *Elements* and Apollonius's *Conics*.[14] A "Completion of the *Conics*" (a reconstruction of the lost Book Eight) possibly also belongs to an early period in his life.[15] A large proportion of his extant writings are devoted to problems of elementary and advanced geometry, including geometrical methodology.[16] In the *Optics*, which belongs to a later period in his career, Ibn al-Haytham solves the problem that has become known since the seventeenth century as "Alhazen's problem": to find the point(s) of reflection on the surface of a spherical mirror, convex or concave, given the positions of the eye and the visible object-point—a problem equivalent to a fourth-degree equation and therefore not solvable by ruler and compass (Sabra, 1982; Hogendijk, 1996). Armed with this solution he is able in *Optics*, Book V to undertake the first systematic investigation of images produced by mirrors of various shapes: spherical, cylindrical, and conical, convex and concave.

The titles and/or descriptions of works mentioned in lists of Ibn al-Haytham's writings also reveal a sustained interest in natural philosophy, which he called physics (*al-ʿulūm al-ṭabīʿiyya*). In his early career, sometime before 417/1027, he wrote a combined "summary" of the *Optics* of Euclid and of Ptolemy, in which he attempted a "completion" of Ptolemy's work by offering a reconstruction of Book I that, apparently, had been missing from the Greek text. This "summary" is not extant, and we can only speculate about the contents of that Arabic addition to Ptolemy's truncated treatise. But it must have been concerned at least in part with what Ibn al-Haytham considered to be the appropriate foundation for Ptolemy's experimental and psychological treatment of a subject which Euclid had dealt with in predominantly geometrical terms. Also non-extant is a treatise on the "Nature of Sight," and another on "Optics According to the Method of Ptolemy."[17] These three compositions, then, and possibly a fourth whose title we do not know (*Optics*, Preface, para. [8]) were preliminary exercises, so to speak, that eventually led to the large "Book of Optics (*Kitāb al-Manāẓir*), in seven treatises," which Ibn al-Haytham composed in mature age (at any rate, after AD 1028). That major work, announcing a decisive break with the basic assumptions of earlier mathematicians, including Euclid and Ptolemy, had the declared ambition of not simply proposing an alternative view, but of building up a new and complete system of optics based on new foundations deliberately planned to combine both "physical" and mathematical doctrines and modes of argument.[18] As is now well known, Ibn al-Haytham's *Book of Optics*, once rendered into Latin (almost certainly in Spain), not long after a Latin version of Ptolemy's treatise had also been made in Sicily from the Arabic, quickly established itself as the chief authority on its subject, a status which it maintained among philosophers and mathematicians in Europe up to the time of Kepler.—It is quite remarkable (perhaps the single most remarkable thing about the history of Arabic optics) that the first mathematician to have fallen under the influence of Ptolemy's *Optics* as the most developed mathematical treatment of vision in antiquity was also the one who wrote the first treatise that superseded it.—This chapter has the immediate aim of identifying and bringing together the main ingredients that went into the making of Ibn al-Haytham's revolutionary project.

II The Substratum: A Phenomenalist Physics of Light

It is clear that for Ibn al-Haytham, as for Euclid and Ptolemy, "optics"/*ʿilm al-manāẓir/hē optikē technē*, was a study essentially concerned with visual perception. Everybody had accepted that the presence of external light (however understood) was a condition of vision; and, apparently, Ptolemy, in the lost

Book I of his *Optics*, had included a discussion that reaffirmed the Platonic doctrine of synaugeia (*ijtimāʿ, ishtirāk*), meeting or cooperation of external light with visual emanation in the production of actual vision (Lejeune, 1948, *première partie*).[19] In the remaining four Books of Ptolemy's treatise, however, external light appears as a factor previously stated but no longer investigated as an independent agency: the visual ray or flux (*visus*, which must have translated *baṣar*, itself corresponding to *opsis*), spreading out from the eye in the form of a continuous cone, constantly figures in Ptolemy's work as the indispensable instrument of vision in all three of its known modes: rectilinear, reflected, and refracted; while illuminated color is asserted to be the primary cause affecting the organ of vision, in such a way as to convey to the perceiver "accidents" or qualifications of color—for example, shape as the outline of a colored area, size as the magnitude of such an area, motion as the changing spatial relations between adjacent colored areas. True vision (*vere videre*) happens when the direct ray is blocked by a dense and shining object (*lucidum spissum*) which thus signals its presence where it is actually located with respect to the viewer (*Ptol. Opt.* II, [4, 26]; see below, sec. IV). Untrue vision occurs when the ray is deflected by a reflecting or refracting surface, thereby giving rise to the perception of an imaginary object lying behind the surface. This account was to be entirely abandoned or drastically reinterpreted when Ibn al-Haytham decided to reject the visual-ray hypothesis altogether, proposing instead a theory of vision based on a coherently articulated theory of light as an independent physical property—a property whose behavior in rectilinear propagation, reflection, and refraction was subject to experimentally verified rules, and whose characteristic physical effect on physiological systems of vision initiated *specified processes* ultimately ending in the visual perception of external objects and of all their visible qualities or properties (*maʿānī/intentiones*) through mental operations of "inference" (*qiyās, istidlāl*) or interpretation.

In the Preface to his *Optics*, Ibn al-Haytham announced a new starting point for his investigation: the mathematical treatment of vision was to be preserved, as he declared, and thus the traditional conception of optics as a mathematical science remained, but it had to be based on a correct understanding of light as the sole agent of vision. As viewed by him, the crisis of the science of vision ("confusion" was his term for the crisis) consisted in what he regarded as an unsatisfactory separation between a rigorous, and to that extent commendable, mathematical approach, and a not-so-thorough approach, to be found in the works of natural philosophers, which nevertheless had physical truth on its side. There was, therefore, need for a new "synthesis" (*tarkīb*) that combined the advantages of these two opposed methods.

It must be emphasized that Ibn al-Haytham was contrasting, not two authorities, but rather two methodologies, and that what he consciously looked for, was not a synthesis by means of juxtaposition, but a new system of explanations to be discovered only by a fresh inquiry based on newly established "principles and premisses." And, as the Preface also announced, the proposed inquiry consisted not in simply conceding what was due to existing doctrines, but first in subjecting the relevant "particulars" and "properties" (of vision and of light) to "inspection" and then, and only then, in "gathering by induction (*istiqrāʾ*) what is found in the manner of vision to be uniform, unchanging and not subject to doubt" (*Optics*, I, 1[6]).[20] Consistently with the declared aim of the book, and in agreement with the proposed procedure outlined in the Preface, the chapter following the Preface, on the "properties of sight," immediately presents the reader with a general, orderly description of the "conditions" of vision as revealed by a series of detailed observations supported by carefully described experiments: the existence of distance between eye and object; the existence of unobstructed straight lines between points on the surface of the eye and points on the object's perceived surface; luminosity of the object; a minimum size of the object that varies with the strength of the viewer's eyesight; opacity and hence color of the seen object; and ascertainable variability of distance with the size of the object, the degree of the object's luminosity and color, and with the strength of eyesight. These were the general, empirically established data of vision to be explained in terms of equally general "properties" of light to which Ibn al-Haytham turns in the next chapter in Book I, with more details to be added in Books IV and VII on reflection and refraction, respectively.

The theory describes how the physical property of light shines or radiates (*ashraqa*) from a self-luminous object (the sun, a star, a flame) in straight lines from every point on the surface of the object in all directions; how it rectilinearly extends (*imtadda/extendere*) through a transparent medium (air, water, glass); how it is reflected from smooth surfaces at a certain angle and in a given plane; and how it is refracted as it passes through surfaces separating media of different transparencies. The theory also takes care to describes how, once "fixed" (*thabata*) in the surfaces of illuminated opaque (nontransparent) objects, or in the body of transparent media considered as always endowed with a certain degree of opacity, the light will shine forth in exactly the way it does from self-luminous objects: that is, from every individual point where the light is fixed in the surface of the opaque object *or* in the incompletely transparent medium, on all straight lines that can be drawn from that point. In all of these statements the theory makes no appeal to any metaphysical entities or doctrines and no reference to anything other than the observable phenomena. The resulting classification of all "lights" is consequently a classification of these phenomena: "essential light" is the observable light in permanently luminous bodies; "accidental light" is that observed in bodies when they are illuminated by external sources; "primary light" is what radiates from essential light; and "secondary light" that which radiates from accidental light. All of these "species" or kinds of light are "found" to behave in exactly the same manner; and what are classified as opaque, smooth, or refracting surfaces are all "found" to behave in identical ways with respect to all kinds of light; and, finally, all lights are said to weaken in strength or intensity as they recede from their respective sources. Thus we are led to characterize the "physical" theory of light underlying Ibn al-Haytham's *Optics* as a *phenomenalist theory*: the theory invokes no hidden entities or properties, and it is merely concerned to establish regular features of the behavior of light by reference to direct observation and experimentation.

Color, in Ibn al-Haytham's theory, is a distinct property of material, opaque bodies—distinct, that is, from light. Self-luminous objects—for instance the sun—are said to have "something that behaves like color," or "of the nature of color," which means that besides their intrinsic luminosity they are also opaque. Colors exist in other opaque bodies whether these are illuminated or not. We do not know (we cannot know) whether colors extend themselves into the adjacent medium in the absence of light. But when illuminated they are "found" to "radiate" in the company of the illuminating light, with which they "mingle;" we can further verify that colors behave exactly as light does in rectilinear transmission, reflection, and refraction; and we know that they are perceived only when they enter the eye mixed with light. Thus all the experimental statements about light in the preceding paragraphs also hold for color, and Ibn al-Haytham's theory of light is at the same time a theory of the distinct property called "color."

In his theory of light and color, as well as in his theory of visual perception, Ibn al-Haytham employs the Aristotelian-Peripatetic term "form" (*eidos/ṣūra*) to refer to the physical properties involved. His experimental, phenomenalist account (in *Optics*, I, 3) of light and color as objective properties gains nothing in explanatory power by substituting "form of light" and "form of color" for "light" and "color." The term "form" does not in fact occur until somewhat late in this account (*Optics*, I, 3[113]), when the discussion turns to color, and it is not always adhered to afterward. But the use of these expressions was of course in keeping with Ibn al-Haytham's declared aim to inject the mathematical study of vision with what he believed to be objective, physical truth. He does, however, introduce a crucial refinement of the concept of form which is essential to the fulfilment of his program for a physical theory of light that is also amenable to mathematical treatment. (And as will be seen presently, "form" was the term inextricably linked to the natural-philosophical doctrine of *minima naturalia* which Ibn al-Haytham accepted.)[21] This refinement appears in his concept of *point-forms* of light and color. The expression itself is not used by Ibn al-Haytham, but he repeatedly speaks of the "forms of points of light" and of color in the surfaces of visible objects, and it is these points of light and color that extend themselves on straight lines from illuminated points on the objects (whether these are self-luminous or not), *and* from points within an illuminated transparent medium like air or water.

The idea or, if we like, the assumption, is that, for any material body or part of a body, opaque or transparent, to carry, receive, repel or transmit the "form" or property *light* (or *color*) it has to be of a certain minimal size. If the body is "divided" further (say, by narrowing the naturally luminous (or illuminated) part on the body's surface, or by narrowing the aperture through which the light passes in a transparent body), the light will vanish. Thus the "point" from which light shines in all directions must have a minimal, finite size; and the "ray" proceeding from this "point" in the adjacent medium, what Ibn al-Haytham calls "the least light" (*aqall al-qalīl min al-ḍawʾ*) or "the smallest light" (*adaqq al-daqīq min al-ḍawʾ, aṣghar al-ṣaghīr min al-ḍawʾ*), must occupy a minimal finite width through the middle of which the *mathematical* ray can be imagined to pass.[22]

It will be remembered that Newton's *Opticks* (1704), which had also deliberately presented a phenomenalist/experimentalist theory of light claiming independence from any hypothesis about the "nature" of light, operated with what he regarded as a neutral concept of the "light ray" as "the least Light or part of Light" passing through a suitably small aperture. He, too, assumed that a minimal width of the passing light beam could be isolated by sufficiently dividing or narrowing the beam's path, while additionally assuming that, by simultaneously allowing a "least part of light" to pass alone in the direction of propagation (e.g., by successively chopping off the beam perpendicularly to the direction of propagation—say, by means of a fastly rotating shutter), a single minimal part could be isolated [*Opticks*, pp. 1–2]. By substituting the concept of ray as "least part of light" for the covert reality of a light corpuscle, Newton thought he could claim to be presenting a purely experimental and nonhypothetical, or phenomenalist theory which he hoped would accommodate a host of light properties including, not only rectilinear propagation, reflection and refraction, but also differential refrangibilities of color, diffraction/inflexion, fits of easy reflection and of easy transmission, and polarity—all of which he believed to be inseparable and indeed "connate" dispositions of the light "ray." Ibn al-Haytham's position was made much easier by the severely limited repertoire of light phenomena that he had to deal with; and thus the simple concept or assumption of a least part, already supported by a widely accepted natural philosophy, was enough for endowing the mathematical ray, strictly an imaginary sraight line or a straight direction of activity, with solid physical reality, that is, the phenomenal reality of a sensible, diffused quality of a material object. And, perhaps even more important, such a simple concept was all that he felt was needed for attacking the specific and basic *problem of vision* he set out to solve in his *Optics*.

A theory of light and color as properties of material objects is not itself a theory of vision. But it is clear that Ibn al-Haytham's theory, whatever the range of its intrinsic possibilities, is utilized in his *Optics* only to the extent required for establishing the new theory of vision. It has been stated above that the foundational theory of light is free from metaphysical presuppositions, for instance such as those associated with the so-called thirteenth-century Latin "perspectivists," such as Roger Bacon, John Pecham, and Witelo, whose enthusiastic appropriation of Ibn al-Haytham's ideas should not be allowed to obscure the fact that their basic commitment was to a very different project. Would it be equally true to assert that the theory also failed to envisage a mechanical structure of matter necessary for or at least enhancing the understanding of the behavior of light? In Books IV and VII of the *Optics* Ibn al-Haytham contemplates, and indeed explores at some length, "explanations" (he calls them *ʿilal*, causes or reasons) of optical reflection and refraction in terms of concepts of motion, speed, density, impact, resistance, repulsion, and the concept of *iʿtimād* (endeavor, pressure), current in earlier and contemporary Muʿtazilite *kalām* discussions; and all of these concepts are directly applied to small, solid, spheres projected against hard or yielding surfaces.[23] What should we make of these *analogies* (for that appears to be their function in the *Optics*) which, we know, later attracted fruitful attention in Europe, for example from Kepler, and especially from Descartes who found some of their mathematical features well suited to his own mechanistic view of the behavior of light? I pose this important question here although it is not essentially relevant to my present concern, since these analogical explanations (or "comparisons," as Descartes called them) do not play a part in Ibn al-Haytham's account of *vision*. Ibn al-Haytham's own opinion seems to be in fact that, strictly speaking, considerations such as those involved in the study of mechanical collision of bodies or the penetration of yielding media did not properly belong in a treatise on "optics" in the restricted sense of a theory of vision.[24]

III An Image-oriented Theory of Vision

Ibn al-Haytham's theory of vision was the only one circulating in Europe, up to the time of the Renaissance, that interposed between the center of vision and the seen object a surface on which a configuration of illuminated points of color directly corresponded to their arrangement in the field of vision. That interposed surface was the slightly flattened spherical surface of the crystalline humor, and the *visually relevant* class of points of light and color existing in it marked intersections of the surface with the straight rays proceeding from points in the field toward the center of the eye, or vertex of the *geometrically defined* "visual cone." The theory maintained that "perception"/*idrāk*/ *comprehensio* of any object in the field, and of all its visual properties (size, shape, distance, and the rest), consisted in a mental reading of this color mosaic (which, alone, is said to be first "sensed" or registered on the crystalline's surface) after it has been transferred as a coherent whole through the humors of the eye and through the optic nerves, and after being ultimately presented to the brain where the final reading process was performed by a sense-faculty understood as a faculty of discrimination and judgment (*tamyīz*).

The analogy between the interposed surface in Ibn al-Haytham's theory and the "picture plane" envisaged by the Renaissance perspectivists is worth noting: the analogy would not hold if Ibn al-Haytham's eye functioned as a pin-hole camera (as in Leonardo) or as a lens-camera (as in Kepler). But the difference between a spherical surface in the one case and a plane surface in the other was of significance to the Renaissance perspectivists for whom there also remained the unique task of artificially constructing on a flat surface a picture which an *external eye* would read as a representation of objects deployed in three-dimensional space. In the present section of this chapter it will be our concern to investigate the status of what Ibn al-Haytham regarded, in Book I of the *Optics*, as a "form of the object" having the same order or arrangement of parts on the surface of the crystalline as on the surface of the external object, and to consider the fate of that concept of *ordered form*, and indeed of the whole theory as we find it in Book I, in the light of experiments reported *only* in Book VII.

Ibn al-Haytham's was not of course the first image- or picture-oriented theory of vision; such had been for example the theory of the ancient atomists in terms of coherent likenesses, or films or idols (*eidōla*), and the theory favored by Aristotle and by the Peripatetic tradition throughout the Middle Ages, in terms of forms received in the eye on the analogy of impressions made by a signet-ring in wax.[25] But the theory proposed by the eleventh-century mathematician was the first to attempt a *mathematical way* of constructing a *physically produced* "form" (*ṣūra*) inside the eye that could serve as the immediate basis for a full mental representation of the seen object. Henceforth, and until the publication of Kepler's *Ad Vitellionem Paralipomena* in 1604, much of the history of optics as a mathematical theory of vision was concerned with problems suggested by Ibn al-Haytham's *attempted construction*, and by problems arising from the *new experiments* fully preserved in the Arabic text and in the medieval Latin translation of Book VII. Am I ultimately contemplating a success story? Yes, but not one that is simple, straightforward, or historically uninstructive. After all, Ibn al-Haytham wrote some nine hundred years after Ptolemy; and four hundred years of rather active interest in vision were to pass between the transmission of Ibn al-Haytham's *Optics* to Western Europe and the publication of Kepler's treatise.

In *Optics*, I, 6 ("On the manner of vision") Ibn al-Haytham clearly formulates his problem as one of identifying the conditions, necessary and sufficient, for obtaining the normal visual perception of the external world as a world of distinct objects and distinctly differentiated colors and shapes (I, 6[5–11])—having previously argued from observation and experiment that we see as a result of an effect (*athar, ta'thīr, infiʿāl/ablatio, operatio, passio*) produced in the eye by the agency of external light (*Optics*, I, 2 & 4) and having experimentally determined the general characteristics of the behavior of light (I, 3) and having provided a general description of the structure of the eye adapted from the current anatomical literature (I, 5). Ibn al-Haytham agreed with the physicists' view that visual perception was a matter of receiving forms, but he insisted that asserting this was not enough to *explain* "distinct" vision, given what he had just established about how forms of light and color arrive at the surface of the eye from points on the facing objects (in accordance with rules of emission and propagation) and what the forms must undergo as they pass through the eye's layers (in accordance with the known rules of refraction). A full explanation, he argued, must provide means of somehow isolating the point-forms which originate at distinct points on the objects, but which must necessarily mingle together on the surface of the eye, and inside the eye itself, where they must make their first impressions.—This was not only Ibn al-Haytham's new problem; it was the real problem of distinct vision.

Now it is important to realize that, both in the way he understood his problem and in the way he went about finding a solution in Book I (and later in Book VII), Ibn al-Haytham simply accepted, *on trust*, the empirical evidence claimed by Galen and by the dominant medical tradition up to Ibn al-Haytham's time—to the effect that the crystalline humor, alone among the parts of the eye, was the "first" or "principal" seat of visual "sensation." The evidence consisted in the reportedly observed fact that vision ceased when the crystalline alone was damaged or obstructed (*Optics*, I, 6[14, 15]; II, 2[10]).[26] I say "important" because this generally received doctrine proved to be a crucial factor, and, as it turned out, a major "obstacle," that seemed persistently to push Ibn al-Haytham's thought into a certain direction. Since forms of light and color will come from all points on the facing objects to all points on the surface of the eye, they will all mingle together as they individually spread over the whole of that surface, and most of them will have been refracted before they have reached the crystalline's surface, thus giving rise to more mingling among them; and further confusion of forms will take place again as a result of further refraction at the crystalline's surface. "*Therefore* [as Ibn al-Haytham concludes in one of many passages to the same effect in Book I], the crystalline cannot perceive the visible object as it is [i.e., truly and distinctly] unless it perceives the color and light of each point on the object by means of the form reaching it through *one point only* on the surface of the eye" (I, 6[16], emphasis added). In other words, for a distinct perception/*idrāk* (read: sensation/*iḥsās* or sense impression) of a given object to be at all possible at the crystalline's surface and within the crystalline's body, only point-forms that have entered both the surface of the eye and the crystalline's front surface at right angles can be considered to be visually effective. And for this necessary condition to be realized, these two surfaces must be considered concentric at the region cut off by the "visual cone" defined by the width of the pupil, namely the cone with vertex at the center of the eye, and base at the object. As for the forms that strike the crystalline's surface after having been refracted at the eye's surface, they are (in Books I & II) rendered ineffective by virtue of two assumptions. The first, expressed briefly in Book I, invokes the superior effect of perpendicular action (I, 6[24]). The other, much more significant and of greater consequence for the theory as a whole, postulates the crystalline's natural disposition to "sense" only forms that go through it perpendicularly, along the lines of the visual cone.[27] As Ibn al-Haytham spells out this second assumption, the crystalline, *as a transparent body*, will refract the forms (or rays) that reach it on lines intersecting the radial lines, and this refraction will take place in accordance with the rules of refraction; but, *as a sensitive body*, it will only take notice of the unrefracted forms that strike its surface at right angles and that subsequently pass through its body along the radial lines. Therefore, what the crystalline *senses* in its surface and throughout its body is a *total* "form of the object" consisting of a configuration of point-forms of illuminated color that correspond one-to-one with all their *distinct* points of origin on the object seen; and it is this *sensed* form that will eventually reach the common nerve where two total forms from both eyes will be united and, finally, perceived by the ultimate sentient faculty (*al-ḥāss al-akhīr/ultimum sentiens*) that resides in the front of the brain.

Ibn al-Haytham has an answer to one who might object to his argument as an example of *ad hoc* conjecturing or hypothesizing, that is, as something perhaps unworthy of the mathematizing natural scientist he claimed to be. Such an objection, he replies, would not be really justifiable, considering that other privileged lines of activity are known to be attested in the natural world: heavy bodies freely fall only in lines normal to the earth's circumference, heavenly bodies move only in circles, and light itself freely travels only in straight lines (*Optics*, I, 6[43]).—He has a point. It is clear, however, that in choosing to privilege one natural mode of receptivity in the crystalline over another he was undoubtedly swayed by the widely received doctrine that assigned sensitivity in the eye to the crystalline.

We should note that while Ibn al-Haytham is certainly concerned to explain *distinct vision*, he does not speak of *distinct forms* (or images) in the eye actually portraying the array of light and color that distinctly *appear* on the object. These forms are *real*, being the effect physically produced in the eye by the incoming forms of light and color. And, in accordance with his physics of light and color, he characterizes that effect in the eye as actual illumination and coloration inside the eye. But as the forms arrive on top of one another at the surface of the eye, and subsequently on the surface of the crystalline, they become confused or "mixed" (*mumtazija/permixtae*) and therefore do not by their combined physical effect inside the eye represent their original distinct distribution on the object. Ibn al-Haytham repeatedly speaks, however, of forms in the eye that have the same arrangement of parts as the forms (or configurations of light and color) existing in the objects (*Optics*, I, 6 [20, 27, 33, 38, 40, 63, 66]). These, as we have just seen, are the forms *geometrically distinguished* by means of the perpendiculars drawn from the object-points to the center of the eye or vertex of the visual cone. But for their *distinct perception*, Ibn al-Haytham simply invokes the power of perception itself, thus by-passing the need for a *segregating optical apparatus* (such as a pin-hole arrangement or a focusing lens), and immediately and precipitously embarking on a psychological explanation of vision.—So, here again, we recognize the extent to which Ibn al-Haytham's theory in Book I of the *Optics* depended on the second, physiological assumption referred to above (p. 98)—an assumption that, we shall now see, Ibn al-Haytham was led to modify in an important passage at the end of chapter 6 in Book VII that explicitly and emphaticaly makes room for refraction as a *constant* factor in vision.[28]

That we see objects outside the cone of vision is a fact that Ibn al-Haytham demonstrates in that passage by a simple experiment (figure 4.1). He places a slender object (a surgeon's "probe"/*mīl*/Gr. *mēlē*) close to the outer corner of one eye while the other eye is closed: the probe will be visible even when it is definitely located outside the limits of the geometrical cone as defined by the

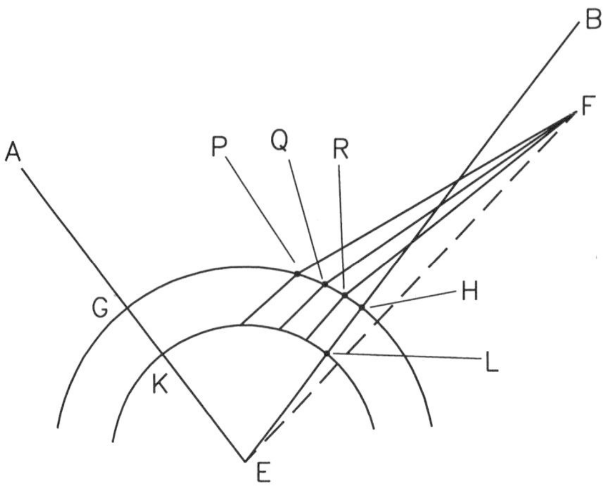

Figure 4.1 (constructed from the text)
Probe experiment. Only refracted forms from F, placed outside the geometrical cone defined by GEH, will reach the crystalline's surface KL. All will be "sensed" by the crystalline, and all will be "perceived" by the "sensitive faculty" as located on the perpendicular FE, itself considered a "radial line" proceeding from E though not one of the lines constituting the cone that is limited by the pupil's width.

narrow pupil. Thus the light reaching the crystalline from points on the probe will have arrived at the crystalline's surface on lines *all of which* intersect the lines of the cone, and, therefore, all are inclined to the eye's surface (i.e., the portion opposite the pupil) where they will have been refracted to points on the crystalline's surface where, we are told, *sensation* will first take place. In these circumstances, then, as Ibn al-Haytham concludes, objects are seen *only* by refracted forms or rays.[29]

Other experiments demonstrate that all objects *within* the cone are also seen by refraction as well as being visible by the agency of the light or forms that arrive unrefracted at the crystalline's surface in the manner explained in Book I (figure 4.2). A needle MN is held close to one eye against a white wall AB. Let E be the center of the eye, GH the surface of the eye, and TV the parallel surface of the crystalline; G, E, H thus define the visual cone. CD, the area on the white background that is concealed by the opaque needle MN, cannot be seen by rays perpendicular to the eye's surface, such as PY, proceeding rectilinearly from points on CD toward E. To be visible, the light from points on CD would have to arrive at the eye's surface on non-perpendicular lines, such as PMS and PNS', and must therefore be refracted through GH before it strikes the crystalline's surface (where it will be refracted again). However, as Ibn al-Haytham observes, the eye will actually perceive the screened area CD, though not as clearly as it does the neighboring areas on AB, but will perceive it as if it were visible through a transparent body much broader than the needle's diameter. The experiment thus shows that while the eye sees the near side of the needle by means of forms reaching it along the radial lines, it can only see CD by lights or forms that have been refracted at the eye's surface. Hence the impression of seeing both the needle and the screened area at the same time, as if the much closer needle had the effect of an interposed transparent body—a conclusion which Ibn al-Haytham confirms further by replacing the thin needle by an opaque object broad enough to prevent the forms/rays from reaching the crystalline's surface from points on the screened area by refraction. He finally concludes that what has been shown with regard to points on CD must be true of all visible points on the white wall and of all points included in the visual cone as previously defined.[30]

Aware that he has just said and done something new, Ibn al-Haytham proudly declares, at the end of chapter 6 in Bk VII, that no one before him, in ancient or in recent times, had recognized the fact that "*all* that sight perceives it perceives by refraction."[31]

Thus we are finally led to ask: How does the introduction of an essential role for refraction as a "universal cause" or explanation (*ʿilla kulliyya/causa universalis*)[32] in Book VII affect the theory of vision presented in Book I,

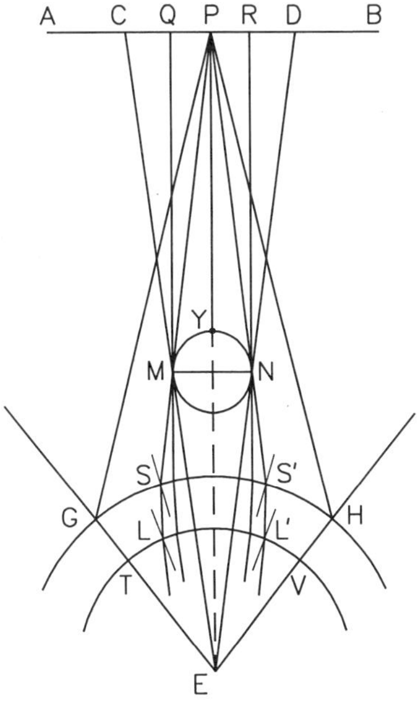

Figure 4.2 (constructed from the text)
Needle Experiment

which the later account obviously makes more problematic? The description of the above experiments in favor of this later addition is preceded by a "general statement" in the course of which we read the following:

> [a] We have shown in the First Book that if the sensitive organ [i.e., crystalline] were to sense, through (*min*) every point on its surface, every form that comes to it, then it would sense the forms of visible objects as mixed with one another and not distinctly. This indicates that the sentient organ does not sense the forms except through (*min*) the perpendiculars to its surface only, and that if it senses the forms through the perpendiculars to its surface then the visible objects will be distinguished and none of their forms will (appear) to it mixed. *And we have shown in the present Book [VII] that refracted forms can be perceived by sight only on ('alā) the perpendiculars drawn from the objects to the surfaces of transparent bodies.* All that being so, the forms refracted through the layers of the eye must be perceived only on the perpendiculars drawn from the objects to the surfaces of the eye's layers.
>
> [b] Now, these perpendiculars are the lines that issue from the center of the eye. Therefore sight will perceive all forms refracted through the eye's layers and sense them, but will perceive them on the straight lines drawn from the eye's center. The forms of all objects facing the portion of the eye's surface that lies opposite the uvea's aperture will thus occur in this portion of the eye's surface, and will be refracted through the transparency of the eye's layers, and reach the sentient organ, viz. the crystalline, and the crystalline will sense them, and the sensitive faculty will perceive them on the straight lines joining the eye's center and those objects.
>
> [c] I mean that the form of every point on every object facing the portion of the eye's surface that lies opposite the uvea's aperture will occur in the whole surface of this portion, and will be refracted from the whole of this portion and reach the crystalline humor, and the crystalline humor will sense all that will reach it of the form of every one of these points, and the sensitive faculty will perceive all that will reach the crystalline of the form of each point on the single line that joins the eye's center and that point.—That is the manner in which sight perceives all visible objects.[33]

In the paradigmatic representation of optical refraction here cited in the underlined sentence in paragraph [a]—a representation transmitted by Ptolemy's *Optics* (in terms of visual rays)[34] and adopted by Ibn al-Haytham and fully stated earlier (in terms of light rays) in the same chapter from which the above passage is quoted—the eye sees an object by refraction on the extension of the refracted ray (Ptolemy's incident ray) at the point where it intersects the perpendicular from the object to the refracting surface. In *that* representation, both the direction of the visible "image" (called *khayāl* by Ibn al-Haytham) and its distance from the eye are determined. Note, however, that Ibn al-Haytham's citation here of the Ptolemaic rule specifies the direction but not the perceived distance of the form/*ṣūra*: the form arriving at the crystalline's surface and refracted through the crystalline's body is said to be perceived "on" (Arabic *ʿalā*/Latin *in*) the perpendicular from the object to the crystalline's surface—period. Thus, neither in Ibn al-Haytham's "general statement," nor in his subsequent experimental reports, is anything said about the meeting or intersection, of refracted rays or their prolongations, with the perpendicular from the object to the crystalline, this perpendicular being one of the radial lines drawn from the eye's center. And there is no mention in these reports of an "image"/*khayāl*, the word consistently used elsewhere for a form/*ṣūra* that is seen behind a reflecting or refracting surface.

What, then, do we finally learn in Book VII with regard to the perception of objects whose light has reached the surface of the eye, by way of rectilinear propagation or by reflection or refraction?—Answer: (1) that these objects are *in all cases* "perceived" by refraction; (2) that some of these objects (namely those inside the radial cone defined by the size of the pupil) will generally, though not always, also make their effect by perpendicular forms or rays; but (3) that all forms, no matter how they have arrived at the eye, will be "sensed" by the crystalline and "perceived" by the "sensitive faculty" as forms of objects situated on the perpendicular from the object to the crystalline's surface.

As a consequence of (1)–(3), we learn further, and *contrary to* Book I, that (4) it is no longer a condition of distinct vision that a given point on a visible object be perceived from a form reaching the crystalline from *a single point only* on the eye's surface (see above, p. 98); rather, the light from an object-point, which always shines in the form of a cone whose base covers the portion on the eye's surface opposite the pupil, will always stimulate the crystalline's sensitivity by means of *all* the forms that may reach it from *any* point on that portion. And, finally, we learn (5) that the concept of "visual cone" has been definitely though quietly re-defined to include lines, such as EF in figure 4.1, which is now called a *radial line* "by way of extension" (*ʿalā ṭarīq al-istiʿāra*) "because it resembles [the traditional radial lines] in that it issues from the center of the eye."[35]

The "obstacle" created by the role traditionally assigned to the crystalline humor as the sensitive part of the eye thus remains. But the experiments described in the *Optics*, Bk VII, Ch. 6, along with their stated conclusions, now point to new problems (including of course the question as to *where on the perpendicular* is the object seen) and to possibilities which some of Ibn al-Haytham's readers in the Muslim world (e.g., Kamāl al-Dīn al-Fārisī) and in Europe (especially Kepler) were led to grapple with.[36]

IV A New Role for Psychology: The Example of Depth Perception

In Book II of the *Optics*, having already argued against the visual-ray doctrine as futile (*ʿabath wa faḍl/superfluus et otiosus, Optics*, I, 6[54]=Risner 14:26) and even "impossible"/*mustaḥīl* (*Optics*, I, 6[59]), Ibn al-Haytham goes on to expose the argument levelled by "mathematicians" [and by Galen] against the view preferred by the "proponents of physical science": ". . . if [so the argument went] vision takes place by means of a form which passes from the visible object to the eye, and if the form occurs within the eye, then why does sight perceive the object in its own place outside the eye while its form exists inside the eye?" He answers: "Those people have ignored the fact that vision is not achieved by pure sensation alone, and that it is accomplished only by means of discernment and prior knowledge (*bi-al-tamyīz wa taqaddum al-maʿrifa/per cognitionem et distinctionem antecedentem*—Risner 39:3–4), and that without discernment and prior knowledge sight would achieve no vision whatever (*lam yatimma li-l-baṣari shay'un min al-ibṣār/ non compleretur in visu visio*), nor would there be perception of what the visible object is at the moment of seeing it" (*Optics*, II, 3[71], emphasis added). This passage, occurring at the beginning of a long discussion of depth perception, points out, clearly and succinctly, the aim and scope of the whole project of the *Optics*. Ibn al-Haytham is saying that no theory of visual perception in terms of the reception of forms or images (which according to him arc analysablc in terms of the effect of light radiation in the eye) can do without the support of a theory of vision which is essentially psychological in character. It is therefore no accident that psychological investigations have come to occupy such a considerable part of the *Optics*: Book II, which is wholly devoted to a general theory of the psychology of vision; Books III and VI, on the errors of rectilinear and of reflected vision respectively; and the parts of Book VII that deal with the errors of vision by refraction, including Ibn al-Haytham's new explanation of the "moon illusion" as a psychological phenomenon (Sabra, 1987, 1991/92, 1995/96).

There is, therefore, a necessary connection between Ibn al-Haytham's theory of vision as an image-oriented theory and the particular emphasis he lays on psychology. More than that, this connection can easily be shown to have substantially influenced the very character of his psychological explanations. If, as Ibn al-Haytham believed, distinct vision required a distinct image as the only accessible message from the object, then the mental activities of "recognition," "comparison," "discernment," "judgment," and "inference," as operations accompanying every act of visual perception beyond the mere reception of sense impressions, must be concerned in the first place with the image itself as the sole optical datum. The significance of this remark can perhaps be best appreciated when we compare Ibn al-Haytham's psychological explanation of depth perception with visual-ray accounts such as we find in Ptolemy and Galen.

Before we do this, let us note briefly that the form/image finally presented in the common nerve to the sense faculty, though still different from the image produced by a pin hole or a lens, can be said to be distinct in a sense that is not applicable to forms/images in the crystalline. The reason is that the form/image in the common nerve has already been disengaged (selected and isolated) from the countless forms with which it was mixed before it was sent off, alone and undistorted, to the common nerve. Ibn al-Haytham thus speaks, in *Optics*, II, 3[47], of the effect exerted by the arriving form on the pneuma that fills the common nerve as one of luminosity and coloration, but now these qualities are understood to represent, distinctly and accurately, their distribution on the object: "The form then reaches the cavity of the common nerve, whereupon that part of the sentient body [= pneuma] in that cavity where the form of the object has arrived becomes colored by the color of that object and illuminated by its light. If the object is of one color, then that part of the sentient body will be of one color; if the parts of the object have different colors, then the colors of the portion of the sentient body that is in the cavity of the common nerve will be different."

As A. Lejeune has already demonstrated in 1948, Ptolemy's explanation of the perception of the distance of a visible object along the stretch of the visual ray, this distance being a factor in discerning the object's localization in space, amounted to nothing more than the assertion that estimation of the distance was simply due to our consciousness of the length of the ray we have sent out to grasp the object (Lejeune, 1948, IV.1, pp. 86–95; *Ptol. Opt.*, II, [26]). Not much of an explanation, perhaps, to a modern mind, but it must have counted as one of the points actually supporting the widely received visual-ray hypothesis, both in antiquity and in the early Arabic tradition, and even later in Islam and in Europe. It will thus be interesting to refer here to one of the most revealing passages in Galen's *De placitis Hippocratis et Platonis* where he defends the view adopted by the "geometers," in words reminiscent of the position and language found in Ptolemy's treatise. Galen states first that "the proper object of sight, which I also call its primary sense-object, is the class of color. For colors are the first thing it perceives, and it perceives them by itself, and it alone of all the sense organs can perceive them, . . . sight alone can discern (*syndiagignōskein*) along with the color of the thing seen its size and shape, . . . There is a detailed discussion of such things in the fifth (book) *On Demonstration*. In any case, sight can discern, in addition to other things, the position (*thesin*) and distance (*diastēma*) of the colored body. . . . I demonstrated in what was said there [in the fifth (book) *On Demonstration*] that everything supports the view that the observed body is seen in the place where it actually is. *This is clearly revealed also through sense-perception itself, and for that reason the geometers give no proof of it but posit it as self-evident;. . . ,*" emphasis added.[37]

Reading Ibn al-Haytham on the question of how an object is seen in its own place (*mawḍiʿ/locus*) we find ourselves in an entirely different framework of explanation. The question is split into others concerning the perception of distance as such or remoteness/*buʿd*/*ex remotione*, direction/*jiha*/*ex parte universi*, and the magnitude or measure of distance/*miqdār* or *kammiyyat al-buʿd*/ *ex quantitate remotionis* (*Optics*, II, 3[67]=Risner 38:42–43). The same distinctions (or some of them) are in Ptolemy (*Ptol. Opt.*, II, [26]; Lejeune, 1948, pp. 87ff). But now we are treated to an incomparably rich discussion the like of which in scope and sophistication is not found in earlier writers in Greek or Arabic. Most important in this discussion is the fact that it is dominated throughout by strict adherence to the various elements of an integrated theory, including especially the all-important element represented by the *newly argued* concept of an optical image in the common nerve. The passage quoted earlier (p. 104) from *Optics*, II, 3[71], continues as follows: "For what the object is is not perceived by pure sensation, but [either] by recognition or by resuming the [original] discernment and inference at the [current] moment of vision. Thus if vision were effected by pure sensation alone, and if all perceptible properties [size, shape, etc.] in the visible objects were perceived only by pure sensation, then the object would not be perceived where it is unless it was reached by something which touched and sensed it. But if vision is not effected by pure sensation alone; and if all perceptible properties of visible objects are not perceived by pure sensation; and if vision is not accomplished without discernment, inference, and recognition; and if many of the visible properties are perceived only by discerment; then to perceive a visible object in its own place there is no need for a sentient [thing] to extend to it and touch it."

In this context, the sentient [thing]/*al-ḥāss* is of course the pneuma or visual flux, and it should be clear against whom this argument is addressed. Other passages in the *Optics* also make it clear that by pure sensation/*mujarrad al-ḥiss* Ibn al-Haytham means nothing more than the light and color excitation in the crystalline and in the common nerve—not even the awareness of light as strong or weak, or the awareness of color as red or blue, these modes of awareness being actually "judgments" involving the various mental operations mentioned (II, 3[49–66]). But the crucial point to be borne in mind (especially in comparisons with Aristotle and Ptolemy) is that these operations do not only presuppose "pure sensation" as "affection" (*infiʿāl/passio*) suffered by the eye. They are applied in the first place to features of the "ordered" form/

image that has been transported to the common nerve. For although the immediate effect of light, as Ibn al-Haytham tells us, is a sensation/excitation "of the nature of pain" and can, as such, share the manner of nervous transmission of painful and tactile excitations (I, 6[67]),[38] it is also a sensation of luminosity and colors and their configuration (I, 6[81]). And, as such, the sensation must be mediated through the visual apparatus by a succession of physiological processes according to definite modes of transmission: that is, along segments of the radial lines within the crystalline, then on suitably refracted lines within the vitreous, and finally along the fibers stretching through the curving optic nerve—all of these differentiated modes of transmission being designed to preserve the integrity of the object's form/image (I, 6[80–81]).[39]

Ibn al-Haytham calls those features in the image "signs" or "cues," *amārāt/signa* (*Optics*, II, 3[22, 24, 45]). Thus the size of the image on the crystalline's surface (equivalent to the visual angle) can be a cue to the object's size, the image's shape a cue to the object's shape, the image's brightness or obscurity a cue to the object's nearness or remoteness, and so on. Hence the characteristic complexity we find in the psychological explanation of estimating distance in the *Optics*. First, we are told that, as with all properties of visible objects other than their luminosity and color, distance is perceived *separately* only by discernment, but as a result of accumulated experience it is perceived by "prior knowledge" [72]. Certain experiences (opening and closing the eyelids, turning our eyes toward or away from the object seen) lead to the judgment that the object is spatially separate from us [73]. With time, this discernment, necessary at first, becomes automatic or, as Ibn al-Haytham puts it, "established in the soul" [74]. While objects are always seen to be at *some* distance from the eye, the magnitude of their distance is not always "ascertained." If an object's distance lies along a sequence of contiguous bodies, then the distance's magnitude will be perceived by estimating the size of the intermediate bodies [76], and this judgement will be correct (or nearly so) when the distance is a moderate one. A correct judgement of distance will thus depend on an accurate (or nearly accurate) estimation of *size* [76–77]. This is supported by observations of clouds below the tops of not very high mountains, and is further confirmed by experiments on the ground [79–80]. Such "inferences," however, are not possible in the absence of intermediate bodies, as when we look at a star high up in the sky [84], and the general conclusion is drawn that distances can be ascertained in magnitude only when they are moderate in size, and only when they stretch along a series of visible objects whose sizes are ascertainable [86].

What happens then when the distances of objects cannot be "ascertained"? The answer is that "the faculty of judgement immediately conjectures their magnitudes by comparing their distances with those of [similar] objects which sight has previously perceived and [the magnitudes of whose distances] it has ascertained" [87]. Strictly speaking, this is a comparison between currently perceived "forms" and "similar forms" previously perceived and ascertained. "That [says Ibn al-Haytham] is the limit of what the discerning faculty is capable of in the process of attaining perception of the magnitudes of the distances of visible objects" [88]. Accompanying the immediate perception of remoteness (distance as such) there is, therefore, also a perception of the magnitude of that remoteness, which is either ascertained or conjectured [90]. And, with regard to familiar objects "recognized" from familiar distances, there will be "no great discrepancy" between the conjectured magnitude of their distances and the real magnitude [93].

It remains to show how the size of an object is accurately estimated by sight, this being a condition for ascertained perception of distance. In his discussion of size (*Optics*, II, 3[135–171]) Ibn al-Haytham begins by accepting Ptolemy's arguments (against "the majority of the mathematicians") in support of the view that size is determined not by the angle of vision alone but also by the object's perceived distance and orientation (*waḍʿ/situs*), citing, among other observations, what is now known as size-distance constancy [135–140]. The angle, or equivalently, the area cut off on the eye's surface by the visual cone, remains a basic criterion/*aṣl/radix* for judging size, which itself varies in magnitude with thc object's distance [142–143]. How, then, is the length of the cone determined? We are back to the question simply answered in the visual-ray theory by postulating immediate consciousness [144–148]. The question is now crucial especially with regard to "ascertained," as distinguished from "conjectural," distances. And it is at this point that Ibn al-Haytham's considerations rise to a still higher level of elaborateness by correlating angles of vision, or forms of objects in the eye, with experiences that furnish non-optical measures of magnitude. He proposes, in effect, a sort of unconscious inference that leads step by step from estimating distances of objects close to our feet to judging distances of farther and farther objects on the ground which subtend progressively decreasing angles at the center of the eye [149–156]. The first *distance-length*, that of a point on the ground next to my foot, is measured by my height; the *size* of an immediately adjacent area on the ground that has been measured (and correlated with corresponding angles) many times by my stepping on it or by stretching my arm to it. This gives me an estimate of the length of the ray reaching my eye from a point displaced by a given amount from the first. And so on, by continually comparing ray-lengths with angles (or corresponding sizes of their images in the crystalline), we get to form notions of the size of terrains separating us from the objects whose distances we come to judge, first by calculation, and subsequently by recognition based on repeated experiences and on repeated and unintended verifications. Ibn al-Haytham finally adds that by "estimation" of distances he does not mean anything quantitatively precise, but simply the process of forming a standard in our imagination for making further estimations.

V Conclusion

The seven books of Ibn al-Haytham's *Optics* constitute a unified project with a single, continuous argument running through them. In the working out of this project the fundamental concept of form is Aristotelian, and so is the basic idea of vision as something initiated in the perceiver by the impression made in the eye by a "form" combining the qualities of luminosity and color. But the *punctiform* analysis of forms (Vasco Ronchi's apt term) as a means of bringing points of the object's light and color into the eye, and as a guide leading them through the eye's tunics and humors, is certainly not. Ptolemy is believed to have presented in the lost Book I of his *Optics* a theory describing the role of external light, as distinguished from the radial emission of visual "flux" from the eye, but we do not have a precise idea of the details of that theory (though possibly more can be learned about it by excavating the Arabic sources). At any rate, Ptolemy's account of the role of external light (whatever its precise nature) was clearly intended to work in company with, and not as a substitute for, the explanations in terms of the visual-ray doctrine that actually dominates the rest of his treatise. Ibn al-Haytham proposed, in contrast, a theory of light radiation whose purpose was to do away with visual radiation altogether by relegating to itself all functions previously assigned to the visual ray as a physical entity, while retaining its use as a geometrical abstraction. His theory of vision further employed the concept of an optical image in the eye, the construction of which according to the established behavior of light had not been part of either the Aristotelian or the Ptolemaic program. And an immediate consequence of his attempts in this direction was his implementation of a persistent psychological approach to the whole problem of visual perception.—It is evident that the synthesis worked out in Ibn al-Haytham's *Optics* incorporates elements of various sorts which it has derived from earlier treatments of vision. It is also evident that those elements are now made to serve the aims of an entirely new project in which they perform new functions.

One inherited element in particular, the role assigned to the crystalline humor as the sensitive part of the eye, had two distinct consequences. The first was that it encouraged (perhaps even pushed) Ibn al-Haytham to postulate a geometrical structure of the eye that retained the basic geometry of the visual cone whose lines alone defined the course of efficacious rays (or point-forms) within the crystalline. Thus he lost sight of a crucial fact of ocular physiology—which is not to say that the "fact" was lying out there for all to see! The experiments described by Ibn al-Haytham in *Optics*, Book VII, chapter 6, some of whose striking conclusions he boldly articulated and accepted, were strictly incompatible with the hypotheses developed in Book I in terms of perpendicular rays. These experiments appear, however, to have come late to Ibn al-Haytham's notice. And although they eventually led him, explicitly, to widen the visual cone in order to admit the experimentally proven efficacy of refracted rays within the eye, he thought he could accommodate them by appealing once more to psychology—which brings us to the second consequence, namely the new emphasis on psychological explanations.

In and of itself the new level of emphasis on psychology in Ibn al-Haytham's *Optics* was historically an important development that has recently attracted the attention of psychologists and historians of psychology. The emphasis allowed him to open a new chapter in the history of the systematic study of vision. But the impressive psychological observations and explorations that we find displayed throughout the book were also frequently overburdened. As an element in a complete explanation of the act of visual perception, the psychological approach was of course appropriate, indeed indispensable; but, in Ibn al-Haytham's treatise, psychology was required to accomplish more than it could possibly achieve by itself. This requirement had a lot to do with the "obstacle" we noted in Section III, and the difficulties involved are already evident in Ibn al-Haytham's theory of direct vision, for example in his treatment of depth perception. But the inherent problems become especially acute in Ibn al-Haytham's treatment of specular images, where we find him following an admirably consistent but ultimately doomed course. I believe that Kepler at one point became aware of these problems which, as I believe, he correctly diagnosed, and which he was able to overcome by removing the "obstacle." It is my intention to pursue these last observations in a sequel to this discussion.

NOTES

1. Prominent examples of mathematicians/astronomers who contributed to medical literature are Thābit ibn Qurra and Ibrāhīm ibn Sinān in the ninth and the tenth century AD, Ibn al-Haytham and al-Bīrūnī in the eleventh, Naṣīr al-Dīn al-Ṭūsī and Quṭb al-Dīn al-Shīrāzī in the thirteenth. Thābit (whose medical writings were considerable) is also known to have practiced medicine, as did other members of the Ibn Qurra family.—For Thābit see Ibn al-Qifṭī, *Ta'rīkh al-ḥukamā'*, ed. J. Lippert, Leipzig, 1903, esp. pp. 116–19; F. Sezgin, *GAS*, III (Medizin, etc.), pp. 260–63.—Work number 44 in the autograph list of writings by Muḥammad ibn al-Ḥasan ibn al-Haytham in the fields of "natural philosophy and metaphysics," all composed before 10 February 1027, consisted of

summaries/epitomes (*jumal wa jawāmi*ʿ) of 30 works of Galen; see IAU/Müller, 1882, II, p. 95.—For al-Bīrūnī see C. Brockelmann, *Geschichte der Arabischen Litteratur*, I (Leiden: E.J. Brill, 1943), p. 627 (no. 27), S II(Leiden, 1937), pp. 874–75; S.H. Nasr, *al-Bīrūnī: An Annotated Bibliography*, Tehran, 1973, pp. 101–03.—The medical activities of Sinān ibn Thābit ibn Qurra and Ibrāhīm ibn Sinān are mentioned in IAU/Müller, 1882, I, pp. 224, 226.—For Ṭūsī and Shīrāzī see Brockelmann, *GAL*, S II(1938), pp. 932-33, and II(1949), pp. 274–75, S II, pp. 296–97.

2. For Ḥunayn's *The Book of the Ten Treatises on the Eye*, see Meyerhof, 1928. The integral text of Aḥmad ibn ʿĪsā remains in manuscript (an edition is being prepared by E. Kheirandish and A.I. Sabra). Two Arabic MSS are: Istanbul, Laleli 2759(2), and Ragip Pasha 934; for dates and discussions of Ibn ʿĪsā, see the Introduction to Ibn al-Haytham, 1989, vol. II; Rashed, 1997; Kheirandish, 1999, vol. I. The surviving optical writings by al-Kindī are edited, translated into French and annotated in Rashed, 1997, which includes a revision by J. Jolivet, H. Sinaceur, and H. Hugonnard-Roche of *Liber Jakob Alkindi De causis diuertitatum aspectus*, first edited by A. Björnbo and S. Vogl—see Björnbo & Vogl, 1912. Also included in the 1997 volume is the work by Qusṭā ibn Lūqā on specular images, *Fī ʿIlal mā yaʿriḍu fī al-marāyā min ikhtilāf al-manāẓir.*

3. Galen/De Lacy, 1980, Second Part, Bk vii, esp. pp. 453ff. See n. 37 below.

4. See, for example, Ḥunayn, *Ten Treatises*, in Meyerhof, 1928, Arabic Text: pp. 91; 77, 80; 79, 81.

5. Ḥunayn, *Ten Treatises*, in Meyerhof, 1928, p. 95 (where *manāẓir* is translated as "glances"); Kindī, "Rectification," in Rashed, 1997, p. 167, line 11, and *passim*; and Qusṭā ibn Lūqā, *ʿIlal*, in Rashed, 1997, p. 583, lines 17–18.

6. A. Björnbo & S. Vogl, 1912, pp. 42ff.

7. Rashed, 1997, p. 173, lines 19–25.

8. Lejeune, 1948, pp. 22–24.—The work of the tenth-century philosopher, Abu 'l-Ḥasan al-ʿĀmirī (d. 382/992) (see al-ʿĀmirī/Khalīfāt, 1988), *Qawl fī al-baṣar wa al-mubṣar* ("On Vision and the Object of Vision"), betrays certain interesting similarities of ideas and terms with Ptolemy's *Optics* which are not easily detectable in the ninth-century writers referred to above. He mentions Ptolemy's name (but not his *Optics*) side by side with Euclid's as two "philosopher-geometricians" (*ḥukamāʾ al-muhandisīn*) who upheld a theory according to which [visual] "light" goes out of the eye in the shape of a cone to the "sensible color;" this light is said to be "strengthened" by the external light before it is "obstructed" (*yaʿūquhu*) by the "colored body," thereby causing the "form of the color" to be fastened (*tanʿaqidu*) to the end or base of the cone, whereupon the "light" turns back (*yankusu munʿakisan*) to the source/origin (*yanbūʿ*) of the cone.—He then attributes this account of *ittiṣāl* (contact between the organ of sight and the object of vision) also to "Galen and his followers among physicians" p. 435.—He offers (p. 419) a list of *seven* "visible properties" (*al-maʿānī al-marʾiyya*), in this order: (1) color, (2) size, (3) shape/figure (*shakl*), (4) number, (5) distance, (6) the state of being at rest or in motion (*hayʾat al-sukūn wa al-ḥaraka*), and (7) substance (*jawhar*). Three of these (here numbered 4, 5, 7) are not included in Ptolemy's formal list of *res videndae* in his *Optica* (II, [2]: "*Dicimus ergo quod uisus cognoscit corpus, magnitudinem, colorem,*

figuram, situm, motum et quietem" (Lejeune, 1989, p. 12, and n. 2), although 4 and 5 are actually discussed by Ptolemy. On the other hand, two visible properties in Ptolemy's list, namely *corpus*/*jism*/corporeity (which should not be confused with ʿĀmirī's no. 7: *jawhar*), and *situs*/*waḍʿ*/local position, are missing from ʿĀmirī's list.

Also according to ʿĀmirī, some of the visible properties are perceived by themselves/ *bi-al-dhāt* (these are: color, size, shape, being at rest or in motion), while others (substance, number, and distance) are perceived "incidentally"/*bi-al-ʿaraḍ*; only color is perceived "firstly" (*mudrak*^{un} *idrāk*^{an} *awwaliyy*^{an}), whereas size, shape, and the state of being at rest or in motion are each perceived "by second intention"/*bi-al-qaṣd al-thānī*. Some of these expressions correspond to similar ones in Ptolemy: compare, for example, ʿĀmirī's *firstly* and *by second intention*, with Ptolemy's *primo*, and *sequenter*, respectively (Ptol. Opt., II, [3, 6].

The above similarities *and* discrepancies immediately pose the question as to what might have been ʿĀmirī's source(s). I hope to discuss this question in more detail in another publication. Here I am inclined to postulate a Peripatetic origin for these terms and concepts in both Ptolemy and ʿĀmirī.

9. For Ibn ʿĪsā, see Ibn al-Haytham, 1989, vol. II, Intro., pp. xxxvi–xxxvii, and lix, n. 85; for al-Kindī's text see Rashed, 1997, pp. 424–427.

10. The relevant text is quoted in Ibn al-Haytham, 1971, pp. 74–77. See also Ibn al-Haytham, 1989, vol. II, Intro., pp. xxxiv–xxxvii.

11. Ibn al-Haytham, 1989, vol. II, Intro., pp. lxix–lxx, n. 110, and p. lxxi, n. 112.

12. Incidentally, this is a fact to be borne in mind when considering the role of optics in Arabic astronomical investigation up to the time when Ibn al-Haytham's *Optics* became *generally* known to astronomers, that is, after the composition of Kamāl al-Dīn al-Fārisī's "Commentary" on it at the end of the 13th century.—On the astronomer Muʾayyad al-Dīn al-ʿUrḍī's knowledge of the *Optics* see Sabra, 1998/99, p. 307, n. 24.

13. Ibn al-Haytham, 1989, II, Intro., pp. lix–lx; Rashed, 1993; Sabra, 1994c.

14. IAU/Müller, 1882, II, pp. 93–94.

15. Hogendijk, 1985, pp. 62–63.

16. For the latter topic see Sabra, 1998, pp. 27–30, and references cited there.

17. IAU/Müller, 1882, II, p. 87 (no. 20), and p. 98 (no. 27); Ibn al-Haytham, 1989, vol. II, Intro., p. xxxiii.

18. Ibn al-Haytham, 1989: *Optics*, I, 1[2]; vol. II, pp. 4–7.

19. For the Arabic translation of *synaugeia* in Plato's account see Daiber, 1980, pp. 202, 550, 590. I suspect that *ishtirāk* in the sense of Plato's term is what underlies *communicant* in Ptolemy's *Optica*, in the Latin translator's Preface, para. [2]—see Lejeune, 1989, p. 5, and facing translation. For translating this verb in the same paragraph, Lejeune has "*s'associent*"; Smith, 1996, has "interact."

20. The reader should note that the Preface and the two following chapters of Book I are missing from the Latin version.

21. See Ibn al-Haytham's *Qawl fī al-Ḍawʾ* ("Discourse on Light") in *Rasāʾil*, pp. 17–18.

22. *Optics*, IV, MS Fatih 3215, fols 62bff.

23. Sabra, 1981, pp. 72–78, 93–98.

24. At the end of his interesting mechanical discussion (in Book IV) of the reflection of vertical free and forced motion, Ibn al-Haytham says that "this not being the place for a thorough discussion of this matter, it will be enough to make do with what we have determined regarding the examination of the motion of reflection" (Fatih MS 3215, fol. 70b).

25. Aristotle, *De anima* II.12: 424a15. A detailed account of what might be described as the most developed "image-oriented" theory of vision up to the time of Ibn al-Haytham can be found in Book III (*al-maqāla al-thālitha*) of the *De anima* composed by his contemporary Avicenna—see Avicenna/Rahman, 1960; Avicenna/Van Riet, 1972. The theory is basically Aristotelian, but it combines Aristotle's view of vision as the reception of visible forms in the organ of sight with physiological and pneumatic doctrines transmitted through Galen's works: The "form" of an object is "conveyed" by an actually transparent medium (illuminated air) to the eye where it makes an impression on the crystalline humor. But this is not perception yet. Visual "perception" takes place only when the two forms in the two crystallines reach the front of the brain, after their uniting together in the optic chiasma where the optic nerves cross each other before diverging again. Only in the front ventricle (*farāgh*) of the brain, which is filled by the pneuma that carries "the power of common sense," will the form finally make the effective impression received by the common sense itself, and vision will be completed.—Avicenna refers to the proponents of this theory as "the upholders of the doctrine of simulacra"/*aṣḥāb al-ashbāḥ*/*auctores sententiae de simulacris*– apparently using *shabaḥ* in this context interchangeably with *ṣūra*/*eidos*. (The Medieval Latin translation alternates between *simulacrum* and *forma*, but also uses other terms including *imago*—see *Lexique Arabo-Latin* in Van Riet's edition.) Avicenna repeatedly contrasts this doctrine with that of *aṣḥāb al-shuʿāʿ*/*auctores sententiae de radiis*, namely the doctrine favored by the mathematicians, which he rejects. But rather than suggest any possible way of reconciling these two positions, Avicenna seems consistently to present them as mutually exclusive. And there is no hint of concern for a mechanism by means of which the form is "conveyed" to the crystalline or from the crystalline to the brain.—There is as yet no satisfactory modern translation of Avicenna's rich discussion; see however the section devoted to Avicenna in Lindberg, 1976, pp. 43–52.

26. See also Ibn al-Haytham, 1989, vol. II, pp. 53–54; and Galen/May, 1968, pp. 463–464, where Galen asserts as a matter of empirical truth "that the crystalline humor is the principal instrument of vision, a fact clearly proved by what physicians call cataracts, which lie between the crystalline humor and the cornea and interfere with vision until they are couched." See also the Index in May's edition for related statements on the crystalline. Ibn al-Haytham does not mention Galen (or any other medical authority) by name, but after describing the experiments which we find reported throughout the medical tradition up to his time and later, remarks: "All this is attested by the art of medicine" *Optics*, I, 6[14].

27. *Optics*, I, 6[65]=Risner 15:7-9; I, 6[90]=Risner 17:1-4; II, 2[11]=Risner 26:1-9; Sabra, 1978, pp. 165–166.

28. There are reasons (which cannot be discussed here for lack of space) that lead me to conclude that the passage in question must represent a later development of Ibn al-Haytham's thought, that is, an expansion of his theory made necessary in light of compelling experiments the force of which he either was not aware of or had not fully realized when he first conceived the arguments expounded in Book I.

29. MS Fatih 3216, fols. 102b-103b=Risner 269:26–46.

30. Ibid., fols 103b–106a=Risner 269:46–270:29.

31. Ibid., fol. 106b:8–9=Risner 270:39–40; emphasis added.

32. Ibid., fol. 106b=Risner 270:37.

33. Ibid., fols. 101a:2–102a:1=Köprülü MS 952, fol. 73a:17-31=Risner 268:61–269:16.

34. Lejeune, 1957, pp. 167ff.

35. MS Fatih 3216, fol. 102b=Risner 279:24.

36. I may briefly mention here two modern comments on Ibn al-Haytham's treatment of refraction in the eye. The first is by Muṣṭafā Naẓīf, who gives a lucid account of the experiments in *Optics* VII.6, points out the serious departure from the theory expounded in Book I, notes how close Ibn al-Haytham came in the final account to the breakthrough which we associate with Kepler, and (in connection with this last observation) ends with the counterfactual: had Ibn al-Haytham proposed that "the image" (*khayāl*) of the visible point was located at the intersection of the perpendicular with the refracted ray itself inside the eye (rather than with its prolongation outside the eye), he would have placed himself "centuries" ahead of his own time (Naẓīf, 1942, pp. 233–239).—True, but the fact that he did not is itself interesting and historically significant.

The second comment is by David Lindberg. He too points out the discrepancy between Books I and VII, and the impossibility (in some cases) of intersection between the perpendicular and the prolongation of the refracted ray (Lindberg, 1976, pp. 76–78).—But, as indicated above, such an intersection was not intended. Both comments seem to me to underestimate the effect of Ibn al-Haytham's initial suppositions and the implications of his precipitous though understandable appeal to psychology (see Section IV below).

My own analysis above is meant to answer two questions: one is about what drove Ibn al-Haytham in the direction he *actually* took, and the other concerns *the nature of the problem he handed down to his successors*. I would however agree with Lindberg's remark that "Kepler's principal innovations were a response to precisely this problem of nonperpendicular rays [i.e., those that must be refracted within the eye] and the necessity of establishing a one-to-one correspondence [between points on the object seen and points on the image]" (p. 78).

37. Galen/De Lacy, 1980, vii.5: 32–40, emphasis added. On the ninth-century Arabic translations of Galen's *De placitis* and of parts of his *On Demonstration*, see

Bergsträsser, 1925, Ḥunayn's Arabic text and German translation, pp. 26–27/21–22 and 47–48/38–39; Sezgin, *GAS*, III, pp. 105–106, no. 37; see Ptol. Opt., II, [22, 23, 26].

38. See also Ibn al-Haytham, 1989, vol II, Commentary, p. 56, and *Optics*, II, 2[14–15]=Risner 26:39ff.

39. Ibn al-Haytham, 1989, vol. II, Commentary, pp. 73–74; Sabra, 1989.

References

Note on abbreviations.—The *Optics* of Ibn al-Haytham is referred to by Book (*maqāla*), chapter, and Paragraph numbers, in this form: I, 2[3]. These numbers are the same for the Arabic text (Ibn al-Haytham, 1983), and the English translation (Ibn al-Haytham, 1989, vol. I).

The abbreviation *Ptol(emaei). Opt(ica).* refers to the second edition and French translation in Lejeune, 1989, by Book and Paragraph: e.g., II, [3].

References to Risner's edition of Ibn al-Haytham's *Optics* in *Opticae thesaurus*, 1572 are by page and line numbers, e.g., 38:42–43.

al-ʿĀmirī/Khalīfāt, 1988: Abu 'l-Ḥasan al-ʿĀmirī, *Rasāʾil*. The Arabic texts, edited by Saḥbān Khalīfāt, and with an Introduction translated into Persian by Mahdī Tadayyun. Tehran: Markaz Nashr Dānishgāhī.

Avicenna/Rahman, 1960: *Avicenna's De Anima* (*Arabic Text*). Being the psychological part of Kitāb al-Shifāʾ. Edited by F. Rahman. London: Oxford University Press. Reprint of 1959 edition by the University of Durham.

Avicenna/Van Riet, 1972: *Avicenna Latinus: Liber de Anima I–II–III*. Édition critique de la traduction latine médiévale par Simone Van Riet. Louvain-Leiden: E. Peeters-E.J. Brill.

Bergsträsser, 1925: Gotthelf B., "Ḥunain ibn Isḥāq über die syrischen und arabischen Galen-Übersetzungen," *Abhandlungen für die Kunde des Morgenlandes*, XVII(2), Leipzig.

Björnbo & Vogl, 1912: A. Björnbo, and S. Vogl (eds), *Alkindi, Tideus und Pseudo-Euclid: Drei optische Werke*. Abhandlungen zur Geschichte der mathematischen Wissenschaften mit Einschluss ihrer Anwendungen, XXVI.3, Leipzig-Berlin.

Daiber, H., 1980: *Aetius Arabus: Die Sokratiker im arabischer Überlieferung*. Wiesbaden: Franz Steiner Verlag.

Galen/May, 1968: Galen, *On the Usefulness of the Parts of the Body (De usu partium)*. Translated from the Greek with an Introduction and Commentary by Margaret Tallmadge May. 2 vols. Ithaca, N.Y.: Cornell University Press.

Galen/De Lacy, 1980: Galen, *De placitis Hippocratis et Platonis*. Edition, Translation and Commentary, by Phillip de Lacy. In three parts. Corpus Medicorum Graecorum V4, 1, 2. Berlin Akademie Verlag.

Hogendijk, 1985: Jan P. H., *Ibn al-Haytham's Completion of the Conics of Apollonius*, New York: Springer.

———, 1996: "Al-Muʾtaman's Simplified Lemmas for Solving 'Alhazen's Problem,'" in *From Baghdad to Barcelona: Studies in Honour of Prof. Juan Vernet*, ed. Josep Casulleras and Julio Samsó. 2 vols. Barcelona: Instituto Millás Vallicrosa de Historia de la Ciencia Arabe. Vol I, pp. 61–101.

Ḥunayn ibn Isḥāq, *see* Bergsträsser; Meyerhof.

IAU=Ibn Abī Uṣaybiʿa/Müller, 1882–84, *ʿUyūn al-anbāʾ fī Ṭabaqāt al-aṭibbāʾ*, ed. August Müller. 2 vols. Cairo-Königsberg. (Reprinted with additions as *Islamic Medicine*, vols 1–5, Frankfurt: Institut für Geschichte der Arabisch-Islamischen Wissenschaften, 1995.)

Ibn al-Haytham, 1971: al-Ḥasan ibn al-Ḥasan ibn al-H., *al-Shukūk ʿalā Baṭlamyūs*, ed. A.I. Sabra and N. Shehaby, Cairo: Dār al-Kutub.

———, 1983: *Kitāb al-Manāẓir, Bks I–III: On Direct Vision*, edition of the Arabic text by A.I. Sabra. Kuwait: al-Majlis al-Waṭanī li-l-Thaqāfa wa al-Ādāb wa al-Funūn.

———, 1989: *The Optics of, Bks I–III: On Direct Vision*. English Translation, Commentary, etc. by A.I. Sabra. 2 vols. London: The Warburg Institute.

———, 1357 A.H.: *Rasāʾil*, Hyderabad, Dn.: Dāʾirat al-Maʿārif al-ʿUthmāniyya.

———, *see* Risner.

Kheirandish, 1999: Elaheh K., *The Arabic Version of Euclid's Optics (Kitāb Uqlīdis fī Ikhtilāf al-manāẓir)*. Edited and translated with Historical Introduction and Commentary. 2 vols. New York: Springer.

Lejeune, 1948: Albert L., *Euclide et Ptolémée: Deux stades de l'optique géométrique grecque*. Louvain: Bibliothèque de l'Université de Louvain.

———, 1957: *Recherches sur la catoptrique grecque d'après les sources antiques et médiévales*. Brussels: Académie Royale de Belgique.

———, 1989: *L'Optique de Claude Ptolémée dans la version latine d'après l'arabe de l'émir Eugène de Sicile. Édition critique et exégétique augmentée d'une traduction française et compléments*. Leiden: E.J. Brill. (First published without the French translation in 1956.)

Lindberg, 1976: David L., *Theories of Vision from Al-Kindī to Kepler*. Chicago & London: University of Chicago Press.

Meyerhof, 1928: Max M., *The Book of the Ten Treatises on the Eye Ascribed to Ḥunain ibn Is-Ḥâq (809–877 AD)*. The Arabic Text, edited from the only two known manuscripts, with an English Translation and Glossary. Cairo: Government Press.

Naẓīf, 1942–43: Muṣṭafā N., *al-Ḥasan ibn al-Haytham: Buḥūthuhu wa kushūfuhu al-baṣariyya*. 2 volumes with continuous pagination. Cairo: Fuʾād I University.

Newton, 1952: Isaac N., *Opticks*. 4th ed., London 1730. Reprint of London 1931, New York: Dover.

Ptolemy: *Ptolemaei Optica*, see Lejeune, 1989.

Rashed, 1993: Roshdi R. (ed. & transl.), *Géométrie et dioptrique au xe siècle. Ibn Sahl, al-Qūhī et Ibn al-Haytham*. Paris: Belles Lettres.

———, 1997: *L'Optique et la catoptrique*. Vol. I of *Œuvres philosophiques et scientifiques d'al-Kindī*. Leiden: E. J. Brill.

Risner, 1572: *Alhazeni arabis libri septem* [*Opticae*]. In *Opticae thesaurus*, ed. Friedrich Risner, Basel. (Reprint, with Introduction by D. C. Lindberg, New York, 1972.)

Sabra, 1972: A.I. S., "Ibn al-Haytham," in *Dictionary of Scientific Biography*, ed. C.C. Gillispie, New York: Scribner, VI, pp. 189–210.

———, 1978: "Sensation and Inference in Alhazen's Theory of Visual Perception," in Peter K. Machamer and Robert G. Turnbull (eds), *Studies in Perception: Interrelations in the History of Philosophy and Science*, Columbus, Ohio: Ohio State University Press, pp. 160–185.

———, 1981: *Theories of Light from Descartes to Newton*. 2nd ed., Cambridge: Cambrigde University Press. (1st ed., London: Oldbourne, 1967).

———, 1984: "The Andalusian Revolt Against Ptolemaic Astronomy," in Everett Mendelsohn (ed.), *Transformation and Tradition in the Sciences: Essays in Honor of I. Bernard Cohen*, Cambridge: Cambridge University Press, pp. 133–153. Reprinted as chapter XV in Sabra, 1994a.

———, 1987: "Psychology versus Mathematics: Ptolemy and Alhazen on the Moon Illusion," in Edward Grant and John Murdoch (eds), *Mathematics and Its Application to Science and Natural Philosophy in the Middle Ages: Essays in Honor of Marshall Clagett*. Cambridge: Cambridge University Press, pp. 217–247.

———, 1989: "Form in Ibn al-Haytham's Theory of Vision." *Zeitschrift für Geschichte der Arabisch-Islamischen Wissenschaften* 5, pp. 115–140.

———, 1991/92: "On Seeing the Stars: Edition and Translation of Ibn al-Haytham's *Risāla fī Ru'yat al-kawākib*," *Zeitschrift für Geschichte der Arabisch-Islamischen Wissenschaften*, 7(1991–92), pp. 31–72. In collaboration with Anton Heinen.

———, 1994a: *Optics, Astronomy and Logic. Studies in Arabic Science and Philosophy*. Aldershot: Ashgate Publishing, Variorum.

———, 1994b: "The Physical and the Mathematical in Ibn al-Haytham's Theory of Light and Vision." Paper first delivered in Jyväskÿlä, Finland, in 1973, first published in Tehran in 1976, and reprinted with corrections as chapter VII (20 pages) in Sabra, 1994a.

———, 1994c: Review of Rashed, 1993. *Isis*, 85 (1994), pp. 685–686.

———, 1995/96: "On Seeing the Stars, II. Ibn al-Haytham's 'Answers' to the 'Doubts' Raised by Ibn Ma'dān," *Zeitschrift für Geschichte der Arabisch-Islamischen Wissenschaften* 10, pp. 1–59.

———, 1998: “One Ibn al-Haytham or Two? An Exercise in Reading the Bio-Bibliographical Sources,” *Zeitschrift für Geschichte der Arabisch-Islamischen Wissenschaften* 12, pp. 1–50.

———, 1998/99: “Configuring the Universe. Aporetic, Problem Solving, and Kinematic Modeling as Themes of Arabic Astronomy,” *Perspectives on Science* (MIT Press), 1998, vol. 6, no. 3, published in 1999.

———, *see* Ibn al-Haytham.

Samsó, 1994: Julio S., “On al-Biṭrūjī and the *hay'a* Tradition in al-Andalus,” in idem, *Islamic Astronomy and Medieval Spain*. Aldershot: Ashgate Publishing, Variorum. Chapter XII, pp. 1–13, first presented at a Symposium held in Granada, 1992.

Sezgin, *GAS*. Fuat S., *Geschichte des arabischen Schrifttums*, Leiden: E.J. Brill, 1967–.

Smith, 1996: A. Mark S., *Ptolemy's Theory of Visual Perception: An English Translation with Introduction and Commentary*. Transactions of the American Philosophical Society, Philadelphia, vol. 86, pt. 2.

III

Mathematics: Philosophy and Practice

5

Mathematics and Philosophy in Medieval Islam

Gerhard Endress

The Andalusian jurist, philosopher, and scientist Ibn Rushd, the great Averroes who died eight hundred years ago in Marrakesh, put his life's work in the service of one project, perceived dimly in his youth, defined ever more clearly in the course of a prolonged struggle with the epistemic paradigm of the religious community, and brought to fruition in his years of maturity in a series of Long Commentaries on five principal works of Aristotle: establishing demonstrative science, the law of reason, as the basis of thought and action in human society.

An ever increasing sense of urgency is pervading all of his approaches to this task, seen as an ultimate duty. The recurring prayer for a last delay that God should grant him to achieve his goal accompanies his praise of Aristotle as a guide and guarantor. It is a goal most difficult to attain where the principles of "natural philosophy" (*al-ʿilm al-ṭabīʿī*) were at variance with the observations described and calculated, precisely and predictably, by the mathematical professions. In his *Commentarium Magnum* on Aristotle's *De Caelo*, the solution of some remaining doubts is referred to the discussion of Metaphysics: "Perhaps, if we shall see the last of our term in life, we may explain this point when devoting a literal commentary to Aristotle's discourse on this science [i.e., Metaphysics]; indeed, this is one of my highest hopes, and perhaps God, in his grace and compassion, will help us to live and see this time and to attain this goal—He is beneficent and generous."[1] But when he finally achieved his Great Commentary on "this science," that is, the Metaphysics or First Philosophy of Aristotle, toward the end of his life, he despaired of his task: the task he had set himself to explain and justify an astronomy true to the principles of Aristotle's cosmology: a valid model for calculation as well as a true representation of reality:

> "We must examine this ancient astronomy from the beginning. It is the true astronomical scheme which is valid in accordance with the natural principles [*al-hayʾatu l-qadīmatu llatī taṣiḥḥu ʿalā l-uṣūli l-ṭabīʿiyya*]. That is, according to my conviction, an astronomy based on [the assumption of] the movement of one and the same sphere around one and the same center, revolving on two different

> poles and more according to what is fitting the appearances. Indeed, motions like these can make a star go faster and slower, forward and backward, and have all the motions for which Ptolemy was unable to find a model (*an yaṣnaʿa lahā l-hayʾata*). . . . In my youth, I hoped to make a complete study of this, but now that I have grown old, I have given up this idea because of the obstacles I found in my way before. But this explanation will perhaps induce somebody to study these things later. In our time, astronomy is not about something real (*laysa minhu shayʾun mawjūdun*); the [model of the] sphere existing in our time is a model conforming to calculation, not to reality (*muwāfiqatun li-l-ḥusbāni lā li-l-wujūdi*)."[2]

This brings us to the heart of the matter at hand: The relation of philosophy and mathematics, and the relation of both to reality. In the final analysis, this concerns the question of what the true and first reality is: the ultimate object of study for the philosopher-scientist.

Knowledge is power. In assuming the prerogative of definition, philosophers and scientists, using different paradigms of concept and method, competed for authority. This authority is based on the belief that true understanding of the being, order and movement of the world will warrant proper action, and will constitute the ultimate good and the felicity of man. This conviction inspired the philosophy of science from Antiquity until the early modern age: from Plato's Academy—μηδείς ἀγεωμέτρητος εἰσίτω[3]—until Kepler, who illumined the first book of his *Harmonice Mundi libri V*[4] with a quotation, in Greek, from Proclus' Commentary on Euclid's *Elements*: "Mathematics makes contributions of the very greatest value to physical science. It reveals the orderliness of the ratios according to which the universe is constructed, and the proportion that binds together all that is in the cosmos."[5]

1 The Philosophical Tradition in Greek Science

Philosophy as a transmitted text and as a system of instruction entered Arabic-Islamic society in the baggage of specific social and professional groups: of scientists and physicians.

It is true that premodern societies did not know the narrow professionalism typical of the modern division of labor. It is nevertheless true that since early Hellenism, philosophy itself competed with the individual sciences for recognition of a professional status in society, and sought to found its claim on the unconditioned knowledge (ἐπιστήμη ἀνυπόθετος) of the principles.[6] On the one hand, the philosophical schools assumed competence, and took charge of education, in the mathematical sciences. The conception of philosophy in Aristotle and the old Peripatetic school had embraced, ideally at least, the applied sciences—these in turn being regarded as elements of a *paideia* in the sense of propaedeutics to philosophy: a stage in philosophical education leading the way to the advanced level of dialectic, the "science of sciences."[7] Neither in the Hellenistic nor in the Roman period, on the other hand, did mathematical studies form part of a general education. Outside the *enkyklios paideia*,[8] the "comprehensive education" of philosophy, such studies were linked up with, and restricted to, the professional training of engineers, architects, geometers, and musicians.[9] But here, even in the individual and practical sciences, the teaching of the leading authorities and their basic texts maintained the intimate connection between applied mathematics and its epistemological and metaphysical background. Beyond the decline of the philosophical schools in the civilization of late Hellenism, the philosophical doctrine of the principles and of the cosmos survived in the gnostic Platonism of the natural sciences, in the Neoplatonism of the mathematicians, in the Peripatetic cosmology of Ptolemy, as also—but this is a matter different and apart—in the elementary logic reading of the Christian *schola* (*uskūl*).[10] This is how Greek philosophy entered the urban and courtly society of Islam: as methodology and ideology of the professional sciences, notably of mathematics and astronomy on the one hand, of medicine on the other. It is a philosophy neither pagan nor Christian nor Islamic, but universal: a rational religion of the intellectuals of Greek erudition, giving an ulterior sense to their activity.

Each scientific tradition carried its own philosophical discourse: a choice of authorities, a methodology, a classification and hierarchy of the sciences, and a general orientation of cosmology and ethics. With the physicians we find Galen's platonism as also Galen's own logic, anthropology and ethics, competing with philosophy in pretending to teach an *ars vitae*. (In consequence, the philosophic or non-philosophic character of medicine, being *technê* or *epistêmê*, was under dispute in apology and polemic from both sides.) The mathematician and astronomer, and the professional astrologer or geometer, pretended to a universal competence no less than the physician, but on a different scale: on the authority of a time-honored tradition, and of an eminent ancestry, in the history of philosophy itself. The mathematicians were Platonists and Pythagoreans in the tradition of Nicomachus, Proclus and Iamblichus. But the astronomers cherished the Aristotelian propaedeutic and, above all, the Aristotelian cosmology conjoined with the authority of Ptolemy. Hence it was Aristotle who came to dominate the system of the physical world, and it was a Peripatetic structure which, since being adopted by Ptolemy, prevailed in the method and epistemology of professional science.[11]

The World Is Number: The Platonic Heritage

The philosophy of mathematics owes to Pythagoras two significant contributions, perennial legacies to the history of thought: firstly, of the attitude and the word of *philosophos*, and secondly, of the concept of *mathêmata* as being everything that can be precisely known and learned. Even Aristotle will use *mathêmata* in this basic sense, and there is a long way leading from those essentials of knowledge to what is being learned as *mathematics* in today's schools. But although the *mathêmata* of Pythagoras were different in subject, method, and the state of conscious approach from the mathematical discipline of the later schools, mathematical principles in the proper sense of number[12] and of mathematical harmony were from the beginning regarded as the very center of Pythagorean thought. Everything knowable has number.[13] If we may trust Aristotle, the Pythagoreans held that numbers have an essential likeness with things of the world, sometimes that things *are* numbers, and that number is "the essence of all."[14] This is, of course, refuted by Aristotle, being assimilated to Plato's doctrine of eternal Ideas.[15] Through Plato, the mathematical disciplines of the Pythagorean canon were introduced as forming the basis of intellectual education. It was the program of the Academy: arithmetic, geometry, astronomy, harmony, and dialectic. This canon is presupposed as a matter of course by Aristotle in the opening of the *Posterior Analytics*. But it was Aristotle who, in the second book of the *Analytica posteriora*, extended the *mathêmata* to include the universe of knowledge, leaving behind the more narrowly mathematical paradigm of the Academy and of its Pythagorean model.[16]

> The programme of the Academy is expounded by Plato as a programme of educating the Guardians of the Republic (*Resp*. VII, 526ff.): The subjects to be taught are arithmetic, geometry (to which is added stereometry as a special subject), astronomy, and music (i.e., the science of harmonical proportions); these are the only disciplines recognized as sciences in the proper sense, yielding *a priori* knowledge of immutable and eternal reality. They are described in *Resp*. book VII with respect to their power of turning the soul's eye from the material world to objects of pure thought. First comes arithmetic, the science of number, the numbers—ideal units—of mathematics being considered "by thought," in the abstract, only;[17] hence "this study is really indispensable for our purpose, since it forces the mind to arrive at pure truth by the exercise of pure thought."[18] Geometry, coming second, is equally "knowledge of the eternally existent," and "will tend to draw the soul toward truth and to direct upwards the philosophic intelligence."[19]

The rise of arithmology on the one hand, and of musical theory on the other, in later Platonism and Neoplatonism is closely connected with the reemergence of Pythagoreanism, the mathematical offshoot, as it were, of Middle Platonism.[20] And as Plato had built upon Pythagoras, Neoplatonism relied on Neopythagoreanism. Not the metaphysics of the One and Intellect of Plotinus, but the gnostic and occult tendencies of the later Neoplatonists drew on this source, leaning toward asceticism, supported by magical practice, contriving the perfection of the soul in view of its ultimate ascent to the world above. The central philosophic message of Pythagoreanism, echoed and supported by many a statement of Plato as well, made it a religion of mathematicians: The world is number; through mathematicals, the transcendental and the divine can be perceived. As mathematicians, some of the leading figures of the Pythagorean school, such as Apollonius of Tyana, were foremost in their age. The "Introduction to Arithmetic" of Nicomachus of Gerasa (first century C.E.), on the other hand, is the work of a philosopher rather than a mathematician, intended as a guide to the late works of Plato and to the Pythagorean treatises, first read by philosophers rather than mathematicians, and still popular at a time when there were no mathematicians left, but only philosophers who incidentally took an interest in arithmology.[21]

> As a true Pythagorean, Nicomachus makes arithmetic—the science of number—the primary object of philosophy, the name of *philosophia* being ascribed to Pythagoras (as was common in the school). This "wisdom," *sophia*, is defined as knowledge of the truth in "real things," things immaterial, unchanging and eternal, among which the subject of arithmetic is foremost, because "it existed before all the others in the mind of the creating God like some universal and exemplary plan" (*Introd. arithm.*, IV.2).[22]

Iamblichus, the 4th century disciple of Porphyry, follower of Plotinus, put forward a program to pythagoreanize Platonic philosophy. As philosophers, the followers of Pythagoras made mathematics, starting with arithmetic, not only the leading propaedeutic art, but also the foremost object of philosophical study in its own right. "One significant result of this is the mathematization of all areas of philosophy that is so striking a feature of later Greek philosophy."[23] In the writings of Proclus, again, Plato supplants Pythagoras as the central authority, and while accepting the pivotal role played by mathematics in the philosophical sciences, Proclus chooses geometry rather than arithmetic as the pre-eminently mediatory mathematical science (as is evident from his commentary on Euclid's *Elements*). Mathematicals are projections by the soul of innate intelligible principles; and it is particularly in geometry, according to Proclus' teacher Syrianus, that the soul projects such innate principles into imagination because in its weakness soul is better able to grasp these principles in the extended forms given in the figures of geometry.[24]

The Transmission of Philosophy through Science

While the principal works of these authors may not have been available to the first generation of Arabic scientists, as al-Kindī and his contemporaries, the general attitude to the mathematical sciences clearly goes back to such authorities, transmitted through the basic manuals and their commentators and translators. Even for astronomy, where the Peripatetic doctrine and attitude determined the method and many of the basic cosmological assumptions, Plato had sanctioned the pursuit of mathematics as a philosophical assignment. Going beyond the mere calculus of an auxiliary model, mathematics established the sympathetic rapport between the higher and lower worlds, man and the universe. The designation as *philosophos* of the savant who in his special field of application evinced this quality, underlined this self-image. Hence the Platonic concept and rôle of the *mathematica* prevails in the system of the mathematical sciences and in the tradition of number theory and of the doctrine of musical harmony.

But long since, philosophy and the sciences had drifted apart with regard to their social status and their rôle in education and intellectual life; philosophy had ceased to pretend to the status of a profession. Even before the decline of pagan Hellenism, philosophy had lost the remarkable role it had played in education and intellectual life as against the applied sciences; it had also lost its social status—philosophy remaining but small farc in the provisions of professional physicians and astrologists. The last Alexandrian commentators earned their living not as professors of philosophy, but—as indicated by explicit hints and by the implicit evidence of their metaphors and their examples given to illustrate a point—as doctors, grammarians, rhetoricians, and astrologers.[25] Instead of the philosopher of universal competence, the authority of definition among the intellectual élite is assumed by the "philosopher" (in Arabic, *ḥakīm*) specialist of the applied arts.

Apart from particular professional features, regional traditions persisted locally from pre-Islamic time before uniting in Baghdad: On the side of the Nestorians working in Sasanid Iran, we find Ābā of Kashgar (c. 600), familiar with astronomical as well as medical sources; on the side of the Monophysites of upper Mesopotamia, Sergius of Reshʿayna translated not only Galen, and books on astrology, but also works of the Christian Neoplatonist known under the name of Dionysius Areopagita, and the translations from the Persian, made by Severus Sebokt of Qinnasrin, provided a remarkable range of astronomical and mathematical works.[26] Most important for our subject is the tradition of the philosopher-scientists from Ḥarrān, the ancient Carrhae, where worship of the heathen star-gods survived until the tenth century: the Ṣābiʾa, claiming the protection due to the *ahl al-kitāb* and rising to high stations in the Abbasid administration. Even though the activity of Thābit ibn Qurra and his descendants, both as translators and as original mathematicians, does not suffice to attribute to these "Sabian" sources every text of Plato and every Neoplatonic or gnostic interpretation of Aristotle leaving traces in the Arabic tradition,[27] the Platonic-Neoplatonic heritage of the mathematicians from Ḥarrān may go a long way to explain the knowledge of Plato, and the influence of Platonism, in the philosophical orientation of early Arabic science. But far beyond this specific transmission of mathematical science, the arithmology of neo-Pythagorean origin pervaded the multiple strands of the popular and practical traditions of Hellenism, from the occult sciences to gnomological wisdom literature; the final triumph of Peripatetic *falsafa* as a school of demonstrative science is due to its reception and adaptation in a new milieu.

Al-Kindī and the Platonic Tradition

The translations which were commissioned by one of the leading philosopher-scientists of the early ninth century, and influenced his own writings, cover a wide range: Abū Yūsuf Yaʿqūb ibn Isḥāq al-Kindī, astronomer, astrologer, versed in mathematics and optics, medicine and pharmacy, a polymath of his age—the age of the caliph al-Maʾmūn (813–833) and his sons and successors—who died in or shortly after 866. Taken altogether, the works of al-Kindī, and the sources made available through his efforts, and translated on his demand, are the most impressive witness to the triumph of Hellenism after an earlier period where the import of Iranian traditions had been prominent both in the political and religious community and in the reception of science. But the different strands, professional and doctrinal, are yet unconnected, even in conflict, different in style and approach, in his vast œuvre.

Al-Kindī's Plato, where he is named, stands for the Platonism of the gnostic, "Hermetic" subculture of popular Hellenism, a religion for intellectuals like the one upheld by the Sabians, of mathematicians and astrologers. An *exposé* of Neoplatonism, transmitted on al-Kindī's authority by Ibn al-Nadīm,[28] is put forward as a doxography of the Ṣābiʿa. It is true that more of Plato's authentic works were available in al-Kindī's generation than were preserved beyond the next century (mainly through the philosophical tradition of medical authors—the tradition of Galen the Platonist). Among the Platonic dialogues available was the *Meno*, the first exposition of the doctrine of recollection (*anamnesis*). But al-Kindī's own treatise on "What the soul remembers of what it had in the world of the intellect," while invoking the Platonic concept, is based on the Neoplatonic tradition of the Arabic Plotinus source and related texts.[29] What is stressed most forcibly here is the existence and nature of *a priori* knowledge: a knowledge of the eternal principles, but brought forth (*tukhrijuhā*) in the individual soul—after its exposure to the secondary intelligibles, the forms-in-matter of the material world—by the autonomous activity of reason only. Not everybody is qualified to attain, to re-collect, this knowledge in its primordial splendor. By stressing the incorporeal substance of the intelligibles and of the rational soul, al-Kindī drives home the ultimate value of his science: Only he who purifies his soul will gain true happiness and the ultimate vision of truth.

> Our main testimony for the Arabic *Meno* is found in the work of a mathematician: Thābit ibn Qurra's Epistle "On the argument ascribed to Socrates on the rectangle and its diameter."[30] This is the well-known problem used in the *Meno* in order to demonstrate how even a mind not trained in mathematics can be guided toward mathematical insight, because the human mind can be made to "remember" what it obtained in its preexistence while viewing the ideas in the world above. But the mathematician Thābit does not touch on *anamnesis*, closing his mathematical analysis with a very general remark on the goal and value of mathematical science. Some of these matters, he says, are more elevated than others; whosoever confines himself to the basic matters of geometry, does this either from incompetence, "like some people in our era," or because he wants to guide learners through gradual stages according to their capacity. This latter was the intention of Socrates, using this problem as a paradigm of his intention, for "mathematics (*ta'ālīm*, μαθήματα) is for the soul what nourishment is for the body—as one of the Ancients said."[31]

The Platonism of Proclus (who had written a long commentary on book I of Euclid's *Elements*),[32] and the Pythagorean attitude of Nicomachus of Gerasa and of the Hellenistic theory of music are obvious in al-Kindī's extant works and in the titles of some which have been lost. As a mathematician, al-Kindī was familiar with the tradition of Iamblichus' treatise "On the Common Mathematical Science," which presented mathematics as the absolute object of contemplation; he knew the Neoplatonic reading of Euclid's *Elements*, and he reworked for his own use the Neoplatonic philosophers' vademecum of number theory, the *Introduction to Arithmetic* by Nicomachus of Gerasa, available in a contemporary Arabic version.[33] Also available was Nicomachus' "Great Book on Music,"[34] and hence, the interest in the actualisation of perfect mathematical relations in musical harmony can be followed up in this same school of thought.

But this is only one side of al-Kindī's philosophical program. On the other hand, he was an heir to, and a conscious continuator of the Academic tradition surviving at the hands of commentators with a Neoplatonic orientation, notably in the school of Alexandria and its Byzantine offshoots. Both sides compete in this early period of translation and adaptation, at times in a striking contrast of style. Al-Kindī's Aristotle is not yet the master of logic and of demonstrative science, styled the First Teacher by al-Fārābī's school in the next century, emancipating philosophy from the applied arts, relegating Plato to an inferior rank restricted to sharing out practical, political wisdom. But already, Aristotle had taken on the rôle of super-philosopher, the "foremost" (*mubarriz*) of the Ancients who, in al-Kindī's words, "for luminous, harmonious souls will lead the way toward the highest spiritual rank."[35] al-Kindī's Aristotle, albeit Platonic in matters of theology and cosmology, is representing the encyclopaedia of the rational sciences; he is the undisputed authority on the physical world.

The system of philosophical (including mathematical) studies is modelled on the Alexandrian curriculum, well-known through the Neoplatonic commentators of Aristotle from the school of Ammonius; and the victory of Hellenism over the Iranian tradition in astronomy and astrology prepared the way for Aristotle to become the First Teacher of Arabic Islamic philosophy. This is evident in the work of al-Kindī's rival, the astrologer Abū Maʿshar (m. 272/886)—his Great Introduction to astrology is the first full-grown handbook to be written in Arabic for any of the ancient sciences—as well as al-Kindī's own.[36]

The result of al-Kindī's reading is, in more than one respect, a compromise between the obvious contradictions apparent in the Corpus Aristotelicum itself, between the "Platonic" and the "Peripatetic" Aristotle, and between the tendencies of the Greek commentators. The synthesis arrived at in al-Kindī's division of the sciences is a case in point, and reflecting a long discussion. The Platonic tripartition of the sciences and of being had been kept by Aristotle in his division of the sciences into physical, mathematical and theological (*Metaph.* K 7, 1064b1), but was in conflict with the ontology of the later Aristotle who—in the final analysis—denied the subsistence of the mathematical entities as well as of soul. In the further attempts of the Academy to reconcile these positions, from Speusippus to Proclus, soul was coordinated ontologically with the mathematicals.[37]

In his introduction to the study of Aristotle,[38] al-Kindī gives the first place to mathematics (*al-riyāḍiyyāt*) as a preparation to the study of philosophy proper (plausible in the context—there were scarcely any mathematical writings ascribed to Aristotle[39]—and in accordance with the information given by some of the commentators, referring to the Platonic curriculum).[40] al-Kindī wrote a treatise confirming this attitude, "That philosophy can be acquired through the science of mathematics only."[41] In the following classification of Aristotelian philosophy, we do not find mathematics as an "intermediate science" ranged between physics (on the motion of sensible substances) and metaphysics (on intelligible being), but psychology: an intermediate discipline concerning objects which are independent of bodily substance, subsistent, but perceived by the senses in conjunction with bodies. These are the topics of psychology, in one instance.—But, in another context, the topic of this "intermediate science" is mathematics. The quadrivium of mathematics is intermediate between the natural sciences on the lower end, and—following the way from the multiplicity of sensual phenomena to the universal simplicity of the principles through abstraction—metaphysics and theology. The soul belongs to the same intermediate realm "in between": the soul as subject, and the mathematicals as object, belong both to the world of the eternal intelligibles and the corruptible sensibles. This realm "in between," *to metaxy*, is the realm of the recollection in the soul of its prenatal view of the universals. Logic, the fourth and propaedeutical discipline beside the former three, is being treated as part of the catalogue, but the final apotheosis of the science of demonstration, based on the *Analytica Posteriora*, was achieved by the Arab logicians of the tenth century. Al-Kindī's starting point for the study of philosophy, as in the Platonic Academy, is mathematics.

Al-Kindī's double esteem of mathematics as a propaedeutic to philosophy, and as a subject worthy of philosophical study in itself, unfolds in great detail, and in the best rhetorical tradition of the Platonic mathematicians, in his treatise "On the string instruments producing sound" (*K. al-Muṣawwitāt al-watariyya*). Introducing *musikē* as a discipline of mathematics, he expounds the position of the mathematical sciences as intermediate between physics and metaphysics:[42]

> "It is a custom with the philosophers to practice the middle science, ranged between a science beneath it and a science above it. The one beneath is the science of nature and what is moulded from nature; the one above is called the science of what is not of nature, albeit its impact is observed in nature. This intermediate science, which leads the way both to the science of what is above and what is below it, is divided into four sections: viz. the sciences of arithmetic, musical harmony (*ʿilm al-taʾlīf wa-huwa l-mūsīqī*), geometry, and astronomy"—

in this order, music being given the second place before the other disciplines according to the "Pythagorean" system of Nicomachus.[43]

After providing some remarks on the precedence of knowledge (*ʿilm*) over action (*ʿamal*), al-Kindī goes on to expound the philosophers' teaching of mathematics in general as a theoretical basis of rational practice:

> "So it was a habit with the philosophers to present the secrets of the science of nature and its manifestations in many of the subjects they treated in books, as in those on arithmetic and the amicable and hostile numbers, on the proportional

> lines, and on the five polyhedra (*mujassamāt* 'bodies') fitting into the sphere. After demonstrating that there is no sensible thing the matter of which is not constituted from the four elements and the fifth nature, viz. fire, air, water, earth, and the sphere, their acumen, intelligence and reflection guided them to establish the stringed instruments of sound, and thus to mediate between the soul on the one hand, and the composition of the elements and the fifth nature on the other, by means of such instruments. They designed many stringed instruments in accordance with the composition of the animal bodies, and brought forth from them the sounds corresponding to the human composition, demonstrating thereby to intelligent minds how noble and excellent this wisdom (*ḥikma*, philosophy) is."[44]

The skilled philosopher-musician will be able to adapt his music to any given situation, creating harmony between the soul and the universe, like the physician diagnosing the humors of his patient and prescribing a treatment inducing the equilibrium of health.[45] The musical instruments, and the string instruments in particular, are constructed so as to present the cosmic structure of the physical and intelligible world. Al-Kindī describes the ethnic and historical varieties of the lute as models of the universe, allocating the number of strings in each case to ontological classes. The four strings of the Greek, as also the earlier Arab lute, are of course in correspondence with the familiar series of cosmic tetrads, the elements, the senses, and the humors, and many others, not forgetting the four cardinal virtues, and the four primary questions put forward by Aristotle in his introduction to the *Analytica Posteriora*.[46] The five-stringed lute, more familiar from the later period of Arab musical practice, receives a similar treatment.[47] In an analogous procedure, the particular characteristics of the four individual strings are associated with the elements, the humors, ecliptic arcs, sections of the zodiac, faculties of the soul and other aspects of physiology and astronomy, the macrocosm and the microcosm—associations which are used to explain the specific reactions and affections evoked by the sound of each.[48]

> A different use, but going back to the same tradition of Neopythagorean arithmetic, is made of mathematical proportions in the composition of drugs. Since al-Kindī's *Kitāb fī Maʿrifat quwā l-adwiya al-murakkaba*[49] and its critique by Ibn Rushd is being discussed by Tzvi Langermann (in another contribution to this volume),[50] I will dispose of it briefly. What al-Kindī does is to apply Galen's doctrine of the "grades" of action in the simple drugs to the compound drugs, and he extends the Galenic model by calculating the effects of the compounds on the basis of geometric proportions and progressions. The arithmology underlying this speculation can be traced back, again, to the *Arithmetic* of Nicomachus of Gerasa, and it is the continuing influence, in twelfth-century Andalusia, of a theory based on "the art of number and the art of music" which exasperates Ibn Rushd and drives him to a torrent of abuse.[51]

The Gnostic Tradition and Mathematics

It is in these various domains that we encounter concurrent offshoots of a tradition which in the work of al-Kindī found its most versatile, serious and influential exponent, and which lived on in his school, but gradually sank to the lower strata of occult practice and heretical obscurantism against the pure rationalism of Peripatetic philosophy.

Cognate traditions of arithmology can be traced in the philosophical encyclopaedia of the Ismāʿīlī Ikhwān al-Ṣafāʾ on the one hand, and in the corpus of alchemy and the occult sciences ascribed to Jābir ibn Ḥayyān on the other. I cannot go into the discussion of age, authenticity and unity of the Jābir corpus of writings; suffice it to say that the transmission of the corpus was carried on in Ismāʿīlī circles, where the *Kitāb Ikhwān al-Ṣafāʾ* was read as well, but goes back—in part at least—to an earlier period of the Arabic reception of Hellenistic thought. It is a tradition of the philosophical and scientific "subculture" of late Hellenism. Al-Kindī knew this tradition and drew upon it, but achieved a first, though incomplete integration with the Peripatetic paradigm. In the Jābir writings, as also in the *Kitāb Ikhwān al-Ṣafāʾ*, this tradition was carried on independently, and in a different context.

For Jābir the role of arithmology is significant. It pervades the "science of the balance," *ʿilm al-mīzān*, a central concept governing the philosophy of nature.[52] This is meant to reduce all domains of human knowledge to a system of quantity and measure, conferring on these the character of an exact science. In particular, it is Jābir's intention to submit nature to measure, and to determine the proportions of the elemental qualities or forces—the hot and the cold, the moist and the dry—represented in the bodies and in their interactions. As in al-Kindī's treatise on the compound drugs, the Galenic physiology, and in particular the Galenic theory of the four degrees of intensity or potency in regard of the elemental qualities, determining the effects of a specific medical or nutritive substance, is at the basis of this theory.[53] Jābir follows closely the Galenic classification. At the same time, the values of the four degrees and their subdivisions are calculated on the basis of arithmetical progressions, familiar from Greek number theory as it was found in Nicomachus, and on the other hand, in the literature on the harmonical proportions.[54] On a larger scale, the rapport between the musical harmony governing the celestial spheres and the harmony found in the physical world is investigated in the Jābirian *Kitāb al-Baḥth*, a philosophical justification of theurgy (the *ʿilm al-ṭilasmāt*). Here again, the author makes reference to Plato's *Timaeus*, or rather the Platonic-Pythagorean tradition elaborated from the cosmology of the Platonic work.[55]

Also based on Greek sources, and put into a systematic framework structured on the lines of the Alexandrian curriculum and system, are the *Rasāʾil*

Ikhwān al-Ṣafāʾ, written in second half of the 10th century.[56] One remarkable trait which springs to the eye is the priority given to mathematics. The book opens with a treatise on arithmetic. The introduction, explaining *falsafa*, *philosophia*, as *maḥabbat al-ḥikma*, is closely dependent on the Pythagorean tradition[57] as transmitted through the *Eisagôgê* of Nicomachus,[58] leading on to the classification of the sciences.

> "The aim and scope of this Epistle is training (*riyāḍa*) the souls of the disciples of philosophy, those who 'choose wisdom,' and who study the real (nature of the) things, and search after the causes of all things; in it, there is explained that the form of number in the souls corresponds to the forms of the beings in matter, being the models from the upper world. Through their knowledge, the novice is led on toward the other propaedeutical [*riyāḍiyyāt*, i.e., mathematical] and physical disciplines. Indeed, the science of number is the root of the sciences, the essence of wisdom, the foundation of knowledge and the (principal) element of all things [*al-maʿānī*, objects of the mind]."[59]

Arithmetic is followed by geometry, astronomy, and music. Apart from the general concept of mathematics, a number of closer parallels to al-Kindī's treatment of musical harmony is found in the *Epistle on Music*, the fourth part of the quadrivium.[60]

The author of the *Risāla* is convinced (following the Pythagorean example)[61] that the science of music is the principal wisdom leading to philosophical thought:

> "Musical harmony in its most exalted and perfect form is embodied in the heavenly spheres and the music that they make, and earthly harmony, including that in the music made by man, is only a pale reflection of that same lofty universal harmony. . . . Since an ordinary mortal cannot hear this music before he is cleansed and purified, he will aspire to be redeemed from the prison of this earthly life, to be prepared for the contemplation of an eternal harmony which is the most real and truthful."[62]

The harmony that governs all celestial and earthly phenomena is explained by means of number. Like al-Kindī's treatise mentioned before, the *Epistle* abounds of arithmetical speculations that spread into many and varied domains of cosmology, physiology and philosophy. The section on the affiliation of the four strings of the lute with various cosmic and physical and physiological tetrads agrees almost verbatim with the third section of al-Kindī's treatise on string instruments.

The general attitude is expressed *in nuce* in one of the aphorisms collected at the end of the *Risāla*, recalling the familiar analogy between soul and number:

> "Since the substance of the soul is of the same nature as that of harmonic numbers (*al-aʿdād al-taʾlīfiyya*) and corresponds to them, when the beats of the rhythms presented by the musicians are measured, when in these rhythms the period of beats and silences are proportionate, human nature takes delight in them, the spirit rejoices and the soul experiences happiness. All this is because of the resemblance, the relation and the kinship which exists between the soul and musical harmony."[63]

The School of al-Kindī

The school of al-Kindī was brought to the East by Abū Zayd al-Balkhī (d. 332/934), who spent in his youth eight years of studies in Baghdad, better known through what has survived of his geographical œuvre, but also bent on a systematical treatment of the sciences (his *Tartīb al-ʿulūm* is now lost): a man who combined competence in the rational sciences with a conservative piety praised by his contemporaries.[64] In Balkh, the meeting place of trade routes from Central Asia, Transoxania and Iran, the tradition of the *Faylasūf al-ʿArab* was passed on to Abu 'l-Ḥasan al-ʿĀmirī who spent some time in Baghdad and at the Buyid court in Rayy before he returned to Nishapur (where he died in 381/992). The most detailed attempt to determine the relation of the religious and the philosophic disciplines in a harmonious symmetry is al-ʿĀmirī's *Iʿlām fī manāqib al-Islām.*[65] The very title signals an apologetic program: the rational sciences are put into the service of Islam, the absolute religion, and of the religious sciences. Both spheres are based on tenets which agree with pure reason and are supported by valid demonstration.

Here again, we find al-Kindī's attitude toward the mathematical sciences: the "science of number" (*ʿilm al-ʿadad*) will "immerse the mind into the intellectual pleasures;"[66] and the science of harmony (*ṣināʿat al-taʾlīf*) will give demonstrations of the harmonious relations, measures and forces in the terrestrial and celestial world, and beyond this, in the corporeal and the spiritual world (*al-ʿālam al-rūḥānī*) in general; without this, the astronomers were not able to verify the states of the celestial bodies. Astronomy, in its turn, will alert the mind to the "doors of felicity."[67] The ethical component of this *ḥikma*, the autonomous ethics of the philosopher who finds in the encyclopaedia of sciences the instruction for educating his soul toward purity and ultimate bliss, is found again in the *Tahdhīb al-akhlāq* of Abū ʿAlī Miskawayh (m. 421/1030). It was al-Kindī's concept of philosophy as an autonomous way of thought and way of life—albeit in the service of the Muslim community and compatible with the Koranic revelation—which stayed alive in the circles of the *ḥukamāʾ*: of scientists, of learned courtiers, and of physicians who in the spirit of Galen's platonism revered in philosophy the healing art of the soul. It was the concept integrated by Ibn Sīnā with al-Fārābī's concept of philosophy as demonstrative science: a universal encyclopedia which in the ranks of scientists and physicians, but also, and increasingly, among the élite of Muslim administration, found an eager readership.

2 Aristotle and the Universal Claim of Philosophy as demonstrative science

From the rise of philosophical schools in late Hellenism to their reception by the Christian transmitters and the Muslim heirs to this tradition, Aristotle was venerated as founder of the paradigms of rational discourse, and of a coherent system of the world. In the course of the hellenization of Arabic science and philosophy, after the initial dominance of Iranian traditions in medicine and astrology (transmitting, it is true, their own brand of hellenism), Aristotle was elevated to the rank of absolute philosopher, *al-ḥakīm* or *al-faylasūf.*[68] At the same time, Plato was shoved gradually aside into the domain of popular wisdom and vulgar gnosticism. Not the philosophers, but the learned doctors of medicine, disciples of Galen, continued to cherish the Platonism of Galen's school, and the little that is extant of the texts still available to the first generation of translators has been preserved not by philosophers but by the dilettanti of philosophy, most of them physicians. The mathematicians, and above all the astrologers, followers of Ptolemy, equally made Aristotle the supreme guide to the "science of sciences,"[69] according to the traditional definition of philosophy.

Plato's dialectic of ideas was replaced by Aristotle's alternative dialectic of discourse: a deductive epistemology. This was first put forward in a radical form in the *Posterior Analytics*: "For we can say goodbye to the *eidê*, for they are nonny-noes, and if there are any, they are nothing to the argument."[70] Even here, Aristotle's closeness to the Platonic model he is replacing, and to the study course of the Academy, is evident in his allusions to Platonist vocabulary and concepts: "All teaching and all intellectual knowledge come about from already existing knowledge. This is evident if we consider it in every case; for the mathematical sciences are acquired in this fashion, and so is each of the other arts."[71] With Aristotle as with Plato, mathematics is the science par excellence, providing both examples and the general problematic. A passing shot at the *Meno* is making it plain, if only implicitly, that Plato's theory of recollection (*anamnesis*), which explained the preexistence of mathematical universals through reference to eternal ideas, has been discarded. In the "aporia of the *Meno*," the partner of Socrates, baffled in his search for virtue, asks: "And in what fashion, Socrates, will you seek that of which you do not even know if it exists?" (*Meno* 80d). Aristotle follows Plato in maintaining that the seeker does in a sense already know what he is looking for. Having explained that the learner must already know the premises, Aristotle adds (*An. Post.* I.1.71a24f.) that in a sense he also knows the conclusion. But only in the end (*An. Post.* II.19), he returns to the implications of his own theory of preexisting knowledge, and proceeds to account for the acquisition of the first, lowest universals by induction (*epagôgê*). Indeed, we cannot demonstrate the principles. Aristotle's primary contention is to expound the universal structure in the acquisition of knowledge based on what is already available to human knowledge from such principles, by expounding the universal system of demonstration, *apodeixis*.

The role of axiomatic mathematics as a background to demonstrative method in philosophy is evident from its very conceptualization and terminology. As early as in the fifth century B.C.E., Greek mathematics had taken the step from simple demonstration, ἀπό-δειξις, from visual evidence, to demonstration from principles: definitions and axioms. Like the science of geometry, logical demonstration had "to rely on principles, which, though unprovable, are nonetheless true and indisputable."[72] In this, Aristotle continued an intellectual tradition which recognized a fundamental affinity between mathematics and dialectic. Even though the mathematical and physical sciences apprehend their principles in a different way, Aristotle regards mathematical procedure—axiomatization, and the use of hypotheses—particularly helpful for the acquisition of all scientific knowledge. Mathematics provided to him a model of deductive-demonstrative science parting from principles (ἀρχαί).[73]

But his noetic concerns are not separable from ontology, and especially from the basic ontological aporia of relating individuals *qua* individuals to individuals as being exemplars of universals. "The newly declared οὐσία, the individual substance, had as individual substance become unknowable except in universal terms, and the abstracted essence took on the detached character of the rejected forms of Plato."[74] For Aristotle himself, this remained the "greatest" aporia (*Metaph.* 1087a13, 999a24–25). It was here that the ways of late Hellenistic metaphysics, of philosophical theology, in Christianity and in Islam as well, parted with the master of demonstrative science: returning over and again to the assumption of preexisting, eternal, hypostatized objects of knowledge. According to Aristotle's theory, presented as a general epistemology, the sciences are to deduce the properties of substances from their essences through syllogisms. Still, in expounding the sciences in a formal axiomatized system, Aristotle proposed for every branch of human knowledge what early Greek mathematics had done for mathematicals (and what Euclid consummated for geometry later on[75]—influencing, in his turn, an axiomatic approach to ontology and cosmology in Neopythagorean and Neoplatonic metaphysics).

In following Aristotle in this overall orientation, all subsequent philosophical systems, notably those of Islam, are essentially Aristotelian—whatever their Platonic or Neoplatonic paradigm in the allegories of the World Above.

The place accorded to mathematics has to be seen in this context.

In c. 2 of the *Metaphysics,* book B, Aristotle discusses the question under which science, or sciences, if any, the "principles of demonstration" will come, that is to say, the "common opinions" or axioms (*axiômata*), which are the starting-point of all demonstrations. The science of these principles cannot be any one of the special sciences, as e. g. geometry. In a later passage (*Metaph.* Γ. 3, 1005a19–b1, cf. K. 1, 1059b14–21), Aristotle gives his solution concerning the position of mathematics:[76]

> "We have now to consider whether it belongs to one science or to different sciences to inquire into what mathematicians call axioms, and into substances. It is manifest that the inquiry into these axioms belongs to one science and that the science of the philosopher; for they hold good of all existing things, and not for some one genus in particular to the exclusion of others. Everyone makes use of them because they belong to being *qua* being, and each genus is (part of) being. . . . This is why none of those who study the special sciences tries to enunciate anything about them, their truth or falsehood; neither the geometer, for instance, nor the mathematician does so, though it is true that some of the physicists have made the attempt, and not unnaturally seeing that they supposed that the inquiry into the whole of nature and into being belonged to them alone. But since there is a class of inquirer above the physicist (nature being only one particular genus of being) it is for the thinker whose inquiry is universal and who investigates primary substance to inquire into these axioms as well. Again, since the mathematician, too, uses the common axioms in a particular application, it must be the business of first philosophy to investigate the principles of mathematics also."

As far as the axioms of mathematics hold of all being, they belong to philosophy, investigating all being so far as each of them *is*.

The physical part of philosophy and mathematics study the same objects, but there is a difference—especially with regard to optics, harmonics, and astronomy, which among the mathematicals are "nearest to the study of nature."[77] In physics, both matter and form are studied: the substances of the physical bodies as well as their shapes—bounded by planes, lines, and points. Mathematics studies these geometrical attributes only, not as attributes of physical bodies, but in abstraction:[78] separate in virtue of cognitive abstraction, not—*pace* Plato—qua being ontologically separate.

Al-Fārābī and Demonstrative Science

By the end of the ninth century, philosophers had gained a readership which had spread considerably beyond the circles of the scientific professions: among littérateurs, among the élite of the secretarial class (the *kuttāb*), and in circles attending the courtly *majālis* of learned and literary exchange. These may also have shared the philosophers' distrust of the infection of Kalām by the growing tide of traditionist orthodoxy. But then, addressing the same readership, a different program was drafted by al-Fārābī (d. 339/950): based on a wider choice of the sources which had become accessible, and envisaging a comprehensive system of knowledge, and integrating the Aristotelian theory of the principles with Neoplatonic cosmology and a Platonic model of the political-religious community.

Encompassing all of Aristotle's logic, physics and metaphysics, the early Fārābī was the first of the *falāsifa* to turn from the compromises of al-Kindī's creationist Platonism in his "Integration of the opinions of the two Sages, Plato and Aristotle." The philosopher realizes that the primary subjects of his inquiry are the universals, not as hypostatized species subsistent in the "world of the intellect," such as ideas—or for that matter, mathematical entities—but *in rebus*, principles of reality subjected to induction and demonstration. He discards with the Platonic concept of *anamnesis* by re-interpreting it on Aristotelian lines: In substance, Aristotle agrees with Plato when he defines the true function of recollection in the beginning of the *Posterior Analytics* (the *Kitāb al-Burhān*).[79] Al-Fārābī added to earlier concepts of philosophy in Islam the radically Aristotelian concept of philosophy as a demonstrative science (*ʿilm al-burhān*) which proves universally what in the particular sciences is deduced by particular "indications" or "signs" (*dalāʾil*), and which perceives absolutely what in the individual religious-linguistic communities is conveyed individually. Philosophy as a science is a method of deduction and demonstration, not an ideology competing with theology: being an independent way toward knowlege, it could be proclaimed as a safeguard for the religious community itself. It is here that Aristotle as being the author of exemplary and encyclopaedic instruction is transformed to become the authority of a method leading to absolute knowledge. The philosopher claims rulership, not only inside the scientific community, but in the religious community as well.[80]

Al-Fārābī integrated the sciences in the framework of a formal axiomatized system, the system of Aristotle's *Posterior Analytics*. Philosophy and religion, the universal, rational sciences and the disciplines specific to the religious and linguistic community, are shown to be complementary parts of the same hierarchical system of cognition and interpretation.

Aristotle's *Posterior Analytics*, the *Kitāb al-Burhān*, provided al-Fārābī with a coherent system of deduction and demonstration, comprising all levels of rational activity, and serving as a guide for the division and hierarchical classification of the sciences, leading up to the First Philosophy, metaphysics. The basic text is the exordium of the *Analytica posteriora* (I.1): "All teaching and all learning come about from already existing knowledge"—by deduction (from the specific), induction (from the particular), and individual "signs" (*dalāʾil*), or, in the practical arts, experience, in descending order of certainty. Al-Fārābī's own summary contains explicit consequences as to the coherence and ranking of the sciences:

> "Of the theoretical sciences, some are universal [sc. the First Philosophy and the universal demonstrative sciences, Topica and Sophistica] and some are particular [sc. mathematics, physics and theology]. The universal sciences have in common the subjects, the objects and most of the premises, but differ in the conditions aforementioned [sc. in the relative status of the principles, subjects and objects used as premises in their specific demonstrations]. The particular sciences are all below the First Philosophy, participating in it in so far as all their subjects are below the Absolute Existent. This science [sc. First Philosophy] will employ universal premises which all the particular sciences employ in the way we have described [i.e., in the mode applied to their particular subject], while the particular sciences employ premises which are demonstrated in that science [sc. in First Philosophy]."[81]

The subordinate, particular sciences and the superior, universal sciences "help each other" in that "the prior sciences provide in the subsequent sciences the knowledge of the causes or of both the causes and the existence, while the subsequent sciences provide in the prior ones the existence." It follows that

> "each art (*ṣināʿa*) which provides the principles of another art is governing (*raʾīsa li-*) that art. Now the governing science in an absolute manner among the sciences which provide the causes, is that which provides the ultimate causes of the beings: and this must be the First Philosophy."[82]

The mathematical sciences are posited between the physical and the metaphysical in being the "abstractive sciences," abstracting immaterial entities from the material substances: separating intellectually what is inseparable from matter in its actual existence.[83] The position of the mathematical sciences (*ʿilm al-taʿālīm*) in al-Fārābī's "Enumeration of the sciences" (*Iḥṣāʾ al-ʿulūm*) follows the same Aristotelian premisses: Theoretical arithmetic (arithmology, *ʿilm al-ʿadad*) examines numbers absolutely (*bi-iṭlāq*), abstracted in the mind (*mujarrada fī l-dhihn*) from the bodies and from anything actually numbered, sensible or not; through this universal application, it enters the realm of the sciences[84] (as against the particular, practical *technai*, like medicine, excluded from the *ʿulūm* in al-Fārābī's *Iḥṣāʾ*).[85] An analogous statement is made about theoretical geometry (*handasa*).[86] This latter is qualified as being "more general" than the one immediately following, optics (*ʿilm al-manāẓir*), but optics requires the status of a science in its own right (*an yufrad*) in examining the "aspects" of what "appears to sight" (*mā yaẓharu li-l-baṣar*) as distinguished from "what (a thing) is in reality" (*mā huwa ʿalayhi bi-l-ḥaqīqa*).[87]

Here the scientific character of optics[88] is pointed out as being a way to establish this difference by means of "certain demonstrations" (*barāhīn yaqīniyya*), that is, proofs yielding certain knowledge.[89] In this, al-Fārābī formally asserts the claim of a mathematical science to demonstrative method, and to the quest for knowing reality as such—the objective of philosophy by definition. The same claim was raised, and had been raised before, by mathematicians who (a) were able to point out Aristotle's use of an optical example, in his *Posterior Analytics,* for elucidating the conclusion from sensible existence (τὸ ὅτι) to cause (τὸ διότι) in scientific demonstration and the discussion of optical phenomena in the Ps.-Aristotelian *Problemata Physica,*[90] (b) took the concept and method of geometrical *apodeixis* from Euclid, and (c) claimed the status of universals for the mathematical "causes" figuring as a middle term in such demonstration, in accordance with Aristotle's own procedure (*An. Post* I.2).

One generation prior to al-Fārābī, Qusṭā ibn Lūqā (died c. 300/912–13)—mathematician, philosopher, and translator of Greek scientific texts—introduces his epistle on catoptrics with a praise of demonstrative science as "the finest of the humaniora," and then continues to commend his own subject, optics, as being "the finest of the demonstrative sciences: the one in which the natural science and the science of geometry partake, since from the natural science it takes the sensual perception, and from the geometrical, the demonstration by means of lines [i.e., linear constructions]"—such, par excellence, is the science of rays (catoptrics).[91]

From here Ibn al-Haytham was able to go on toward establishing mathematical astronomy and optics as the noblest of sciences about *universalia in rebus.*

Ibn Sīnā: the New Encyclopædia

Avicenna united and integrated the early traditions of *falsafa*, both in respect to groups of readership and professional circles, and also in uniting the Platonic and Peripatetic fundamentals. Taking up and completing the work of al-Fārābī, he projected the conceptual framework of the Arabic *Posterior Analytics* onto all domains of scientific and philosophical knowledge, conceiving all strata of cognition—including the highest degrees of discursive and intuitive thought (the latter being the *ḥads*, bereft altogether of its mystical connotation)—as applications of the syllogism.

In his "Division of the Intellectual Sciences" (*Taqsīm al-ʿulūm al-ʿaqliyya*),[92] the mathematical quadrivium is dealt with in a basic and straightforward manner, dependent upon the manuals of the Hellenistic tradition, and repeating the classical *topos* of the "intermediate position" of the mathematicals: Mathematics, as a part of the theoretical philosophy (*al-ḥikma al-naẓariyya*) is intermediate, *al-ʿilm al-awsaṭ,* between physics and metaphysics ("theology"); its objects, regarding their existence, are bound up with matter and motion, but their concepts—their "definitions" (*ḥudūd*)—are not, since they can be understood without reference to any bodily substrate.

> Mathematics is duly mentioned in his "Autobiography," which is an idealized curriculum of the accomplished philosopher: In his youth, preceding systematic studies, he learned some practical geometry and "Indian calculation." Then he studied the *Isagoge* and elementary logic with his first teacher in philosophy, al-Nātilī, going on to Euclid and the *Almagest*, and then to physics and metaphysics. After learning the practical art of medicine, he took up the systematical study of theoretical philosophy, to be crowned by a deepened understanding of metaphysics.[93]

The question of the place of mathematics and its objects in the philosophical sciences is dealt with in some more detail in the Metaphysics (*al-Ilāhiyyāt*) of the philosophical encyclopædia *al-Shifāʾ*.[94] The subject of mathematical science is measure (*al-miqdār*), *qua* being abstracted, in the mind, from matter.[95] Number may be found both in sensible and in non-sensible objects; measure, whether said of a corporeal dimension or of a limited quantity taken from a continuous extension, is never separate from matter, although in the first sense (of dimension), it is a principle in the existence of all natural bodies, hence it is prior in essence to the sensible beings.[96]

He goes on to discuss the subject-matter of metaphysics, that is, "what is beyond nature," *Mā baʿd al-ṭabīʿa*.[97] It might be called properly "the science of what is before nature," because its objects of study are essentially and generally before the natural bodies. Now someone might object—someone, we might add, in the tradition of al-Kindī or the Ikhwān al-Ṣafāʾ—"that the questions of pure mathematics, studied in arithmetic and in geometry, are equally 'before nature,' and especially number," because this can exist independently of a material substrate. As for the lines and surfaces treated in geometry they subsist in bodies. The measure (*miqdār*) treated in geometry is not an absolute principle or form, as of prime matter, but accidental, as of bodies possessing the three dimensions width, breadth and height. As for arithmetic, it does not study absolute number. It is true that number can be found in the separate beings, but it is also found in the natural things. It may arrive that in the imagination (*fī l-wahm*) a number abstracted from any adventitious substrate be conceived. But number cannot exist unless it adheres to a thing in existence (*lā yumkinu an yakūna l-ʿadadu mawjūdan illā ʿāriḍan li-shayʾin fī l-wujūd*). As far as being among the separate existents (*al-ashāʾ al-mufāriqa*), on the other hand, number cannot be subjected to any quantitative relation, augmentation or reduction, but subsists as it is. As an object of quantitative relation, augmentation or reduction, it must be in matter. This, however, is the subject of arithmetic: studying number in respect of it being in bodily nature, albeit abstracted from its natural states in the imagination (*fa-idhan ʿilmu l-ḥisābi min ḥaythu yanẓuru fī l-ʿadadi innamā yanẓuru fīhi wa-qad ḥaṣala lahū l-iʿtibāru lladhī innamā yakūnu lahū ʿinda kawnihī fī l-ṭabīʿa*).[98] Arithmetic, in consequence, is not a study of the essence of number, or of absolute number, but of number in its accidental inherence in matter. The essence of number, however, is an object of metaphysics indeed.

Book VIII of the *Shifāʾ* "On unity and multiplicity"[99] brings an extended refutation of the Platonic doctrine of ideas, and especially—on the lines of Aristotle's *Metaphysics,* A.5 (985b23ff.)—of the Pythagorean notion of subsistent "numeric forms" or "numeric numbers" (313.11 *al-ʿadad al-ʿadadī*, 314.3 *al-ṣuwar al-ʿadadiyya*), viz. of "numeric numbers from which they constitute the forms of the natural existents" (319.10 *al-qāʾilīna bi-l-ʿadadi l-ʿadadiyyi l-murakkibūna minhā ṣuwara l-ṭabīʿiyyiyāt).*

On the one hand, Ibn Sīnā insists on the primacy of philosophy—of the First Philosophy: on the rank of its subject matter, and the universal control of demonstrative method. On the other hand, he integrates the fundamentals of the quadrivium—geometry, arithmetic (including numerology), spherical astronomy (the science of the *Almagest*), and musical harmony—into the encyclopædia of the sciences established under the aegis of philosophy. But his is not a mathematicians' philosophy; and contrary to all of his predecessors, he leaves out all aspects of mathematical science where observational practice meets demonstrative method.

3 Ibn al-Haytham: Mathematics as demonstrative science

The enormous success of Ibn Sīnā's encyclopedia was not only due to his new metaphysics, which promised to solve the antinomies of metaphysics and religious thought, but also to the systematic coherence of his logic and epistemology. Scientists and physicians were his most eager students and readers. It was not an Avicennian scientist, however, who at first, and uncompromisingly, made demonstrative science his very own:

"Truth is sought for itself; and in seeking that which is sought for itself one is only concerned to find it. To find the truth is hard and the way to it is rough. . . . But God has not protected scientists from error. . . . The seeker after the truth is, therefore, not he who studies the writings of the ancients and, following his natural disposition, puts his trust in them, but rather the one who suspects his faith in them and questions what he gathers from them, the one who submits to argument (*al-ḥujja*) and demonstration (*burhān*) and not to the arguments of a human being whose nature is fraught with all kinds of imperfection and deficiency."[100]

The famous introduction to the *Shukūk ʿalā Baṭlamyūs* of Avicenna's contemporary Ibn al-Haytham (c. 354/965–432/1040) is like a radical restatement of Aristotle's frequent proposal, before studying a problem, first to consider the opinions of the Ancients. It echoes in spirit, if not in its wording, the statement of Avicenna in the introduction to his *Manṭiq al-mashriqiyyīn*, acknowledging the merit of "the most excellent of their [the Peripatetics'] predecessors," Aristotle, but scoffing at "the common philosophasters who are infatuated with the Peripatetics and who think that no one else was ever guided by God or attained to his mercy."[101] But the approach and method of Ibn al-Haytham, although evolving from a mathematical science embedded in the traditional system of ancient cosmology, end up in a radical rejection of transmitted authority where it contradicts the results established by the very method of Aristotle. Aristotle remains a vague symbol of the authority of any philosophy whatsoever, while mathematical science becomes the foremost of the demonstrative sciences.

Evincing the principles of his science, Ibn al-Haytham enjoins the true scientist to be a true philosopher, following the rules of demonstration. Remarks on method are frequent. For the general principles of physics, Ibn al-Haytham turns to the opinions of "all the philosophers" or "those of the philosophers who arrived at the truth" (*al-muḥaqqiqūn min al-falāsifa*).[102] Aristotle "laid down the principles from which the way to the truth will be found, its nature and substance be attained, and its essence and quiddity be found" (*aḥkama l-uṣula llatī fīhā yuslaku ilā l-ḥaqqi fa-yudraku ṭabīʿatuhū wa-jawharuhū wa-tūjadu dhātuhū wa-māhiyyatuhū*).[103] Aristotle's physical philosophy was, as a matter of course, his point of departure, an authority invoked frequently, and the subject of summaries and commentaries listed among his early writings. But in the end, Ibn al-Haytham remained an Aristotelian only in the sense of a general methodological orientation. In an earlier treatise "On the Configuration of the World" (*fī Hayʾat al-ʿālam*),[104] he expounds, in a separate appendix, the principles of celestial movement, all of which can be traced up to Aristotelian physics.[105] In the later treatise "On the Light of the Moon" (*fī Ḍawʾ al-qamar*), he spurns all mention of Aristotle's celestial physics, such as the nature of the fifth body, *aithêr*, to be used as premisses for his theory. Instead of metaphysical doctrines, such general principles as can be observed behind his argument are specific theorems, developed from physical theory, but closer to the facts under discussion.[106] Aristotle—the only philosopher actually named—remains but a symbolic authority of demonstrative method: a virtual text, while his own writings fall into oblivion.

The observance of demonstrative method by itself had become the passport of competence for the pursuit of knowledge in the epistemic community. When in his "Solution of the Aporias in Euclid's *Elements*," Ibn al-Haytham raises his own apodeictic method above the time-honored authority of the master of demonstration in geometry, he still refers to the principles of science pronounced by Proclus and Aristotle, but claiming their ultimate perfection: "The causes in matters scientific are the premises employed in the geometrical proofs—these are the proximate causes; but what we seek in each construction is the remote and first cause—and this has not been pointed out by any of the earlier nor any of the later authorities."[107]

In Ibn al-Haytham's remarks on his method of inquiry, the use of *istiqrāʾ* (*epagôgê*, "induction")[108] is an explicit pointer to the logical procedure described in the final chapter of Aristotle's *Posterior Analytics* as the way to detect the principles or universals used as premisses in a valid demonstration. It is true that the word is used somewhat loosely by Ibn al-Haytham in many instances.[109] According to al-Fārābī's reading of Aristotle, induction (*istiqrāʾ*) aims at establishing a universally affirmative or negative proposition. As a procedure, he understands induction as the act of surveying all or most of the particular cases falling under a given universal to see whether a certain predicate applies or does not apply to the particulars surveyed. If complete, the induction is called "perfect," if incomplete, "imperfect." al-Fārābī's understanding of induction in terms of a one by one examination of the particulars does not correspond to the meaning of this term in the relevant Aristotelian passages; there, it is not attending to the particular cases, bur rather the advance from these particular cases to the corresponding universal which is known as induction. "It seems that al-Fārābī's understanding of the matter is a consequence of the fact that in the Arabic *Prior Analytics*, *epagôgê* was rendered as *istiqrāʾ*, a term that must in this case be taken to refer to the act of "collecting" the individual cases."[110] The mathematician Ibn al-Haytham goes on from here to check the limits of the theoretical model by means of systematic observation (*iʿtibār*, "experience").[111]

But Ibn al-Haytham, starting from the familiar concepts of Aristotelian epistemology and from the traditional models of astronomy and optics, transformed both. In his hands, the objective of induction, instead of a collection of universals from the particulars of any observation whatsoever, became focused on the refinement of complex procedures, apt to provide criteria for the validity of the models and hypotheses they were to yield. The true progress, owing to a true revolution in method, was based on a new conception of the use of mathematics for the description of those particulars, collected and surveyed in order to support a perfect inductive inquiry yielding valid results. While mathematical models are based on the data of observation, the philosopher-mathematician is convinced of the essential coherence between valid models and the plan—the *logos*—of nature. This conception of mathematical relations in natural science is founded on the basic assumption that physical theory, in order to be valid, must deal with real bodies, and not with imaginary hypotheses, and must be developed through a process of observation, experiment, and induction. It is thus possible to conclude that the most simple of mathematical relationships, arrived at under the most excellent conditions of scientific observation, can be supposed to correspond to the structure of the physical world.[112]

Through induction from the phenomena, the Aristotelian-mathematician grasps for demonstrable evidence of the absolute—forms-in-matter, but nonetheless universal. The only medieval dissertation on the aesthetically beautiful not bound up with ethical instruction about the morally good is found in Ibn al-Haytham's *Optics*: the beautiful (*al-mustaḥsan,* "what is regarded as beautiful") as harmonious proportion.[113] The ultimate object of contemplation of the Pythagorean and Platonic philosophy of mathematics reappears in demonstrative mathematics: not assimilation to the First Good, but absolute form.

Philosophy and Spherical Astronomy

The rationalism of Hellenistic philosophy, in Islam as before, is made visible through the reality of the cosmos. The order of the spheres, the eternal, circular motion of the heavenly bodies, the progression from the one and first cause to its manifestations in the celestial hierarchy and to the changeable and corruptible substances of the sublunar region evolved into an increasingly differentiated system (in Arabic, the *hay'a*, "shape" or "configuration," of the world). Plato, according to a well-known tradition, had enjoined the astronomers to find out "by the assumption of what uniform and orderly movements the apparent motions of the planets can be acccounted for"—to "save the phenomena" through mathematics: both were founded on the eternal Ideas.[114]

This was achieved, or very nearly so, at an early age of Greek science. But Aristotle's assumption that Eudoxus' geometrical model of concentric spheres was a physical reality, a mechanical system in conformity with pure mathematics, and obeying the laws of natural movement established in his

Physics, created new problems. The more precise the astronomical observation, and the more intricate the mathematical calculus of the celestial revolutions grew, the more difficult it became to reduce all phenomena of the heavens to a coherent system of uniform, circular motion: the natural movement of *aithêr*, the celestial substance. Based on two theorems of Apollonius, Ptolemy crowned the achievement of his predecessors, with an elaborate construction of epicyclic and eccentric vectors explaining (in the *Almagest*) the phenomena of the planetary cycles. And indeed, he took this both as a mathematical solution of Plato's assignment and as a true model of the physical cosmos (a system of contiguous nested shells, as sketched in his *Hypotheses*).[115] But this quantitative conception had to deal with the variations in angular velocity of the eccentric deferent, with the variances of precession and trepidation in the sphere of the fixed stars, and related difficulties, which violated the principle of uniform circular movement. Consequently, the application of Aristotle's physical theory to the Ptolemaic system required a new planetary theory.[116]

The problematic was known, and was seen as a fundamental aporia, from the times of late Antiquity. Various doubts are raised and refuted by the commentators of Aristotle with reference to early astronomers as well as the philosophers of the Academy.[117] While the main bulk of our literature is concerned with the conflict between the mathematical astronomy of Ptolemy and the physical philosophy of Aristotle, the Platonists had a difficult stand as well. Proclus the philosopher, having to give an appropriate placc to astronomy among the four disciplines leading the way to the Good, had to defend Plato the astronomer against the "modern" astronomy of his time, devoting many pages to the task of refuting the assumption of eccentric spheres and epicyclic motion. The main fault of the astronomers was "to pass from the domain of the physical bodies to mathematical considerations," which are imaginary, "and to give an account of the natural movements on the basis of things which do not exist in nature"—repeating, as it were, the mistake of the Pythagoreans criticized by Aristotle in the *Metaphysics*, in making mathematical realities account for natural processes.[118]

This is, in principle, the very same criticism which was raised against Ptolemy by Ibn al-Haytham, and by the Aristotelian philosopher-scientists of Andalusia. The astronomy received by the Arab authors represented the state of the art of professional science, and was adopted as well in al-Fārābī's philosophical cosmology. But the Ptolemaic system, while valid as a purely mathematical—geometrical—model to serve as a basis for hypothetical calculus, was interpreted as physical reality, and hence, got into conflict with the principles of celestial physics.

Ibn al-Haytham's Critique of Ptolemy

Convinced of the power of inductive method, Ibn al-Haytham leads a vehement critique against those models of Ptolemaic astronomy which under the scrutiny of the mathematician proved insufficient to suit both the universal principles prevalent in the cosmos and the observation of the celestial motions. Ptolemy's celestial model (*hay'a*) consists of magnitudes supposed to move in epicycles—imaginary vectors "which cannot have by [themselves] a sensible movement so as to produce something real in the world" (*laysa yataḥarraku bi-dhātihī ḥarakatan maḥsūsatan tuḥdithu maʿnan mawjūdan fī l-ʿālam*).[119] What is more, "the assumptions made in Ptolemy's astronomy (*hay'a*, i.e., a model of celestial mechanics) for the movements of the five [lower] planets are invalid, because they are contrary to theory (*khārija ʿan al-qiyās,* i.e., to the theory demonstrable from valid premisses) and to sound principles."[120]

> "It is not possible that the movement of the stars, being eternal, homogeneous, following a single order, unchangeable and incorruptible, should be against the principles of theory (*khārij ʿan al-qiyās, παρὰ τὸν λόγον*). It has become evident from all that has been said that the configuration (*hay'a*) established by Ptolemy for the motions of the five planets is invalid, and that a valid model to be constructed for the motions of the planets based (on the assumption) of bodies in homogeneous, eternal, and uniform movement, implying neither absurdity nor admitting doubt, will be different from the model established by Ptolemy."[121]

While admitting that some of the contradictions found in the *Almagest* may be excused, being due to inadvertence, others were admitted consciously; indeed, Ptolemy admits that at some points, he was "compelled by the nature of our subject to use a procedure not in strict accordance with theory" (*ashyā' khārija ʿan al-qiyās*).[122] Once the imaginary circles and lines (*dawā'ir wa-khuṭūṭ mutakhayyala*) posited for the celestial model were assumed to apply to existing bodies, contradiction was bound to follow, but Ptolemy accepted this consciously, being fully aware of the methodological implications:[123]

> "If someone posits a line in his imagination and moves it in his imagination, a line analogous to this line in the heaven will not move in the same way, and neither will a star, if somebody imagines a circle in the heaven and imagines a star to move along this circle, move on this same imaginary circle. If this is the case, then the models which Ptolemy imposed on the five planets are futile, and he posited them although being aware that they are futile, because he was not able to find others. But . . . it is not valid to assume a sensible, eternal, orderly movement unless it conforms to a valid model to be found in existing bodies."[124]

The philosopher-scientist takes the mathematician to task, and calls for the repair of a system branded as being *alogon*: contrary to the universals of "reasoned theory" on which demonstrative science rests.[125] But he stays within the mathematical paradigm of explaining the physical phenomena: he does not call for the abolishment of the Ptolemaic system.[126] Contrary to the spokesmen of the "Andalusian revolt"—we shall come back to this—he was content, in the final analysis, to replace the Aristotelian doctrine of the aetherial body with a physical theory of solid nested shells obeying the cinematic laws established and calculated by Ptolemy—a celestial mechanic in accordance with observational data. But while in his optics, he evolved a sophisticated methodology of experimental control of the mathematical models, the division between the celestial and sublunar realms of the cosmos—both in cosmological theory and in factual experience—forbade an inductive, experimental control in the light of the results obtained.[127]

> Among mathematicians, criticism of Aristotle the mathematician was growing, and was raised explicitely. Abu 'l-Futūḥ Ibn al-Sarī (Ibn al-Ṣalāḥ, died in 548/1153) wrote a treatise refuting Aristotle's assumption, put forth in *De Caelo* III. 8: 306b3–8, that there are two regular solids which can fill up three-dimensional space, the pyramid and the cube, and proving that cubes only can fill a space. See the text ed. by Mubahat Türker, "İbnu'ṣ-Ṣalaḥ'ın De Coelo ve onun şerhleri hakkındaki tenkitleri," *Araştırma* 2 (Ankara, 1964): 1–79, p. 71f.

4 The Primacy of doctrine: the Andalusion reaction

"Once such imaginary assumptions were applied to existing bodies, contradiction followed as a consequence."[128] Ibn al-Haytham's point of criticism against Ptolemy was at the basis of the attacks against Ptolemy rising in the Andalus. But the philosophers of Muslim Spain would deny the mathematicians sufficient competence to grasp the intelligible reality of the cosmos. Starting with Ibn Bājja, they would attack Ibn al-Haytham for trespassing into foreign territory:

> "Al-Zarqālluh did not cease to criticize Ptolemy in most of his opinions. Others before him have held this opinion, and I do not wonder that Ibn al-Haytham detected his [Ptolemy's] evident errors; and if you want to study what I referred to, read Ibn al-Haytham's treatise «On the Doubts Raised Against Ptolemy. . . .» But if you should look at this treatise, you will detect that Ibn al-Haytham read this discipline only in the most elementary manner (*innahū lam yaqra'i l-ṣināʿata illā min ashali l-ṭuruq*), and perhaps he could not decide in his time if he should give a firm judgement in view of its refutation, or if he should just disregard it. Nay, he was not one of true experts of his science, and much farther off the goal than al-Zarqālluh."[129]

There is more to this attack on Ibn al-Haytham than a subtle point of mathematical-physical analysis. It is the start of a fierce competition for the prerogative of universal, rational knowledge. Before going into the notorious problem of the planetary motions, I would like to recapitulate a more specific criticism raised by Ibn Rushd in his Commentary-Paraphrase (*Talkhīṣ*) of Aristotle's *Meteorologica*, referring to Ibn al-Haytham's explanation of the halo surrounding the moon.

In his commentary, he looks down upon the doubts of the earlier commentators, and exalts Aristotle whom God distinguished among all mankind. In a comment on *Meteorologica* III.3, he refers to Ibn al-Haytham's explanation of the shape of the halo, "why it is a circle and why it appears round the sun or the moon or one of the other stars" (*Meteor.* 372b12–13). At the outset, he makes a distinction between the methods of "this science" and mathematical optics:

> "Et quia subiectum istorum [*sc.* signorum *e.g.* halonis etc.] sunt corpora naturalia, et cum hoc ipsa accidunt in situ determinato et in figuris determinatis, necessarium est, ut sit investigatio de eis secundum unum modum naturalis, secundum alium mathematica. Nos autem consideremus hoc de dispositione istorum de eis, de quibus considerat naturalis, utendo illis rebus quae declaratae sunt in mathematicis tanquam suppositionibus et fundamentis positis, et maxime eis, de quorum consuetudine est ut accipiantur hic principia directionis."[130]

In his exposition, Ibn al-Haytham had made an explicit statement on the role of mathematics: the substratum of the object under consideration is a physical body, hence the method of investigation must be physical; but since these objects have a round shape, they must also be investigated mathematically.[131] "This is why the inquiry by means of which the nature of these two effects is investigated comes to be composed of a physical and a mathematical (examination)."[132] This is precisely Ibn Rushd's point of criticism—or rather, apology of Aristotle's apparent omission: Physical science, that is to say: the philosophical theory of physical phenomena, and the mathematical science of optics are two disciplines of different orders. What Ibn al-Haytham explained, in his "famous treatise found in the hands of everybody," "does not belong to this science" (*laysa min hādhā l-ʿilm*). In his own *Epitome*, Ibn Rushd continues, he had taken the mathematical principles as postulates. But mathematics and physical science differ with regard to the causes they look into: while the philosopher studies the proximate causes, essential and evident, of the phenomena *qua* physical bodies, the mathematical science of optics studies the remote causes: geometrical models accidental to the physical substances.[133] Those who confound the two sciences will make mistakes such as Ibn al-Haytham committed. The principles of optics cannot serve as premises in a valid demonstration brought forward in physical philosophy.[134] "Praise to the Lord," he continues, "who distinguished him [Aristotle] with the human perfection. What he understood easily, is understood by the common man only after prolonged study and many difficulties, and what others understand easily is contrary to what is understood [to be correct] by him."[135]

Is he singing in the dark? Averroes takes care not to engage in a discussion of Ibn al-Haytham's advanced mathematics against Aristotle's. Indeed, Aristotle had evidently omitted to give a comprehensive explanation which would "save the phenomena" in the context of mathematical optics. Hence Averroes felt the need for an apology—an apology of his own philosophy which, in order to be universally valid, was dependent on the binding authority of Aristotle's method.

On the authority of Aristotle, only the physical part of philosophy studies real substances; the abstraction of mathematics is but a tool fur the purpose of analogous description, and cannot penetrate into reality in its own right.[136] Interpreting Plato's program of intellectual education, Averroes foists upon him a decisive shift in view of the rank of the mathematicals:

> "The intelligibles [of the mathematical sciences] are defective intelligibles since they are not conceived of in any particular objects but in what imitates them. Hence Plato divides the intelligibles of things into two parts. One of them he calls direct; these are the intelligibles of things that truly are. And the second [he calls thought]; these are the intelligibles of the appearances of existing things—and they are the mathematical sciences [Plato, *Resp.* VII, 533E4–5]. . . . Plato asserts of them that they are not of the rank of the other theoretical sciences as regards human perfection. Hence he says of them that they are sciences whose beginnings are unknown and whose ends are unknown; and [only] what is between the beginnings and the ends is known [*Resp.* VII, 533B6–C8]. This being so, the mathematical sciences are not intended [*mekhuwwanot*] initially and essentially for human perfection, as is the case with physics and metaphsics. Although they differ in this respect—and particularly in what these two sciences take from them [sc., the mathematical sciences] by way of principles for the investigation of the end (as when the divine science [i.e., metaphysics] accepts the number of movements from astronomy)—this difference is not only with respect to their kinds [i.e., arithmetic, geometry, astronomy, and music], but also exists with respect to the parts of the particular science."[137]

The mathematician gives a descriptive model of the phenomena, apt to yield correct calculations, but abstracted from reality; the physical philosopher looks into the proximate causes governing the reality of the world. In cosmology, the noblest object of both physics and mathematical astronomy, the philosopher proves his competence to grasp the highest, yet remote object of intellectual study.[138]

The Andalusian Restoration of Aristotle's Cosmos

The relation between the "physical science" (*al-ʿilm al-ṭabīʿī*) of philosophy and mathematical astronomy, and the relation of either to the cosmic reality, was discussed from the side of natural philosophy throughout the Andalusian school of *falsafa*: Ibn Bājja, Ibn Ṭufayl, Ibn Rushd, al-Biṭrūjī, Maimonides. Starting from the same principles of cosmological theory, Arabic astronomers of the West put forward solutions based on the Aristotelian models of homocentric spheres. Al-Zarqālluh, Ibn Ṭufayl (Ibn Rushd's predecessor as physician to the Almohad court) and the latter's disciple, the astronomer Abū Isḥāq al-Biṭrūjī were the most prominent advocats of such theories. The point of criticism turned against Ptolemy and his followers remains the same as before: The calculus of the mathematicians may fit, albeit imperfectly, the observations, but does not account for the actual processes governed by uniform principles and eternal laws; indeed, it is in evident contradiction to the principles established by Aristotle. Going further than previous critics of Ptolemy, they made bold to build a new configuration which should conform both to the principles and to the calculus matched with observation. Previous failure to achieve this was due not to the principles set up by the philosophers but to the imperfections of observation based on sensual data. [139]

Ibn Bājja made this clear in the terms of logical methodology in his short treatise *fī ʿIlm al-hay'a*, starting out from Aristotle's *Posterior Analytics*: The astronomers have to fall back on data arrived at indirectly, through observation and calculation. In trying to build a universal proof of their *hay'a*, they set up a syllogism in which the inferences of such findings will form the middle term of a syllogism; hence, their results must be at variance with the principles of physical science.[140]

Ibn Rushd was not enough of an astronomer to evaluate the theories of his predecessors and contemporaries mathematically. He used to say so himself.[141] His approach is dogmatical: Aristotle's physical and metaphysical doctrine, seen and interpreted as a closed system, advancing from the evidence and induction of existence in physics to the causes and principles of essence in metaphysics, required the astronomers—as in Plato's assignment—to provide calculable models which would link the data of observation with the essential, unchanging principles of the eternal circular movement: principles "pointing" in their turn to the cosmic essences of the spheres, immaterial soul-intellects moved by the desire to emulate the First Mover.

Aristotle's theory of homocentric spheres,[142] and Averroes' defense of the Aristotelian model against the Ptolemaic system, proceed from two basic assumptions: The eternal movement of the celestial bodies must be absolutely regular;[143] and the theories explaining their apparent movements should not regard the spheres and planets as mere mathematical entities, but as animate substances, "enjoying life and action."[144] Ibn Rushd's attempt to reconstruct Aristotle's true system—an attempt hampered by the overwhelming success of the Ptolemaic system in mathematical astronomy, and compounded by errors of translation in the Arabic version of the *Metaphysics*[145]—, is accompanied by constant attacks against Ptolemy's use of epicycles and excentric cycles in his interpretation of the planetary movements, starting from his *Epitome* of Aristotle's *De Caelo*:

> "The apparent advancing and receding motion of the (sphere of the fixed) stars cannot exist in their actual motion. . . . This precession and recession was not observed by the ancient Greeks except in the case of the planets, nor were many of the multiple motions established by Ptolemy observed by the Babylonians, such as the movement of the epicycles."[146]

The movement of the spheres must be homocentric, because the center of the earth must coincide with the centre of the universe.[147] Ptolemy deviated from this principle, because in his model, the center of the deferent axis (carrying the epicycles) must be excentric against the centre of the earth, and the centre of the epicycle moves on the eccentric with varying velocity, in such a way that only when seen from the *punctum aequans*—that is, the point on the line of apsides whose distance from the earth is the double of the linear eccentricity—the motion in the eccentric appears to be uniform.[148] Averroes objects:

> "That the earth is in the center and at rest is attested by the demonstrations which the mathematicians are accustomed to apply to these matters. If it were not in the centre, as Ptolemy claimed, three possibilities would obtain, all of which would result in absurdities and are in evident contradiction with the evidence of observation."[149]

The problem accompanied Averroes all his life and is prominent in the Long Commentaries on both the *Metaphysics* and the *De Caelo,* the magisterial works of his last years, containing numerous references to the problems of celestial mechanics where the *Almagest* was at variance with Aristotle.

> In his early work, the project of Ibn Rushd was limited to a "state of the art" summary of the encyclopaedia of the rational sciences, limited to "what is necessary for the human perfection" as a rational being, among the basic methods of juris-

prudence, of logic, and of natural philosophy.[150] The same is explicitely indicated for his "Summary of the *Almagest*,"[151] pointing to Ibn al-Haytham as an authority on this science, while denying the astronomers the philosophers' competence in the methods of demonstrative science.[152] In his early Compendium (*Jawāmiʿ*) of the *Metaphysics* (written in the early fifties of the 6th century A.H.), following closely the cosmology of al-Fārābī (and revised in a second version), Averroes seems to find epicycles acceptable. But attacks on Ptolemy start about this time already, in the physical *Compendia* as well as in the Summary of the *Almagest*: The method followed by the mathematicians offers neither "signs" (*dalāʾil*) nor demonstrations; most of what they attribute to the celestial bodies is impossible, notably the epicycle.[153] The most detailed discussion of the celestial movements, based on *Metaph.* XII.8, is found in his *Great Commentary* (*Tafsīr*) of the *Metaph.*, completed a few years before his death c. 590/1194. The commentator founds his attack on the principle of simple, circular movement found in the celestial bodies, a principle in conflict with the system of eccentric circles and epicycles, to be replaced by a system of homocentric circles of each planet where the poles of one circle rotate in the plane of the adjacent one.[154]

Leaving aside the intricacies of these discussions with regard to the technicalities of astronomy and mathematics[155]—what is striking, and relevant for a final perspective of Averroes' scientific approach, is the apparent subordination of applied science to physical theory: Aristotle's true philosophy was founded upon true science; if this science of the Ancients could be restored, all pieces in the cosmic puzzle would fall into place. Basically, the issue under discussion is sound philosophic method.

Referring to Aristotle's *De Caelo*, chapter II.3, Ibn Rushd elaborates Aristotle's note that "we have to pursue our inquiries at a distance—a distance created not so much by our spatial position as by the fact that our senses enable us to perceive very few of the attributes of the heavenly bodies" (268a4–7): With regard to these things, only certain premises are available to human induction,

> "and the things from which are acquired the premises, by which man scrutinizes many of the things concerning the heavenly body and through which he aspires to know their causes, are derived from the things which are closest to those in resemblance, viz. the animate bodies, and especially man, since it has been made clear that this body is animate. However, it is evident that this (kind of statement) is ambiguous (*yuqāl bil-tashkīk*) [being ambiguous] about what is prior and posterior, and therefore this kind of concept and judgment is weak."[156]

The systematic reasoning behind Averroes' cross-references to the other parts of philosophy is based on rules of the *burhān*: The "demonstration of existence" will be provided in the various natural sciences, as, for example, the parts of the soul in the psychological part of natural philosophy; the "demonstration of causes" is the privilege of a discipline higher in rank with regard to the lower levels, and for the very highest can be given only by induction from *dalāʾil* perceived in the posterior ones.[157]

This systematic relationship between natural philosophy and metaphysics is underlined in many statements from the very earliest phase of Averroes' activity, as in the *Compendia* of the *Ṭabīʿiyyāt* and of the *Metaphysica*:

> "Demonstration in an absolute way (*demonstratio simpliciter, al-burhān al-muṭlaq*)[158] in this science is founded on the propositions accepted from natural science and theological science: It has been explained in the *Physics* that the mover of the celestial bodies is not in matter, and in the book *De Anima* that what is of this kind is intellect, and in the first treatise [of *De Anima*] that the intelligible form [i.e., soul] is moved only through desire coming from its intellect; hence this must have its object in imagination—it is a celestial body exercising desire."[159]

The fundamentals of epistemology, where the demonstration of existence, essence and cause constitutes a hierarchy and interdependence of the sciences, of physics and metaphysics, not only justify, but require systematic cross-references between the disciplines; these principles were established by Aristotle in *Metaph.* VI.1, and are constantly called upon by Averroes as a guideline for philosophic method.[160]

Another case in point is a remark on the order of the planets: Contrary to the doctrine of Ptolemaeic astronomy, the sphere of the Sun must be assumed between the Moon and the remaining planets in order to conform with the principles established by Aristotele. In fact, however, Aristotle had not discussed the relative positions, distances, and velocities of the stars and planets in detail; these topics empirical astronomy was left to deal with adequately (*De Caelo* II. 10, 291a29–32). The astronomer, Averroes explains (*Comm mag. De Caelo*, II c. 57, fol. 64ra33–56 ad locum), demonstrates the existence of data apparent from or indicated by sense perception, regarded as mathematical entities abstracted from matter, while the natural philosopher gives the causes of the same subjects regarded as natural substances. But the philosopher refers to astronomy, as Aristotle does in *De Caelo* II.10, because he considers the causes of those things whose existence has been established in astronomy:

> "Now both the natural philosopher and the astronomer engage in the study of these questions; however, the astronomer mainly gives the existence [*quia*; Arabic, *anna l-shayʾ*] while the natural philosopher gives the cause [*propter quid*; Arabic, *li-ma l-shayʾ*]. What the astronomer mainly gives is based only on those things that appear to the senses . . . , the natural philosopher, however, endeavours

to give the cause why this is so [*propter quam hoc est supra ipsam*; Arabic, *li-ma huwa ʿalā hādhā*]."[161]

The relation between the angular velocity of the planets and their distances with regard to the first heaven underlies a general principle stated by Aristotle: The absolute speed of the planet nearest to the first revolution (the circle of the fixed stars)—that is, Saturn—is greatest, while the others are slower, the decrease in velocity being in proportion with their distance (*De Caelo* II 10, 291b6–10).[162] While recognizing this general principle, Ptolemy tried to establish the precise relative order of the planets with respect to the Sun.[163] Based on observation and computation of the relative distance, and on the apparent eccentricity of the spheres of Mercury and the Moon, he concluded that the spheres of Moon, Mercury, and Venus lie below the Sun while Mars, Jupiter and Saturn lie above.[164]

Unabashed by the reference of the Arabic version to the *aṣḥāb al-Majisṭī*—Aristotle's *mathêmatikoi* (291b9)—Ibn Rushd declares that the conditions underlying Aristotle's exposition are reconcilable only with the "opinion of those who say that the Sun is above Mercury and Venus, and not below—here the astronomers are at variance, and the truth of the matter has not yet been established."[165] However, Aristotle's statement on the connection between the planet's velocity and its distance from the first heaven does not imply a mathematical, proportional ratio; even though the Sun's motion may be quicker than the motions of Mercury and Venus, it may nevertheless move in a sphere above, "because its potency surpasses theirs . . . because there, local proximity is similar to mutual proximity of the essences, and this is proximity in knowledge and in rational cognition: the stronger the cognition of the first movement, the more perfect the desire toward it will be, and the stronger the desire, the quicker its motion will be."[166]

In linking the cosmic order with Aristotle's metaphysics—the cosmic motion originates from the conscious desire of rational souls toward the Unmoved Mover—Averroes puts forward the reasoning of philosophy against the celestial mechanics of the astronomers: *isti enim motus quos ponit Tholomeus fundantur super . . . fundamenta quae non conveniunt scientiae naturali.*[167] But does he betray his own principles, making the natural phaenomena and the facts established by natural science subordinate to a-priori postulates of metaphysics? The doubt and caution expressed, again and again, in view of the difficult and controversial field of astronomy may convince us that he is not: The very passion of his aporia between astronomy and metaphysics betrays his awareness that only a true understanding of the astronomical cosmos will yield true answers to the ultimate questions of metaphysics. Indeed, it is the task of metaphysics to "save the phenomena" of observation.

However, the Commentator deplores the astronomers' (and his own) inability to reconstruct the true Aristotelian cosmos, that is, to provide not just a mathematical emulation which does not contradict his physics, but a true physical astronomy. Regarding the order of thc planets, as also the models contradicting Aristotle's doctrine of homocentric spheres, "the necessary movements in these things have not yet been demonstrated in this science: for the movements assumed by Ptolemy are based on premisses not reconcilable with natural science, sc. eccentrics and epicycles, which are both false."[168] He felt the truth to be near at hand:

> "Maybe, if God will grant me life, I shall investigate the [science of] the sphere of Aristotle's age (*al-hay'atu llatī kānat fī 'ahdi Arisṭū*); and it will turn out that it did not contain any [such] absurdities with respect to physical science. This is [the system founded upon] what Aristotle called 'spiral motions' (*al-ḥarakāt al-lawlabiyya*). It is, as I think, a movement where the poles of one sphere turn about the poles of another sphere, so that its own movement will proceed along a spiral line, as for example the motion of the sun with regard to the diurnal motion. On the basis of [the assumption of] this motion we can give [an explanation of] what happens to the star, such as the differences in movement, backward movement and movement in a straight line. We shall make a study of these motions, for it is impossible that there should be irregularity in the celestial bodies unless [explained] in this way."[169]

It is a tragic irony that this "spiral motion" (*ḥaraka lawlabiyya*), the term behind which Averroes suspected the final solution, goes back to a mistake of the translator.[170] But in his old age he despaired of this hope. A few years before his death, he returned to the problem of the planetary movements in his *Commentarium magnum* of the *Metaphysics* (in the context of Aristotle's discussion of the number of the eternal moving principles of the heavens, *Metaph.* Λ. 8.1073a14ff.).[171]

Averroes is ready to concede to Aristotle that "the specialist of this science" must accept from astronomy the information it gives about the number of the motions, but he insists that he need not accept "the other matters it comprises."[172] For the discipline of astronomy which inquires into the celestial motions "cannot establish, on the basis of the spherical motions apparent to us, the course of the causes *unless it does not contradict the principles of physics* (*illā mā laysa yalḥaqu min waḍ'ihī muḥālun fī l-'ilmi l-ṭabī'ī* [i.e., physical philosophy])."[173] The apparent irregularities in the planetary motions "do not conform to the nature of the motions of the celestial bodies; that is, it has become evident in the physical science that all their motions are regular."[174] The astronomer must therefore postulate a model from which the phenomena (*al-aḥwāl al-ẓāhira*) would result without violating the principles of natural philosophy (*min ghayri an yalzama minhā muḥālun fī l-ʿilmi l-ṭabīʿī*).[175] The complexities involved—composite movements, excentricities, epicycles—result in disagreeement among the "modern" mathematicians (*al-ḥadathu mina l-taʿālīmiyyīn*) as to the number of these motions: an undisguised allusion to Ibn al-Haythams's *Shukūk*; and here, Ibn Rushd is eager to note that the calculus based on such mathematical models is, in certain cases at least, in blatant disagreement with observed phenomena.[176] The mathematicians of Aristotle's time, who provided him with a tentative solution, are commended for being in closer agreement with physical philosophy than Ptolemy and the "moderns." The theory of epicycles, positing centres of motion beside the centre of the world, and the model of excentric spheres, involving "superfluous bodies in heaven with no purpose but as filling (*ḥashwan*)" are both "contrary to nature."[177]

The only solution Averroes can propose is a closer examination of the mathematical astronomy Aristotle himself relied upon, that is, the theories of Eudoxus and Callippus; perhaps the model of "spiral motion" he ascribes to Aristotle "would allow us to do without these two things [i.e., epicycles and eccentric spheres]. . . . Ptolemy failed to notice what had compelled the Ancients to accept spiral motions, namely the impossibility of the epicycle and the eccentric sphere." But—and here we return to the starting point of our inquiry—"in our time, astronomy is no longer something real; the model existing in our time is a model conforming to calculation, not to reality."[178]

In this, Ibn Rushd not only observed the letter of Aristotle's doctrine, but also the spirit of his science, where metaphysics investigated a cosmic reality, not just "units with a serial order," but "enjoying life and action"—in the final analysis, Plato's heritage in the Aristotelian encyclopaedia.

5 In the Face of the Almighty: Theology and Science

The models of concentric planetary spheres, postulated by Ibn Bājja, understood but vaguely by Ibn Rushd, and constructed, however imperfectly, by al-Biṭrūjī, had little influence beyond al-Andalus and the oncoming decline of Western Muslim civilization. Neither here, nor indeed in the coming bloom of the mathematical sciences in the East, the philosophical Weltbild played a rôle in the improvement of the old approaches, let alone a renewal apt to launch a new paradigm. Science thrived at the hands of theologians who left philosophy to the *falāsifa* who in vain hat whetted their steel to tackle the apories of the *hay'a*, and discarded with principles which had proved useless for the practical tasks of the art.

*

The apories admitted, however unwillingly, by Ibn Rushd, were stated bluntly and uncompromisingly, in a remarkable passage of the "Guide of the Perplexed," *Dalālat al-ḥā'irīn,* of the Jewish jurist, physician and philosopher Maimonides, on the authority of Ibn Bājja:[179]

> "As far as the action of ordering the motions and making the course of the stars conform to what is seen is concerned, everything depends on two principles: either that of the epicycles or that of the eccentric spheres or on both of them. . . . Both those principles are entirely outside the bounds of *qiyās* [*logos*, the method of demonstrative science] and opposed to all that has been made clear in natural science [*al-ʿilm al-ṭabīʿī*, i.e., the part of philosophy demonstrating rationally the processes of the physical world]."

While explaining the difficulties involved, controverting the principles established by Aristotle, and after going into the problems of planetary motion in detail, Maimonides concludes:

> "Consider how great these difficulties are. If what Aristotle has stated with regard to natural science is true, there are no epicycles or eccentric circles and everything revolves around the center of the earth. But in that case how can the various motions of the stars come about? Is it in any way possible that motion should be on the one hand circular, uniform, and perfect, and on the other hand the things that are observable should be observed in consequence of it, unless this be accounted for by making use of one of the two principles [sc. epicycles or eccentric circles] or of both of them? This consideration is all the stronger because what is calculated on the hypotheses of the two principles is not at fault even by a minute. . . . This is the true perplexity. However, . . . *this does not affect the astronomer*. For his purpose is not to tell us in which way the spheres truly are [*laysa maqṣūduhū an yukhbiranā bi-ṣūrati wujūdi l-aflāki kayfa hiya*], but to posit an astronomical system in which it would be possible for the motions to be circular and uniform and to correspond to what is apprehended through sight [*an yafriḍa hay'atan yumkinu bihā an takūna l-ḥarakātu dawriyyatan wa-mustawi-yatan wa-tuṭābiqa mā yudraku ʿiyānan*], regardless of whether or not things are thus in fact [*kāna l-amru ka-dhālika aw lam yakun*]."

Ibn Bājja already had expressed doubts whether Aristotle was aware of some of the intricacies of planetary motion, sc. the eccentricity of the sun: indeed, he was not, and would have been baffled if he had been—resorting, as in other cases, to "guessing and conjecturing."

> "However, regarding all that is in heaven, man grasps nothing but a small measure *of what is mathematical* [*illā bi-hādhā l-qadri l-taʿlīmiyyi l-yasīr*, i.e., the little that can be established mathematically]. I shall accordingly say in the manner of poetical preciousness: 'The heavens are the heavens of the Lord, but

the earth hath He given to the sons of man' [Ps. 115:16]. I mean thereby that the deity alone knows the true reality [*ḥaqīqat al-samā'*], the nature, the substance, the form, the motions, and the causes of the heavens."[180]

The decision taken by the great Jewish thinker is in tune as well with the overall development of Muslim intellectual attitudes in the later Middle Ages: while theology becomes scientific, science becomes theological. The agencies of this transition are manifold, and comport a far-reaching shift in the social and intellectual milieu of science in Islam. From the perspective of philosophy, the main factors are the Hellenization of Kalām in the school of al-Ghazālī, notably in the attendance of Fakr-al-Dīn al-Rāzī in Sunnī, and of Naṣīr-al-Dīn al-Ṭūsī in Shīʿī Islam, the adoption of Avicenna's encyclopaedia of intellectual knowledge in the religious community, and at the same the elimination of the stumbling stones in Aristotelian physics—the eternity of the world, and the laws of physical causality submitting God to a necessity imposed by reason. In consequence, the defense of the rational sciences was undertaken by members of the same religious community who regarded the methods of demonstration as an indispensable basis of sound argument in the service of Islam. As a further consequence of this development in the social and intellectual communities of rational science, the teaching traditions of *kalām, falsafa, and riyāḍiyyāt* grow from diverse branches to become parts of an integrated curriculum of learning in the Iranian, and later on in the Mughal and Ottoman *madrasas.*[181]

Following al-Ghazālī, the Ashʿarite interpretation of scientific knowledge[182] defended logical method as a "just scale" (*al-qisṭās al-mustaqīm*) and "vessel of knowledge" (*miʿyār al-ʿilm*), and in the same vein, the teachers of the sciences—many of whom were powerful authorities in theology as well—referred to the epistemology of the *Analytica Posteriora*, as restated by al-Fārābī and realized in his system of the sciences. As al-Ṭūsī states in his epochal *Tadhkira* on theoretic astronomy, "every science has a subject which is investigated in that discipline, and principles, either self-evident, or else obscure—in which case they are proved in another science and are taken for granted in this science," and in the case of astronomy, "its principles that need proof are demonstrated in three sciences: metaphysics, geometry, and natural philosophy (*al-ṭabīʿiyyāt*)."[183] This is closely reminiscent of al-Fārābī's *Kitāb al-Burhān.*[184]

In the same vein, ʿAḍud-al-Dīn al-Ījī (d. 1355), in his monumental *Summa* of Ashʿarite Kalām, presents the principles of astronomy as neutral with respect to religious law, "prohibition does not extend to them, being neither an object of belief nor subject to affirmation or negation." But this compromise is reached at a price: "These [sc. the hypotheses of mathematical astronomy] are imaginary things that have no external existence, . . . mere imaginings which are more tenuous than a spider's web." Like in the apories stated by Averroes and admitted by Maimonides, we come back to the basic conflict between philosophy and astronomy: the conflict about the reality of their object of study.[185] There was, however, an alternative open to the theologian-astronomer, assuming full responsibility for his science while discarding with the principles taken from *falsafa.* In a commentary on al-Ṭūsī's dogmatic, *Tajrīd al-ʿaqāʾid,* al-Qūshjī—an astronomer from the circle of Ulugh-Bek, active in Istanbul until his death in 879/1474—seeks to establish his science in a creation not governed by the physical laws of causality, but by the will of God.[186] "What is stated in the science of astronomy does not depend upon physical (*ṭabīʿiyya*) and metaphysical (*ilāhiyya*) premises. . . . For of what is stated in this science, some things are geometrical premises, which are not open to doubt; others are premises arrived at through intuition (*muqaddamāt ḥadsiyya*),[187] as we have mentioned; others are premises determined by reason in accordance with the apprehension of what is most suitable and appropriate. On this basis, "it is sufficient for the scientist to conceive, from among the possible models, the one by which the circumstances of the planets with their manifold irregularities may be put in order in such a way as to facilitate their determination of the positions and conjunctions of these planets for any time they might wish and so as to conform with perception and sight, this in a way that the intellect and the mind find wondrous."[188] Bent to describe adequately the *opificium mundi*, the theologian-scientist felt free to discard the causes proclaimed by the philosophers as universal principles, and going further, felt free to explore possibilities contradicting the principles of Aristotelian physics.[189]

It was a proud and competent astronomer who took Ptolemy himself to task for an *metabasis eis allo genos*—trespassing into the foreign domain of cosmology: al-Bīrūnī who in his *Qānūn al-Masʿūdī* upbraided Ptolemy for having, in his *Planetary Hypotheses*, "deviated from the path he had followed in the *Almagest,* [having taken up] that which related to opinions outside of this science, that is in the belief of people that the celestial bodies have life, perception, sensation, and the choice of the noblest motions."[190]

The alternative, chosen by many generations of creative mathematicians and astronomers, was to follow the principles of demonstrative science as they had been established by Aristotle, reformulated by al-Fārābī as supreme scientific method and restated by Ibn Sīnā, and at the same time, leaving cosmology aside: Conceding that every science has its own principles; that one science will build upon the other; and that metaphysics comes not first, but next (if at all). The winds of change rose not from the side of philosophers, but from mathematicians who observed, described and calculated the phenomena, and contemplated the wonder of creation in the eternal splendor of the cosmic order.

Notes

1. Ibn Rushd, *Sharḥ k. al-Samāʾ wa-l-ʿālam*, I, c. 90; ms. Tunis, Aḥmadiyya, no. 5538, fol. 70a = ed. in facsimile by G. Endress, *Commentary on Aristotle's Book On the Heaven and the Universe*, by Ibn Rushd (Frankfurt am Main: Institute for the History of Arabic-Islamic Science, 1994), 47 (passage missing in the Latin version of Michael Scot, and perhaps added by the author at a later date); cf. another passage in the same spirit, ibid., II c. 35, fol. 38b22–24 = facs. ed., p. 210 (*in ansaʾa* [MS. *in shāʾa*] *Llāhu fī l-ʿumr* [see below, p. 267]; *v.* G. Endress, "Averroes *De Caelo*," *Arabic Sciences and Philosophy*, 5 (Cambridge, 1995): 9–49, esp. p. 45, and also Jamāl-al-Dīn al-ʿAlawī, *al-Matn al-rushdī* (Casablanca, 1987), 107.

2. Ibn Rushd, *Tafsīr Mā baʿd al-ṭabīʿa*, XII c. 45, ed. M. Bouyges, Bibliotheca Arabica Scholasticorum, V–VII (Beyrouth, 1938–42), p. 1663.11–12, 1664.2–7; cf. English trans. by Ch. Genequand, *Ibn Rushd's Metaphysics* (Leiden: Brill, 1984), 179.

3. "Nobody who is ignorant of geometry shall enter"—according to Hellenistic tradition, the inscription of the Platonic Academy at Athens; see, e.g., Elias, *Comm. in Arist. Cat.*, ed. A. Busse (Berlin, 1900), 118.19.

4. Facs. of the first edition (*Ioannis Keppleri Harmonices mundi libri V*, Lincii Austriae: Plancus, 1619) in Johannes Kepler, *Gesammelte Werke*, Bd VI: Harmonice mundi, ed. Max Caspar (München: Beck, 1940), 13 (facsimile of the original title).

5. Proclus, *In I Euclidis Elem.*, ed. G. Friedlein (Lipsiae: Teubner, 1873): prologus I, p. 22, 17–20: πρὸς δὲ τὴν φυσικὴν θεωρίαν τὰ μέγιστα συμβάλλεται, τὴν τῶν λόγων εὐταξίαν ἀναφαίνουσα, καθ' ἣν δεδημιούργηται τὸ πᾶν, καὶ ἀναλογίαν τὴν πάντα τὰ ἐν τῷ κόσμῳ συνδήσασαν. English translation by Glenn R. Morrow, Proclus, *A Commentary on the First Book of Euclid's Elements* (Princeton, N.J.: Princeton University Press, 1970), 19.

6. See Albert Dihle, *Philosophie als Lebenskunst*, Rheinisch-Westfälische Akademie der Wissenschaften, Geisteswissenschaften, Vorträge, G 304 (Opladen: Westdeutscher Verlag,1990), 12, on the superior claim of the Platonists for the study of physical philosophy—building upon mathematics, but transcending from the mathematicals, verified empirically, to the purely intelligible principles (Proclus, *In Tim.*, 23.9 ff.); Pierre Hadot, *Qu'est-ce que la philosophie antique?* (Paris: Gallimard, 1995), 100 f.

7. See Ilsetraut Hadot, "Les aspects sociaux et institutionels des sciences et de la médecine dans l'Antiquité tardive," *Antiquité tardive*, 6 (Turnhout, 1998): 233–250; *ead.*, *Arts libéraux et philosophie dans la pensée antique* (Paris: Etudes Augustiniennes, 1984).

8. "Encyclopaedia," that is, a general, "all-round" education encompassing the fields of knowledge preparing the way to higher learning.

9. I. Hadot, ibid., 242–244.

10. Ḥunayn b. Isḥāq, *Risāla ilā ʿAlī b. Yaḥyā*, ed. G. Bergsträsser: *Ḥunain b. Isḥāq über die syrischen und arabischen Galenübersetzungen* (Leipzig, 1925), Arabic text, 18; German trans., 15. The earliest translations of Aristotelian logic were commissioned to

scholars of the Christian churches, who had kept up the teaching of logic in Syriac, and of the isagogic tradition of the Alexandrian school; most prominent—but quite independent from those of the astrologers and physicians—are the activities of the patriarch Timothy, working by commission for the caliph al-Mahdī; *v.* Henri Hugonnard-Roche, "Les traductions du grec au syriaque et du syriaque à l'arabe (à propos de l'Organon d'Aristote)," in *Rencontre de cultures dans la philosophie médiévale: traductions et traducteurs de l'Antiquité tardive au XIVe siècle* (Louvain-la-Neuve; Cassino, 1990), 133–147; Sebastian Brock, "Two letters of the Patriarch Timothy from the late eighth century on translations from Greek," *Arabic Sciences and Philosophy*, 9 (1999): 233–246.

11. Cf. Albrecht Dihle, "Philosophie—Fachwissenschaft—Allgemeinbildung," in *Aspects de la philosophie hellénistique*, Entretiens sur l'Antiquité classique, 32 (Vandœuvres-Genève: Fondation Hardt, 1986), 188–232.

12. τὰ μαθήματα καὶ οἱ ἀριθμοί, Pythag. fr. 14 A 17 Diels/Kranz, from Apollon, *Mirabilia* 6 Keller.

13. Philolaos, fr. 44 B 4.

14. ἀριθμὸν εἶναι τὴν οὐσίαν πάντων, Arist. *Metaph.* A 5, 987a13–19. Cf. the sources presented by C. J. de Vogel, *Greek philosophy: a collection of texts*, vol. 1 (Leiden, [3]1963), 10ff., esp. p. 16f. (nos. 37–42).

15. *Metaph.* A.6, 987b10; e 8, 1073a18: ἀριθμοὺς γὰρ λέγουσι τὰς ἰδέας οἱ λέγοντες ἰδέας.

16. Cf. the introduction of H. G. Zekl to his translation of the *Analytica posteriora*, Aristoteles, *Erste Analytik; Zweite Analytik*, in *Organon/Aristoteles*; Bd 3/4, griechisch-deutsch, Philosophische Bibliothek, Bd. 494/495 (Hamburg: Meiner, 1998), lviii–lxxxvi.

17. *Resp.* VII, 526a6: ὧν διανοηθῆναι μόνον ἐγχωρεῖ.

18. *Resp.* VII, 526a1–3: προσαναγκάζον αὐτῇ τῇ νοήσει χρῆσθαι τὴν ψυχὴν ἐπ' αὐτὴν τὴν ἀλήθειαν.

19. *Resp.* VII, 527B7, 9–10: τοῦ ἀεὶ ὄντος.

20. On mathematics in the history of Platonism, see Heinrich Dörrie and Matthias Baltes, *Der Platonismus im 2. und 3. Jahrhundert nach Christus*, Der Platonismus in der Antike, Bd 3 (Stuttgart-Bad Cannstatt: Frommann-Holzboog, 1993), 68–71, 266–279.

21. Thomas L. Heath, *A history of Greek mathematics*, I, 98; Dörrie and Baltes, *op. cit.*, 267f.

22. On the Hellenistic reading and transmission of Nicomachus, *v.* Leendert Gerrit Westerink, "Deux commentaires sur Nicomaque: Asclépius et Jean Philopon, *Revue des études grecques,* 77 (1964): 526–535; Étienne Evrard, Jean Philopon, son «Commentaire sur Nicomaque» et ses rapports avec Ammonius," ibid. 78 (1965): 592–598; Leonardo Tarán [ed.], Asclepius of Tralles, *Commentary to Nicomachus' Introduction to Arithmetic* (Philadelphia: American Philosophical Society, 1969). A close contempo-

rary was Theon of Smyrna, who wrote an "Exposition of mathematical matters useful for the study of Plato" (ed. E. Hiller, Leipzig 1878).

23. Dominic J. O'Meara, *Pythagoras Revisited: mathematics and philosophy in late Antiquity* (Oxford: Clarendon Press, 1989), 212.

24. Ibid., 166–169; see also Ian Mueller, "Mathematics and philosophy in Proclus' commentary on book I of Euclid's *Elements*," in *Proclus: lecteur et interprète des anciens*, Actes du colloque international du CNRS (2–4 October 1985) publiés par Jean Pépin et H. D. Saffrey (Paris: CNRS, 1987), 305–318.

25. L. G. Westerink, "Philosophy and medicine in Late Antiquity," *Janus*, 51 (1964): 169–177; *id.*, "Ein astrologisches Kolleg aus dem Jahre 564," *Byzantinische Zeitschrift*, 64 (1971): 6–21.

26. Cf. G. Endress, "Die wissenschaftliche Literatur," in *Grundriss der Arabischen Philologie*, 2 (Wiesbaden: Reichert, 1987), 407–409.

27. The textual evidence does not support the thesis proposed by Michel Tardieu that Simplicius, on his return from the court of Khosrow Anūshīrwān, retired to Ḥarrān, and that the *Ṣābiʾat al-Yūnāniyyīn* mentioned by al-Masʿūdī should have kept up a Platonic Academy until the 10th century; *v.* Tardieu, "Ṣābiens coraniques et «Ṣabiens» de Ḥarran," *Journal asiatique*, 274 (1986): 1–44; *idem*, "Simplicius et les calendriers de Ḥarrān d'après les sources arabes et le commentaire de Simplicius à la Physique d'Aristote," in *Simplicius: sa vie, son oeuvre, sa survie* (Berlin: de Gruyter, 1987), 40–51; cf. Dimitri Gutas, "Plato's *Symposion* in the Arabic tradition," *Oriens*, 31 (1988): 36–60 (p. 44); Concetta Luna, review of: Rainer Thiel, *Simplikios und das Ende der neuplatonischen Schule in Athen* (Stuttgart: Steiner, 1999), in: *Mnemosyne*, ser. 4, vol. 54 (Leiden, 2002): 482–504.

28. Through his pupil al-Sarakhsī; *v.* Franz Rosenthal, *Aḥmad b. aṭ-Ṭayyib as-Saraḫsī* (New Haven: American Oriental Society, 1943), 41–51.

29. See G. Endress, "Al-Kindī über die Wiedererinnerung der Seele: arabischer Platonismus und die Legitimation der Wissenschaften im Islam," *Oriens*, 34 (Leiden, 1994): 174–221.

30. Ed., with a Turkish translation, by Aydın Sayılı, "Sâbit ibn Kurra'nin Pitagor teoremini tamimi," *Belleten*, 22 (Ankara: Türk Tarih Kurumu, 1958): 526–549.

31. Ibid., p. 541.

32. Qutotations are found in the Arabic scholia of al-Nayrīzī, who in his turn relied not on the full commentary of Proclus—whom he does not mention by name—but rather on the scholia found in his manuscript, containing explicit references to Heron and —transmitting the doctrine of Proclus—Simplicius; *v.* Rüdiger Arnzen, *Abu 'l-ʿAbbās an-Nayrīzīs Exzerpte aus (Ps.-?) Simplicius' Kommentar zu den Definitionen, Postulaten und Axiomen in Euclids Elementa I*, eingeleitet, ediert und mit arabischen und lateinischen Glossaren versehen (Köln, Essen, 2002), xxvi–xxxvii and Index nominum. These concern specific problems, but convey scarcely any traces of the philosophic prolegomena. Proclus' *Elementatio physica* was also known; a selection of the

propositions, in Arabic translation, is quoted in a treatise of Yaḥyā ibn ʿAdī on the refutation of atomism; *v.* G. Endress, "Yaḥyā ibn ʿAdī's Critique of Atomism: three treatises on the indivisible part, ed. with introduction and notes," *Zeitschrift für Geschichte der Arabisch-Islamischen Wissenschaften*, 1 (Frankfurt a. M., 1984): 155–179.

33. Translated from the Syriac by the Nestorian Ḥabīb (ʿAbdyashūʿ) ibn Bihrīz; *v.* M. Steinschneider, *Die hebräischen Übersetzungen des Mittelalters und die Juden als Dolmetscher* (Berlin, 1893; repr. Graz: Akad. Druck- und Verlagsanstalt, 1956), § 320, p. 516–519; this version, which is no longer extant, was replaced by that of Thābit ibn Qurra (d. 288/901), ed. W. Kutsch: *Ṯābit b. Qurras arabische Übersetzung der 'Αριθμητικὴ Εἰσαγωγή des Nikomachus von Gerasa* (Beyrouth: Imprimerie Catholique, 1959). A Hebrew commentary-paraphrase, based on al-Kindī's redaction, is extant in manuscript (*v.* Steinschneider, *loc. cit.*), and is being prepared for publication by Gad Freudenthal.

34. *Kitāb al-Mūsīqī al-kabīr*, *v.* Ibn al-Nadīm, *al-Fihrist*, ed. Flügel, 269.23 ("*wa-li-hādhā l-kitābi mukhtaṣarāt*").

35. al-Kindī, *K. ilā l-Muʿtaṣim bi-Llāh fī l-falsafa al-ūlā*, ed. M. ʿAbd-al-Hādī Abū Rīda (Cairo, 1950–53), 1:97–162, p. 103.1; *Risāla fī kammiyyat kutub Ariṣṭāṭālīs*, ed. M. Guidi, R. Walzer, *Studi su al-Kindī I: Uno scritto introduttivo allo studio di Aristotele* (Roma: Accademia Nazionale dei Lincei, 1940),§ 6, p. 393 = ed. Abū Rīda, *Rasāʾil*, 372f.

36. Abū Maʿshar, *K. al-Madkhal al-kabīr ilā ʿilm aḥkām al-nujūm: Liber introductorius maior ad scientiam iudiciorum astrorum*, ed. Richard Lemay (Napoli: Istituto Universitario Orientale, 1995), 1:23–32 (introd.).

37. On the position of the mathematicals, esp. on the doctrine of the three kinds of being, see Heinrich Dörrie and Matthias Baltes, *Die philosophische Lehre des Platonismus: einige grundlegende Axiome; platonische Physik (im antiken Verständnis)*, I, Der Platonismus in der Antike, Bd 4 (Stuttgart—Bad Cannstatt: Frommann-Holzboog, 1996), 48–66 (texts), 266–90 (commentary); Philip Merlan, *From Platonism to Neoplatonism* (Den Haag 1953; [2]1960), 220–25; G. Endress, "al-Kindī über die Wiedererinnerung" [*supra*, n. 29], 182.

38. al-Kindī, *Risāla fī kammiyyat kutub Ariṣṭāṭālīs wa-mā yuḥṭāj ilayhi fī taḥṣīl al-falsafa*, edd. M. Guidi and R. Walzer: *Uno scritto introduttivo allo studio di Aristotele*, Studi su al-Kindī I (Roma: Accademia Nazionale dei Lincei, 1940); *v.* Christel Hein, *Definition und Einteilung der Philosophie: von der spätantiken Einleitungsliteratur zur arabischen Enzyklopädie*, Europäische Hochschulschriften, 177 (Frankfurt a.M. [etc.]: Lang, 1985), 174–177.

39. Cf. Ilsetraut Hadot, "La division néoplatonicienne des écrits d'Aristote," in Simplicius, *Commentaire aux Catégories d'Aristote,* trad. commentée, (Leiden: Brill, 1990), fasc. 1:63–93, p. 91.

40. *Op. cit.*, p. 391, followed by a detailed explanation of the quadrivium, p. 394. See the references to the Greek commentators given by R. Walzer, ibid., p. 377.

41. Ibn al-Nadīm, *al-Fihrist*, ed. Flügel, 255.28.

42. Ed. by Zakariyyā Yūsuf, *Mu'allafāt al-Kindī al-mūsīqiyya* (Baghdād: al-Majmaʿ al-ʿIlmī al-ʿIrāqī, 1962), 67–92; cf. Amnon Shiloah, *The theory of music in Arabic writings (c. 900–1900)* (München: Henle, 1979), 254f.

43. al-Kindī, *al-Muṣawwitāt al-watariyya*, l. c., p. 70; on the classification of mathematics, see C. Hein, *op. cit.* [above, n. 38], p. 182ff.

44. *al-Muṣawwitāt,* 71.10–20.

45. Ibid. 72.

46. Ibid. 77.5–10.

47. Ibid. 78.10–79.20.

48. Third *maqāla*, p. 85ff.: *fī mushākalat al-awtār.*

49. Ed., with introduction and French translation, by Léon Gauthier, *Antécédents gréco-arabes de la psychophysique* (Beyrouth: Imprimerie catholique, 1938); see also M. Ullmann, *Die Medizin im Islam* (Leiden: Brill, 1970), 302, with further references.

50. Tzvi Langermann, "Another Andalusian revolt? Ibn Rushd's critique of al-Kindī's pharmacological computus," *infra,* chapter 12.

51. Ibn Rushd, *al-Kulliyyāt fī l-ṭibb*, edd. Saʿīd Shaybān, ʿAmmār al-Ṭālibī (Cairo: al-Majlis al-Aʿlā li-l-Thaqāfa, 1989), 309–313, quotation from p. 312.8; corresponding to ed. J. M. Fórneas Besteiro, C. Álvarez de Morales (Madrid: Consejo Superiór de Investigaciónes Científicas; Granada: Escuela des Estudios Árabes, 1987), 389–392 (391.–6).

52. A detailed exposition was given by Paul Kraus, *Jābir ibn Ḥayyān: contribution à l'histoire des idées scientifiques dans l'Islam; t. 2: Jābir et la science grecque* (Le Caire, 1942; repr. Paris: Les Belles Lettres, 1986), ch. V, "La théorie de la balance," 187–303.

53. For these "orders" (τάξεις) and "degrees" (ἀποστάσεις) cf. Galen, *De comp. med. per genera*, II.1 (*Opera*, ed. Kuehn, vol. 13: 464f.); III.2 (p. 572f.); *De simpl. med.*, V. 27 (vol. 11: 786–788); *v.* quotations in Kraus, *Jābir*, 189, nn. 2–4.

54. See the excerpt from *K. al-Baḥth*, edited by Paul Kraus, *Jābir ibn Ḥayyān: textes choisis* (Paris, Le Caire: IFAO, 1935), 510–513, and the relevant passage in Kraus, *Jābir*, II:194f.

55. Cf. Kraus, *op. cit.*, 20ff., adding abundant references to Greek philosophical and mathematical literature, and also to the use of arithmetical symbolism in Christian Greek authors.

56. Or *Kitāb Ikhwān al-Ṣafā'*: ed. Friedrich Dieterici, *Die Abhandlungen der Ichwan es-Safa* (Leipzig, 1883); ed. Khayr-al-Dīn al-Ziriklī (Cairo, 1928); repr. with an introd. by Buṭrus al-Bustānī (Bayrūt: Dār Ṣādir, 1957), and other editions; cf. Amnon Shiloah, *The theory of music in Arabic writings (c. 900–1900)* (München: Henle, 1979), 230–233.

57. Explicitly: *mithlamā kāna yafʿaluhū l-ḥukamā' al-Fīthāghūriyyūn*, I:48.11 (*Risāla* I, 1).

58. Nicomachus Gerasenus, *Introductio arithmetica*, ed. Hoche (Lipsiae: Teubner, 1866), 1–2, referring to Pythagoras; cf. Martin Luther D'Ooge (trans.), Nicomachus of Gerasa, *Introduction to Arithmetic* (New York: Macmillan, 1926), 181.

59. *Rasāʾil*, Preface, I:21.10–22.2. Cf. Nicomachus, *Introd. Arithm.*, p. 1 (trans. D'Ooge, p. 182): "This wisdom he [sc. Pythagoras] defined as the knowledge, or science, of the truth in real things, conceiving science to be a steadfast and firm apprehension of the underlying substance, and real things to be those which continue uniformly and the same in the universe [etc.];" of the four mathematical sciences, one should start with the one "which naturally exists before them all, is superior and takes the place of origin and root;" "this is arithmetic, not solely bcause we said that it existed before all the others in the mind of the creating God like some universal and exemplary plan, relying upon which as a design and archetypal example the creator of the universe sets in order his material creations and makes them attain to their proper ends, but also because it is naturally prior in birth, inasmuch as it abolishes other sciences with itself" (p. 9; trans. D'Ooge, p. 187).

60. *The Epistle on Music of the Ikhwān al-Ṣafāʾ* (Baghdad, 10th century), transl. Amnon Shiloah, Documentation and Studies: Publications of the Department of Musicology and the Chaim Rosenberg School of Jewish Studies, Tel Aviv University (Tel-Aviv: Tel Aviv University, Faculty of Fine Arts, School of Jewish Studies, 1978). References to the Arabic text are to the edition printed at Beirut, Dār Bayrūt & Dār Ṣādir, 1957.

61. The Hellenistic antecedents of this attitude are found—apart from Plato himself and older Pythagoreanism—in the ps.-Platonic *Epinomis*, in the Neoplatonic commentaries on Euclid's *Elementa* and on Nicomachus—Iamblichus, Proclus, Philoponus—and in Porphyry's *In Ptolem. Harm.* (quoting, like Nicomachus, the treatise *On Harmony* by Archytas of Tarentum); cf. the documentation by F. E. Robbins and L. Ch. Karpinski in M. L. D'Ooge [trans.], Nicomachus of Gerasa, *Introduction to Arithmetic* [*supra*, n. 58].

62. Shiloah, *The Epistle on Music*, introd., 7.

63. *Rasāʾil,* I:237.2–6; tr. Shiloah, p. 68 (aphorism 18); cf. also pp. 234–39 = Shiloah, 66–9, aphorisms no. 7, 9, 11, 20, 21.

64. A detailed bibliography has been presented by Everett K. Rowson in his analysis of the teaching tradition leading from al-Kindī to the later heirs of Balkhī's school: *A Muslim philosopher on the soul and its fate: al-ʿĀmirī's Kitāb al-Amad ʿalā l-abad* (New Haven, Conn.: AOS, 1988), 1–19; see also the same, "The philosopher as littérateur," *Zeitschrift für Geschichte der Arabisch-Islamischen Wissenschaften* 6 (1990): 40–92, esp. pp. 1–7 on Abū Zayd al-Balkhī.

65. Abu 'l-Ḥasan M. b. Yūsuf al-ʿĀmirī, *K. al-Iʿlām bi-manāqib al-Islām*, ed. A. ʿAbdalḥamīd Ghurāb (Cairo: Dār al-Kitāb al-ʿArabī, 1967).

66. *Manāqib,* 88.5.

67. *Manāqib,* 90.3, 90.8–10.

68. The epithet *al-muʿallim al-awwal* does not seem to occur before al-Fārābī, but al-Kindī already calls him *mubarriz al-Yūnāniyyīn fī l-falsafa* (*al-Falsafa al-ūlā*, ed.

Abū Rīda, 102f.) For the stages of Arabic Aristotelianism, cf. G. Endress, "L'Aristote arabe: réception, autorité et transformation du Premier Maître," *Medioevo,* 24 (Padova, 1998): 1–42.

69. ἐπιστήμη ἐπιστημῶν, *ʿilm al-ʿulūm*; *v*. Hein, *Definition*, p. 39.

70. *An. post.*, I.22, 83a33–34; cf. J. Barnes [trans.], *Aristotle's Posterior Analytics* (Oxford: Clarendon Press, 1975), 34, 169.

71. Ibid., I.1, 71a1–4.

72. Cf. Kurt von Fritz, "Die ἀρχαί in der griechischen Mathematik," *Archiv für Begriffsgeschichte* 1 (Bonn, 1955): 13–103, repr. in von Fritz, *Grundprobleme der Geschichte der antiken Wissenschaft* (Berlin, 1971), Árpád Szabó, "The transformation of mathematics into deductive science and the beginnings of its foundation on definitions and axioms," *Scripta Mathematica,* 27 (1964), 1:27–48, 2:113–39 (quotation from p. 137).

73. Walter Leszl, "Mathematics, axiomatization, and the hypotheses," in *Aristotle on science: the «Posterior Analytics»,* ed. by Enrico Berti, Studia Aristotelica, 9 (Padova: Antenore, 1983), 270–328 (v. conclusions pp. 304f., 313f., 326f.); Wolfgang Kullmann, "Die Funktion der mathematischen Beispiele in Aristoteles' *Analytica posteriora,*" ibid., 245–70 (conclusions p. 267f.).

74. Edward Booth, *Aristotelian aporetic ontology in Islamic and Christian thinkers* (Cambridge [etc.]: Cambridge University Press, 1983), 2.

75. J. Barnes, *Aristotle's Posterior Analytics*, xi.

76. Trans. Thomas Heath, *Mathematics in Aristotle* (Oxford: Oxford University Press, 1949), 8f.

77. Aristotle, *Phys.* II.2, 193b22–194b15 (194a8: τὰ φυσικώτερα τῶν μαθημάτων).

78. Ibid. 193b34: χωριστὰ γὰρ τῇ νοήσει κινήσεως.

79. al-Fārābī, *K. al-Jamʿ bayna raʾyay al-ḥakīmayn*, ed. Dieterici/Nādir (Beirut 1960), 97.19–98.8.

80. Cf. G. Endress, "The defense of reason: the plea for philosophy in the religious community," *Zeitschrift für Geschichte der Arabisch-Islamischen Wissenschaften,* 6 (Frankfurt a. M., 1990): 1–49 (esp. 16–23).

81. al-Fārābī, *K. al-Burhān*, ed. Mājid Fakhrī, *al-Manṭiq ʿind al-Fārābī* (Bayrūt: Dār al-Mashriq, 1987), 65.18–22.

82. Ibid. 70.12–14.

83. al-Fārābī, *K. al-Burhān*, 68.ult.–70.11.

84. al-Fārābī, *Iḥṣāʾ al-ʿulūm*, ed. ʿUthmān Amīn (Cairo, ^{2}1949), 74.

85. But *v.* Sarah Stroumsa, "Al-Fārābī and Maimonides on medicine as a science," *Arabic Sciences and Philosophy*, 3 (1993): 235–249.

86. *Iḥṣāʾ*, 77–78.

87. Ibid., 79–83, esp. 80.1–13.

88. This is analyzed on the basis of al-Fārābī's *Iḥṣāʾ*, and interpreted in the light of related texts from the philosophical and the scientific traditions, in the enlightening contribution to the present volume by Elaheh Kheirandish, "The many «aspects» of Appearances: Arabic optics to 950 A.D.," *supra*, pp. 53–81. I owe to her article the references given below, notes 90–91.

89. *v.* al-Fārābī, *K. al-Burhān*, 25f. Cf. Aristotle, *An. Post.*, I.2.71b18: ἀπόδειξιν δὲ λέγω συλλογισμὸν ἐπιστημονικόν.

90. *An. Post.* I.13.78a30ff.; *Probl. Phys.* XV.6.911b19–34.—The Arabic version of *Probl. Phys.* XV.6, referred to in the original Greek by E. Kheirandish (*infra*, p. 56, text P), is now available in the edition of L. S. Filius, *The Problemata Physica attributed to Aristotle: the Arabic version of Ḥunain ibn Isḥāq and the Hebrew version of Moses ibn Tibbon*, Aristoteles Semitico-Latinus, 11 (Leiden: Brill, 1999), 658/659.

91. Quoted by E. Kheirandish, *infra*, 59 (text L), ed. Roshdi Rashed, *Œuvres philosophiques et scientifiques d'al-Kindī*, vol. 1 (Leiden: Brill, 1997), appendice II: "La catoptrique de Qusṭā ibn Lūqā," 572/573. The addressee is the Abbasid regent al-Muwaffaq (regn. 843–91).

92. Ibn Sīnā, *Tisʿ rasāʾil fī l-ḥikma wa-l-ṭabīʿiyyāt li-l-shaykh al-raʾīs*, ed. Ḥasan ʿĀṣī (Beirūt: Dār Qābis, 1986) [a reprinting of one of several early Cairo editions of the "nine epistles"], no. 5, 83–94, on the mathematical science: pp. 84 and 89.

93. William E. Gohlmann, *The life of Ibn Sina*, Studies in Islamic Philosophy and Science (Albany: The University of New York Press, 1974), 20–27; Dimitri Gutas, *Avicenna and the Aristotelian tradition*, Islamic Philosophy and Theology, 4 (Leiden: Brill, 1988), 26.

94. Ibn Sīnā, *al-Shifāʾ, al-Ilāhiyyāt*, ed. [George Sheḥāta] Qanawātī [Anawati], Saʿīd Zāyid (Cairo: al-Hayʾa al-ʿāmma li-shuʾūn al-maṭābiʿ al-amīriyya, 1960), 10f.; French translation by Georges C. Anawati: Avicenne, *La Métaphysique du Shifāʾ*, Etudes musulmanes, 21 (Paris: Vrin, 1978), 91ff.

95. *al-Ilāhiyyāt*, 10.10–14.

96. Ibid., 11.12–12.2.

97. Ibid., 21–24; cf. Anawati, 100–102.

98. Ibid., 24.1–2.

99. ed. Anawati and Zāyid, 303–324.

100. Quoted from the translation of A. I. Sabra in *The Optics of Ibn al-Haytham*, books I–III (London: Warburg Institute, 1989), II: Commentary, 3; cf. also Shlomo Pines, "Ibn al-Haytham's critique of Ptolemy," in *Proceedings of the Xth International Congress of the History of Science* (Paris, 1964), 547–50 = *Studies in Arabic versions of Greek texts and in mediaeval science*, The Collected works of Shlomo Pines, vol. 2 (Jerusalem: Magnes Press; Leiden: Brill, 1986), 436–39.

101. Fom the introduction to Ibn Sīnā's *Manṭiq al-mashriqiyyīn*, translated by Dimitri Gutas, *Avicenna and the Aristotelian tradition* (Leiden [etc.]: Brill, 1988), 44f.

102. al-Ḥasan Ibn al-Haytham, *al-Shukūk ʿalā Baṭlamyūs*, edd. ʿAbd-al-Ḥamīd Ṣabra, Nabīl al-Shahbānī (Cairo: Dār al-Kutub, 1971), 19.12, 37.6.

103. Ibn al-Haytham as quoted from his autobiography by Ibn Abī Uṣaybiʿa, *ʿUyūn al-anbāʾ fī ṭabaqāt al-aṭibbāʾ*, ed. A. Müller (Cairo, Königsberg, 1884), 2:93.4; cf. Matthias Schramm, *Ibn al-Haythams Weg zur Physik,* Boethius, 1 (Wiesbaden: Steiner, 1963), 141, cf. 285.

104. Ibn al-Haytham, *On the Configuration of the World*, edition, translation, and commentary by Y. Tzvi Langermann (New York: Garland; 1990).

105. Cf. M. Schramm, *Ibn al-Haytham*, 136f.

106. M. Schramm, *Ibn al-Haytham*, 136–138.

107. Ibn al-Haytham, *Kitāb fī ḥall shukūk kitāb Uqlīdis fī l-uṣūl*, facsimile ed. by F. Sezgin, *Ibn al-Haytham, On the Resolution of Doubts in Euclid's Elements and Interpretation of Its Special Meanings* (Frankfurt a.M., 1985), 4.4–7. Cf. Proclus, *In I Eucl.*, 31.2–3, 31.22–32.2, trans. Morrow, 26: "Every form of knowledge which apprehends the logos, or cause, of the thing it knows is a science. . . . For genuine science is the one, the science by which we are able to know all things, the science from which come the principles of all other sciences, some immediately and some at further remove." And cf. Aristotle, *Phys.* II.3.195b21; *Metaph. H*.4.1044b1–2: the physicist looks for the proximate, most pertinent cause.

108. Ibn al-Haytham, *K. al-Manāẓir*, al-maqālāt 1–3, ed. ʿAbdalḥamīd Ṣabra (Kuwait, 1983), [6], p. 62.7–9: *wa-nabtadiʾu fī l-baḥthi bi-stiqrāʾi l-mawjūdāti wa-taṣaffuḥi aḥwāli l-mubṣarāti, wa-numayyizu khawāṣṣa l-juzʾiyyāti wa-naltaqiṭu bi-stiqrāʾi mā yakhuṣṣu l-baṣara fī ḥāli l-ibṣār*, etc.; cf. Sabra, *The Optics of Ibn al-Haytham*, I (Translation): quoted by Sabra in this volume, p. 85.

109. Cf. A. I. Sabra, *loc. cit*., II, 12f.: "inspection or examination of particulars," referring to al-Fārābī's *al-Alfāẓ al-mustaʿmala fī l-manṭiq* (ed. M. Mahdī, 93) for the use of *istiqrāʾ* in conjunction with *taṣaffuḥ* "to review, to survey."

110. Joep Lameer, *Al-Fārābī and Aristotelian syllogistics: Greek theory and Islamic practice*, Islamic Philosophy, Theology and Science, 20 (Leiden [etc.]: Brill, 1994), 144—confirming A. I. Sabra's analysis of the terminology. It is true that this understanding may be claimed for the Aristotelian *epagôgê* as well.

111. M. Schramm, *Ibn al-Haytham*, 260–264.

112. Ibid., 285–289.

113. Ibn al-Haytham, *al-Manāẓir* [*supra*, n.108], book II, section 3, [200–232] pp. 307– 316; *v.* A. I. Sabra, *The Optics of Ibn al-Haytham* [*supra*, n. 100], I (trans.), pp. 200–206; II (commentary), pp. 97–201.

114. Σῴζειν τὰ φαινόμενα, Simplicius, *In De Caelo comm.*, ed. Heiberg, 488.18–24,

492.25ff.; Proclus, *Hypotyposis astronomicarum positionum*, ed. C. Manitius (Leipzig, 1909), 5.10; cf. Harold Cherniss, "The philosophical economy of the theory of ideas," *American Journal of Philology,* 57 (1936): 445–56 = Cherniss, *Selected papers* (Leiden, 1977), 121–132; Shmuel Sambursky, *The physical world of the Greeks* (London, 1956), 59; idem, *The physical world of late Antiquity* (London, 1962); Jürgen Mittelstrass, *Die Rettung der Phänomene: Ursprung und Geschichte eines antiken Forschungsprinzips* (Berlin, 1962).

115. Willy Hartner, "Mediæval views on cosmic dimensions and Ptolemy's *Kitāb al-Manshūrāt*," in *Mélanges Alexandre Koyré*, 1 (Paris: Hermann, 1964), 254–282 = W. Hartner *Oriens—Occidens* (Hildesheim: Olms, 1968), 319–348; Bernard R. Goldstein, *The Arabic version of Ptolemy's Planetary Hypotheses,* Transactions of the American Philosophical Society, new series, vol. 57.4 (Philadelphia, 1967).

116. Willy Hartner, "Falak," in *Encyclopædia of Islam,* new ed., 2 (1963): 761–763.

117. The position of Sosigenes, the teacher of Alexander of Aphrodisias, was analyzed, as an introduction to his study of Ibn al-Haytham's theory, by M. Schramm, *Ibn al-Hayt͟ham*, 15–63.

118. Alain Philippe Segonds, "Proclus: astronomie et philosophie," in *Proclus, lecteur et interprète des anciens* [as quoted *supra,* note 24], 317–334; quotation p. 332 from Proclus, *Hypotyposis astronomicarum positionum*, ed. C. Manitius (Leipzig: Teubner, 1909), 236. See also W. Hartner, "Mediæval views," quoted *supra*, n. 115; Lucas Siorvanes, *Proclus: neo-Platonic philosophy and science* (Edinburgh: Edinburgh University Press, 1996), 266.

119. Ibn al-Haytham, *Shukūk* [*supra*, n. 102], 16.2

120. Ibid., 33.2–4.

121. Ibid., 34.3–8.

122. Ibid. 37.14–15, cf. also 33.5, 63.14ff.; the reference to the *Almagest*, book IX.2, is found in *Claudii Ptolemaei opera quae exstant omnia*, vol. 1: *Syntaxis mathematica*, ed. J. L. Heiberg (Lipsiae: Teubner, 1898–1903), 2:211.22–24 (καταχρήσεσθαί τινι παρὰ τὸν λόγον); *Ptolemy's Almagest*, trans. and annotated by G. J. Toomer (London: Duckworth, 1984), 422; Régis Morelon, 'La version arabe du Livre des Hypothèses de Ptolémée,' *Institut Dominicain d'Études Orientales du Caire: Mélanges (MIDEO)*, 21 (1993): 7–85 [Bk I].

123. Ibid. 38.2ff.

124. Ibid. 41.ult.–42.8.

125. See the detailed and penetrating discussion of these passages by A. I. Sabra, "Configuring the universe: aporetic, problem solving, and kinematic modeling as themes of Arabic astronomy," *Perspectives on Science*, 6:1998 (Cambridge, MA: MIT, 1999): 288–330, esp. pp. 298–305: "Ibn al-Haytham and the aporetic argument: Ptolemy's dilemma," preceded by a review of Ibn Haytham's *Maqāla fī Hayʾat al-ʿālam*, ed. and trans. by Y. Tzvi Langermann, *Ibn al-Haytham's On the Configuration of the World,* Harvard Dissertations in the History of Science (New York: Garland, 1990).

126. A. I. Sabra, "Configuring the universe" [see preceding note], 304.

127. M. Schramm, *Ibn al-Haytham*, 143–46.

128. Ibn al-Haytham, *Shukūk,* 38.5.

129. Ibn Bājja, *Kalām baʿatha bihī li-Abī Jaʿfar Yūsuf ibn Ḥasdāy*, in *Rasāʾil falsafiyya li-Ibn Bājja*, ed. Jamāl-al-Dīn al-ʿAlawī (Beirut, Casablanca, 1983), 78. On the 5th/11th century Andalusian astronomer (Ibn) al-Zarqālluh, *v.* J. Vernet in *Dictionary of Scientific Biography,* XIV:592 *s.n.*

130. *Comm. med. Meteor.*, b. III <latine, ex hebraico Kalonymi> (Venetiis: apud Iunctas, 1562), t. IV: fol. 448v L. The Latin differs from the Arabic, Ibn Rushd, *Talkhīṣ al-Āthār al-ʿulwiyya*, ed. Jamāl-al-Dīn al-ʿAlawī (Bayrūt: Dār al-Gharb al-Islāmī, 1994), p. 141.10–12, but this is corrupted, being in evident contradiction to p. 145.1–10 (cf. the comments of the editor, ibid., p. 144, note 4, and p. 145, note 4).

131. This was pointed out by A. I. Sabra, *The Optics of Ibn al-Haytham*, books I-III, 2: Commentary (London: Warburg Institute, 1989), p. 6, quoting the mediæval Latin translation of Averroes' commentary. The reference in Ibn al-Haytham, as quoted by Sabra, ibid., p. 4, is his treatise *On the Rainbow and the Halo* (completed in Rajab, 419/1028), as referred to by Kamāl-al-Dīn al-Fārisī, *Tanqīḥ al-Manāẓir* (Ḥaydarābād: Dāʾirat al-Maʿārif al-ʿUthmāniyya, 1347–48/1928–30), II, 258–279. The text identified by the editor, Jamāl-al-Dīn al-ʿAlawī, as being a possible source of Ibn Rushd's reference (*Maqālat al-Ḥasan b. al-Ḥasan Ibn al-Haytham fī l-Aṯar al-ẓāhir fī wajh al-qamar*, ed. ʿAbd-al-Ḥamīd Ṣabra, *Journal of the History of Arabic Science*, 1 [Aleppo, 1977]: 5–19) treats a different topic, and does not contain specific statements on method.

132. Translated by A. I. Sabra, *op. cit.*, 4, from the quotation found in Kamāl-al-Dīn al-Fārisī, *Tanqīḥ al-Manāẓir*, ed. Muṣṭafā Ḥijāzī, II, p. 259, 7–8.

133. Ibn Rushd, *Talkhīṣ al-Āthār al-ʿulwiyya*, ed. ʿAlawī, p. 145.1–10 (l. 4–5: *wa-laysa ḥālu ʿilmi l-manāẓiri maʿa* [sic leg. pro *al-munāẓirīn*] *hādhā l-ʿilmi fī iʿṭāʾi hāḍihi l-asbābi ka-ḥāli ʿilmi l-manāẓiri maʿa ʿilmi l-handasa*). For Ibn al-Haytham's conception of the causes determined in mathematical demonstrations, cf. the text quoted above, note 107).

134. *Talkhīṣ al-Āthār al-ʿulwiyya*, p. 144.

135. *Talkhīṣ al-Āthār al-ʿulwiyya*, p. 145.ult.–146.6.

136. *V. supra*, n. 78.

137. *Epitome of Plato's Republic and its Scientific Arguments* [written, perhaps, in the same period as the *Middle Commentary on the Nicomachean Ethics*, finished in 572 H.], Hebrew text ed. Erwin I. J. Rosenthal, *Averroes' Commentary on Plato's Republic,* Cambridge Oriental Publications, 1 (Cambridge: Cambridge University Press, 1956; repr. 1966, 1969), b. II, ch. XV: 75.16–35; trans. Ralph Lerner, *Averroes on Plato's "Republic"* (Ithaca: Cornell, 1974), 96f.

138. Cf. Aristotle, *Phys.* II.2, quoted *supra*, note 77. On the different "causes" determined in physics and astronomy, see the passage from Averroes' commentary on *De Caelo*, quoted *infra*, n. 161.

139. al-Biṭrūjī, *De motibus celorum:* critical edition of the Latin translation by Michael Scot, ed. by Francis J. Carmody (Berkeley and Los Angeles, 1952); id.: *On the Principles of Astronomy*, an ed. of the Arabic and Hebrew versions with translation, analysis, and an Arabic-Hebrew-English glossary by Bernard R. Goldstein (New Haven, 1971). A superb analysis of the Andalusian "revolt" against Ptolemy—or should we say "restoration" of Aristotle?—in planetary theory has been given by ʿAbdalḥamīd I. Ṣabra in his article "The Andalusian revolt against Ptolemaic astronomy: Averroes and al-Biṭrūjī," in *Transformation and Tradition in the Sciences: essays in honor of I. Bernard Cohen*, ed. by E. Mendelsohn (Cambridge [etc.]: Cambridge University Press, 1984), 133–153. The following synopsis of Ibn Rushd's positions is expounded in more detail in my article "Averroes *De Caelo*: Ibn Rushd's cosmology in his commentaries on Aristotle's On the Heavens," *Arabic Sciences and Philosophy*, 5 (1995): 1–41.

140. *Kalām li-Abī Bakr ibn Yaḥyā fī l-hayʾa*, MS. Berlin: Staatsbibliothek, Wetzstein, 87 (Ahlwardt, *Verzeichniß*, no. 5060; now at Cracow, Biblioteka Jagiellońska), fol. 203a–204b.

141. Ibn Rushd, *K. Mā baʿd al-ṭabīʿa,* IV 13, ed. Quirós, 133 = ed. Amīn, 130; see also below, § 21.

142. Fully developed in *Metaph., Λ* 8.

143. *De Caelo* II.6.

144. *De Caelo* II.12, 292a21.

145. See Ch. Genequand, *Ibn Rushd's Metaphysics: a translation, with introduction, of Ibn Rushd's commentary on Aristotle's Metaphysics, book Lām* (Leiden, 1984), 54f.

146. Ibn Rushd, *Jawāmiʿ al-Samāʾ wa-l-ʿālam* (Ḥaydarābād, 1365/1946), p. 47.

147. See also Henri Hugonnard-Roche: "Remarques sur l'évolution doctrinale d'Averroès dans les commentaires au *De caelo*: le problème du mouvement de la terre," *Melanges de la Casa de Velazquez* 13 (Madrid, 1977): pp. 103–117 (pp. 105–108: I. Que la terre est au centre; pp. 108–111: II. Que le centre de la terre est le centre du monde; pp. 111–113: III. Cause du mouvement et du repos de la terre; pp. 113–115: IV. Sur la théorie du équilibre par indifférence).

148. Willy Hartner, *Oriens—Occidens* (Hildesheim, 1968), 461.

149. Ibn Rushd, *Talkhīṣ al-Samāʾ wa-l-ʿālam*, ed. Jamāl-al-Dīn al-ʿAlawī (Fes, 1984), 272.5–15 on *De Caelo* II 14.

150. On the development of Averroes, and on the indications given with regard to this primary scope of his early work, *v.* Jamāl-al-Dīn al-ʿAlawī, *al-Matn al-Rushdī* (Casablanca: Tūbqāl, 1986), 49ff., 140–143, 159f.; G. Endress, "Le projet d'Averroès," in *Averroes and the Aristotelian Tradition*, ed. by G. Endress and J. A. Aertsen, in the series "Islamic Philosophy, Theology and Science," vol. 31 (Leiden: Brill, 1999), 3–31; *id.*, "*Law ansaʾa Llāhu fī l-ʿumr:* the project of Ibn Rushd," in *La Obra de Averroes: el pensamiento filosófico y científico de Averroes en su tiempo,* Proceedings, VIII Centenario de Averroes, Córdoba, 1998 (forthcoming).

151. Juliane Lay, "*L'Abrégé de l'Almageste*: un inédit d'Averroès en version hébraïque," *Arabic Sciences and Philosophy*, 6 (1996): 23–61, p. 26. Only the Hebrew version of Ibn Rushd's *Mukhtaṣar al-Majisṭī* (*Qiṣṣūr al-Magisṭī*) is extant.

152. Ibid., 42f., 53.

153. Ibid., p. 54f.; p. 54 n. 87 on the terminology: *demonstratio per signum (demonstratio quia)* and *demonstratio propter quid*; see also p. 154.

154. E. g., *Comm mag. De Caelo* II c. 49, fol. 62ra23–26: *et si ponantur centra diversa, accidit impossibile quod diximus; et hoc quod faciunt mathematici, quod ponunt eccentricos, numquam dixit hoc Aristoteles, sed causa diversitatis apud ipsum sunt motus leulab*. On *lawlab*, see p. 156 and note 178.

155. They have been dealt with in some detail by F. E. Carmody [ed.], Al-Biṭrūjī, *De motibus celorum* [see above, n. 139]; idem, *Innovations in Averroes De Caelo* (Berkeley, Cal., 1982). See also the article of A. I. Sabra, referred to above, n. 139.

156. Ibn Rushd, *Jawāmiʿ al-Samāʾ wa-l-ʿālam* (Ḥaydarābād 1365/1946), p. 42.4–10.

157. Cf. also *Jawāmiʿ al-Samāʾ wa-l-ʿālam*, p. 10: *sa-nubayyin hādhā fī l-falsafa al-ūlā* [. . .], *hāhunā innamā huwa ʿalā jihat al-muṣādara ʿalā mā tabayyana fī kitāb al-Burhān*. For a systematic presentation of the epistemological principles underlying this procedure—which go back ultimately to Aristotle's *Analytica posteriora* and al-Fārābī's *Kitāb al-Burhān*—see Ibn Rushd, *Sharḥ al-Burhān*, ed. Badawī (al-Kuwayt, 1984), 348ff., on *Analytica Posteriora*, I, 13.

158. For the Arabic terminology see Ibn Rushd: *Tafsīr Mā baʿd al-ṭabīʿa*, ed. Bouyges [*infra*, n. 171], 703.11; on *barāhīn muṭlaqa* cf. Ibn Rushd, *Sharḥ al-Burhān, l.c.* [see preceding note].

159. For the Latin text, see Averroes, *Commentarium Magnum in Aristotelis De Caelo*, in *Aristotelis Opera cum Averrois Cordubensis Commentariis* (Venice: apud Junctas, 1550), vol. 5: b. II, c. 61, fol. 65v34–42. For the reference to *De Anima*, cf. *Commentarium Magnum in Aristotelis De Anima Libros*, ed. F. S. Crawford (Cambridge 1953), I, c. 89, p. 119, and II, c. 20, p. 159.

160. Cf. Jamāl-al-Dīn al-ʿAlawī: "Ishkāl al-ʿalāqa bayn al-ʿilm al-ṭabīʿī wa-Mā baʿd al-ṭabīʿa ʿind Ibn Rushd," in: *Majallat Kulliyyat al-ādāb wa-l-ʿulūm al-insāniyya*; ʿadad khāṣṣ 3 (Fes: Jāmiʿat Sīdī Muḥammad b. ʿAbdallāh, 1988), 7–51.

161. *Commentarium magnum in De Caelo* [Latin version], c. 57 ad *De Caelo* II.10, 291a29-b11 (text according to the unpublished ed. of F. E. Carmody, corresponding to *Aristotelis opera cum Averrois commentariis* [Venetiis: apud Junctas, 1562, repr. Frankfurt am Main.: Minerva, 1962], vol. V, f. 136rE–F: naturalis vero et astrologus communicant in consideratione istarum quaestionum, tamen astrologus in maiori parte dat quia, naturalis vero dat propter quid, et quod dat astrologus in maiori parte non est nisi eis quae sensui apparent . . ., naturalis autem laborat in dando causam propter quam hoc est supra ipsam . . . . Astrologus enim considerat de causis abstractis sermone idest ratione a materia scilicet causis doctrinalibus [Arabic, *taʿlīmiyya* "mathematical causes"], Naturalis vero de causis que sunt cum materia; verbi gratia in utraque scientia queritur

quare celum est sphericum: Naturalis dicit quia est corpus neque grave neque leve, Astrologus dicit quia linee exeuntes a centro ad circumferentiam sunt equales.—For the principles of method observed by the philosopher, cf. al-Fārābī, *Kitāb al-Burhān*, ed. M. Fakhrī (Beirut, 1987), p. 66.13–17. Consider also Aristotle, *Analytica priora*, I 30, 46a17–22: "It falls to experience to provide the principles of any subject. In astronomy, for instance, it was astronomical experience that provided the principles of the science, for it was only when the *phainomena* were adequately grasped that the proofs in astronomy were discovered," and see G. E. L. Owen, "«Tithenai ta Phainomena»," *Aristote et les problèmes de méthode*, éd. par S. Mansion (Louvain, 1961), 83–103.

162. See Elders, *Aristotle's cosmology*, p. 227f.; Moraux [ed.], Aristote, *Du ciel,* introd., p. civ f.

163. Ptolemy, *Planetary Hypotheses*, ed. B. Goldstein, p. 31 = fol. 90a3–6.

164. Ibid. 27–29 = fol. 88b24–90a6; for the Arabic tradition, see Ibn Sīnā, *al-Shifā'*, [III, 4]: *'Ilm al-hay'a*, edd. Muḥammad Mudawwar, Imām Ibrāhīm Aḥmad (Cairo, 1980), p. 463 (following Ptolemy); also F. J. Carmody [ed.] Al-Biṭrūjī, *De motibus celorum*, introd. p. 62f., Latin text p. 127ff. (Venus above the sun, Mercury below).

165. *Comm. mag. in De Caelo*, II, c. 58, fol. 64r45–48, cf. 64va30.

166. Ibid., fol. 64va48–59. We may compare this with Alexander, apud Simplicium: *In De Caelo*, p. 472.10ff.: The planetary sphere is not moved unwillingly, but in accordance with its »purpose and desire«; it may be necessity, but is also recognized as good.

167. Ibid., II c. 62, fol. 66ra57.

168. Ibid., fol. 66ra55–9 ad *De Caelo* II 12.

169. *Tafsīr al-Samā' wa-l-'ālam,* II c. 35, fol. 38b22–39a4.

170. See Charles Genequand, op. cit., *supra*, n. 145, pp. 54f.

171. Averroès, *Tafsīr Mā ba'd al-ṭabī'at*, ed. Maurice Bouyges, Bibliotheca Arabica Scholasticorum, série arabe, V–VII (Beyrouth: Imprimerie catholique, 1938–52), vol. 3:1639ff. (comm. 42ff.); trans. Ch. Genequand, *Ibn Rushd's Metaphysics* (*v*. n. 145), 168ff.—On the problem, *v.* W. Ross, *Aristotle's Metaphysics* (Oxford, 1924), II: 382f.)

172. *Tafsīr Mā ba'd al-ṭabī'a*, 1654.3–6 ed. Bouyges.

173. Ibid., 1655.11–13.

174. Ibid., 1655.15–1656.1.

175. Ibid., 1656.3–4.

176. Ibid., 1656.8–13.

177. Ibid., 1661.8–1662.7.

178. *Tafsīr Mā ba'd al-ṭabī'a* XII c. 45 (on *Metaph.* XII.8), ed. Bouyges p. 1662–1664 (cf. above, note 2); trans. Ch. Genequand, *Ibn Rushd's Metaphysics,* p. 178f. (correct

178.–6 "Ptolemy was free from" : "P. failed to notice" [1663.3 *dhahaba ʿalayhi*]).—The word in question is *lawlabī*, meaning circular, and more especially heliocoidal or spiral motion. The underlying Greek word, however, is ἀνελιττοῦσα "backward-rolling'. In *Metaph.* Λ 8 (1074a2), putting the purely geometrical account of Eudoxus and Callippus into a mechanical model, Aristotle postulated for each planet except the moon the existence of certain spheres which "roll back" the outer sphere of the planet, preventing the deferent sphere of one planet from affecting the next. Now the Arabic translation Averroes had before him translated σφαίρας ἀνελιττούσας by *ukarun tadūru bi-dawrin lawlabī* (*Tafsīr*, p. 1669.7, 1670.4–5). This gave him a quite different idea of the model Aristotle had in mind (cf. the passage from his commentary on *De Caelo*, quoted above, p. 245)—and deluded him with false hopes. Cf. Ch. Genequand, *Ibn Rushd's Metaphysics*, 55; W. D. Ross, *Aristotle's Metaphysics*, II: 391f.; Th. Heath, *Mathematics in Aristotle,* 218.

179. The passage has been discussed, and compared with relevant statements in Maimonides' *Mishneh Torah*, by Y. Tzvi Langermann, "The True Perplexity: the *Guide of the Perplexed*, part II, chapter 24," in *Perspectives on Maimonides: philosophical and historical studies*, ed. by Joel L. Kraemer, The Littman Library of Jewish Civilization (Oxford: Oxford University Press, for the Littman Library, 1991), 159–174.

180. Maimonides, *Dalālat al-ḥāʾirīn*, book II, ch. 24, ed. S. Munk, 51b–54b = ed. S. Munk, Y. Yoel (Jerusalem: Junovitch, 1931), 225–229; ed. Hüseyin Atay (Ankara: Ankara Üniversitesi, 1974), 349–353; trans. S. Pines, *The Guide of the Perplexed* (Chicago: The University of Chicago Press, 1963), 322–327.

181. Cf. G. Endress, "Philosophische Ein-Band-Bibliotheken aus Isfahan," *Oriens,* 36 (2001): 10–58.

182. This is treated, in the context of our present subject, by A. I. Sabra in his groundbreaking article, "Science and philosophy in medieval Islamic theology: the evidence of the fourteenth century," *Zeitschrift für Geschichte der Arabisch-Islamischen Wissenschaften*, 9 (1994): 1–42.

183. *Naṣīr al-Dīn al-Ṭūsī's Memoir on Astronomy (al-Tadhkira fī ʿilm al-hayʾa)*, ed., trans. F. J. Ragep, in Sources in the History of Mathematics and Physical Sciences, 12 (New York [etc.]: Springer, 1993), 1:90/91.

184. *V. supra*, p. 167 (notes 81 and 82).

185. Al-Ījī, *al-Mawāqif fī ʿilm al-kalām*, quoted from A. I. Sabra, "Science and philosophy in medieval Islamic theology," *supra*, n. 182, 37.

186. First presented, and put into context, by F. Jamil Ragep, "Freeing astronomy from philosophy: an aspect of Islamic influence on science," *Osiris* 2nd series, 16 (2001): 49–71. I am grateful to Jamil Ragep for letting me read his article before publication.

187. On *ḥads*, *v.* D. Gutas, *Avicenna* [as in n. 93], 159–177.

188. Quoted from the excerpts edited and translated by Ragep, *l.c.*, 68, 70.

189. Including the Earth's rotation; *v.* Ragep, ibid.

190. al-Bīrūnī, *al-Qānūn al-Masʿūdī* (Hyderabad 1954–1956), I, p. 27, translated by F. J. Ragep in *Naṣīr al-Dīn al-Ṭūsī's Memoir on Astronomy,* 1:40 (see above n. 183), see p. 260 and p. 302 note 144 in this work for the reference to Aristotle's *De Caelo*, II.12.

6

Tenth-Century Mathematics through the Eyes of Abū Sahl al-Kūhī

J. Lennart Berggren

I Introduction

Abū Sahl Wījan ibn Rustam al-Kūhī[1] was a mathematician from Ṭabaristan[2] who flourished in the second half of the tenth century C.E. and whose work was well-known among the mathematicians of his age working in the Būyid domains. He had as patrons at least three kings of the Būyid Dynasty: ʿAḍud al-Dawlah, Ṣamṣām al-Dawlah, and Sharaf al-Dawlah, whose combined reigns cover the period 962–989. In the times immediately following his, Ibn al-Haytham and al-Bīrūnī knew of several of his works, and ʿUmar al-Khayyāmī cites him as one of the "distinguished mathematicians of Iraq."[3] Later, some of his works were studied by Muḥammad ibn Sirṭāq in the first half of the fourteenth century, and the eighteenth century Egyptian scholar Muṣṭafā Ṣidqī not only copied several of his works but was sufficiently interested in his treatise on the volume of the paraboloid to make a shorter version of it.

Al-Kūhī had, however, not only undoubted ability but the good fortune to be a geometer in a century that was one of the most active periods of geometrical research in medieval Islam. The tenth century included most of the working lives of geometers such as Ibrāhīm ibn Sinān (d. 946) Aḥmad al-Ṣaghānī, Abū Saʿd al-ʿAlāʾ ibn Sahl, Aḥmad al-Sijzī (fl. 970), Abū Naṣr ibn ʿIrāq and Abu ʾl-Wafāʾ al-Būzjānī, geometers whose works variously cite,[4] complement, and contrast with those of Abū Sahl. (We know that al-Kūhī was in personal contact with at least three of these, who worked with him on observations of the sun, which he supervised, during the reign of Sharaf al-Dawlah in 988. And he was in correspondence with at least one more.) In view of this rich intellectual working environment we are fortunate in having available not only some thirty-two works by the great Būyid geometer, as well as excerpts and quotations from not-yet-discovered works of his, but letters that he wrote and prefaces to his works, in which he makes a number of comments on the mathematics of his time. Taken together these provide a view of the work of the better part of a century through the eyes of one of its major figures, and it is some elements of this view that I want to sketch out in this chapter.

We have already spoken of Abū Sahl as a geometer, rather than a mathematician generally, and something that stands out as one reads his treatises is that all of them are devoted to geometry. Indeed, al-Kūhī was known as "Master of his age in the art of geometry" by his two younger contemporaries Abu'l-Jūd and al-Shannī.[5] In the preface to his treatise on the regular heptagon that is preserved in MS. Cairo MR 40 Abū Sahl writes of the benefit of geometry to the mathematical sciences and describes it as "the example which ought to be imitated in the pursuit of truth (in the theoretical sciences) and the leader which is to be followed when it comes to honesty."

Perhaps the reason for his devotion to geometry lay in what he saw as the surety of the knowledge gained from it. In the preface to his treatise on the regular heptagon just cited, he writes, "Its (geometry's) foundation is firm and its rules are consistent and unchanging. It can be affected neither by refutation nor can it be afflicted by infirmity." In contrast to this certainty, he writes in his treatise on the distance to the shooting stars of Galen's account of the variety of explanations offered for shooting stars, and the controversies these opinions occasioned. He then says, "But the other group, whom neither Galen nor anyone else could criticize . . . because they depended on proofs . . . were the mathematicians," and he strikes a similar note in his correspondence with al-Ṣābī where he contrasts "opinion" in physics with "demonstrative science."

It was, however, geometry in a wide sense that interested him, not only the works of Euclid, Apollonius, and Archimedes. For example in his treatise on the possibility of infinite motion in a finite amount of time he studies the movement of the shadow of a sundial's gnomon, and in the aforementioned *On Shooting Stars* he is interested in finding the distance to and size of the shooting stars. Moreover, he tells us in his treatise on the computation of rising times that he has investigated astronomy as well as centers of gravity and optics, and his treatise on the complete compass was, as he himself notes, intended to describe a mathematical instrument useful for drawing conic sections on sundials and astrolabes.

Al-Kūhī and Problems from Hellenistic Geometry

However, these latter treatises are details against an overwhelming background of the rich geometrical heritage from the Hellenistic world, and the intellectual presuppositions of that heritage, as found in and implied by the works of Euclid, Archimedes and Apollonius. Indeed, no fewer than fifteen of his works deal directly with problems discussed by these three mathematicians and many of the others are either very much in the spirit of their works generally or are specific extensions of classical problems. Two examples of such extensions are:

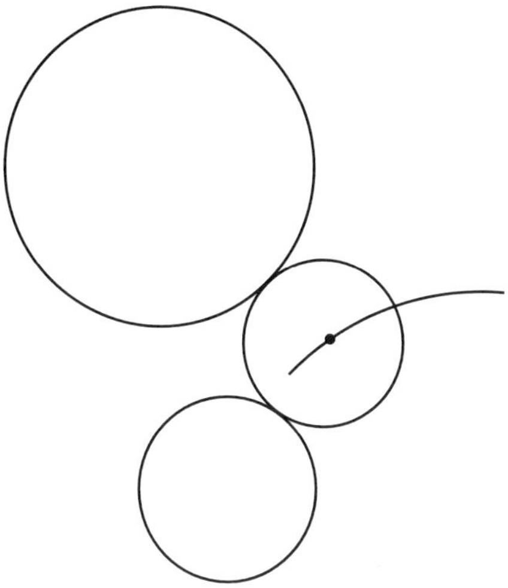

Figure 6.1

1. Al-Kūhī's problem of constructing circles "tangent" to two given objects (any one of which may be a point, a straight line, or circle)[6] and having its center on a line given in position.[7] This is just Apollonius's problem on constructing a circle "tangent" to three given objects with the three objects replaced by two but the requirement added that the center of the circle to be constructed must lie on a given (possibly curved) line (as in figure 6.1 above).

2. Al-Kūhī's extension of two problems Archimedes solves in *On the Sphere and Cylinder II*, namely:

a) To construct a segment of a sphere equal in volume to a segment of a given sphere and similar to another segment of that sphere, and

b) Do the same as in (a) with the word "surface" substituted for "volume,"

Al-Kūhī then solved the following problem:

c) To construct a sphere having a segment equal in surface area to a given segment of one sphere and in volume to a given segment of another sphere.

Other examples are his extensions (not all successful, as indicated below) of Archimedes' results on centers of gravity of figures having an axis of symmetry.

Of course it is possible to find other tenth-century geometrical works that might be described as extending Greek problems. For example, the work of Abu 'l-Wafā' on inscribing and circumscribing figures with some symmetry properties in other such figures comes to mind, a problem also treated by Abū Kāmil and al-Kūhī,[8] but al-Kūhī seemed to be unique in finding and (generally) solving problems of some real depth.

However, other Islamic geometers frequently made creative contributions in extending the Greek tradition, as Hogendijk (1984) points out in his survey of Islamic work on the regular heptagon. Although this work peaked in the latter half of the tenth century, when it involved a half-dozen well-known geometers,[9] it also stimulated five solutions of the problem by Ibn al-Haytham, whose most important work fell in the eleventh century, and even work by Kamal al-Dīn ibn Yūnus at least 200 years later.

The problem of constructing the side of a regular heptagon in a given circle seems to have arisen because Abu'l-Jūd, whose flattering opinion of al-Kūhī we quoted above, criticized the solution found in the Arabic treatise titled *Book of the Construction of the Circle Divided into Seven Equal Parts*. In its incipit the treatise is said to have been translated from an original of Archimedes by Thābit ibn Qurra. However there are a number of perplexing features of the treatise, not the least of which are the facts that of its eighteen propositions only two concern the subject announced in its title, the regular heptagon, and the fact that in these two propositions the author used a particularly vexing verging construction to divide a line segment into three segments satisfying certain proportions. These proportions, in turn, guaranteed that the resulting segments would, as sides of a triangle, produce a regular heptagon in the circumscribed circle.[10]

Almost as damning as Abu'l-Jūd, who described the solution given in the *Construction of the Circle Divided into Seven Equal Parts* as something that appeared more difficult than the original problem, al-Kūhī wrote in the introduction to his first treatise on the heptagon (found in the Cairo ms. MR40) that "it appeared from the book Archimedes authored on the subject—a subtle book—that he did not fulfill his aim and achieve his desire in his solution in any way." Some, al-Kūhī continued, believed that Archimedes' treatment was incomplete because it was impossible for him to do more, others because he filled-in the gap in a treatise that had not come down to the Arabic writers. (Abu'l-Jūd went further and suggested that Archimedes had made a mistake, a suggestion tantamount to heresy among tenth-century geometers, and on for which al-Sijzī roundly criticized him.)

The question was, however, what to do next. Some held—according to al-Kūhī—that "there was no way in which and no method by which one could find"[11] what Archimedes had failed to find. Others believed it could be done, and it appears that both al-Ṣaghānī and al-Kūhī solved the problem at close to the same time. (Hogendijk (1984) argues that al-Ṣaghānī's solution was earlier, and this may very well be the case since his solution bears clear traces of the verging found in the *Construction of the Circle Divided into Seven Equal Parts*, whereas al-Kūhī's solution removed all trace of the verging and divided the line segment directly by means of conic sections.)

Two other interesting features of tenth-century geometry that emerge from the various works on the heptagon are:

1. The importance of the patronage of the Būyid kings in supporting geometrical research, and, something that A. Anbouba has pointed out,
2. The role correspondence played in the development of mathematics at that time.[12]

As for the importance of patronage, the initial satisfactory solutions of the heptagon problem, that of Abū Sahl as well as that of al-Ṣaghānī,[13] were both dedicated to the king ʿAḍud al-Daula, and Abū Sahl wrote another one dedicated to his son, Prince Abu 'l-Fawāris, who later took the title Sharaf al-Dawla. Moreover, his *Treatise on the Ratio Between Three Lines* is dedicated to Sharaf al-Dawla, and the introduction to his treatise on rising times makes it clear it was written in the context of activity at the Royal Palace.

As for the matter of correspondence, one is struck by how much of what has survived just on the topic of the heptagon is the subject of letters, for example: Letters by Abu'l-Jūd ibn al-Layth to one al-Ḥāsib (in Bukhara) in which the former describes the methods of al-Ṣaghānī and al-Kūhī as well as his own method, letters to Abu'l-Ḥasan al-Ghādī, and to al-Bīrūnī, as well as correspon dence between al-Sijzī and Abū Saʿd al-ʿAlāʾ ibn Sahl. The topic is, in addition, one of those treated in the letter of al-Sijzī to "the people of Khorasan."

Al-Sijzī was particularly active as correspondent. Twenty-two of the thirty-five treatises listed under his name in Sezgin (1974) are letters (or "answers"), and in several the correspondent is named.

Al-Kūhī himself participated in the scientific correspondence of his epoch, as is witnessed by the partially extant correspondence he had with al-Ṣābī. This seems to have had the benefit for al-Kūhī of providing him a forum in which to try out ideas, and evidently saved him from the embarrassment of publishing his alleged discovery on the center of gravity of a semicircle and its consequence that the value of π was $3\frac{1}{9}$.

This topic, centers of gravity, is another example of one which interested the Greeks, and in particular Archimedes, and also elicited interest from some of the tenth-century geometers, including al-Kūhī, although the level of interest hardly rivaled that surrounding the regular heptagon. Thus al-Kūhī gives us the precious information that he possessed a book attributed to Archimedes' "On Centers of Gravity," but one that did not contain any proof of the law of the lever, since he refers to it as a "premiss" for Archimedes' treatment of barycentric questions. We also learn that Abū Saʿd al-ʿAlāʾ ibn Sahl wrote on centers of gravity and also inquired on the status of the law of the lever. Finally, it appears from al-Kūhī's treatise on the volume of a segment of a paraboloid that his work on centers of gravity motivated his work on that question in mensurational geometry.

Two versions of his treatise on the volume of a segment of a paraboloid (a third topic in which the tenth-century geometers knew they were following in Archimedes' footsteps) have come down to us, and it is interesting to note al-Kūhī's almost apologetic tone in his introduction to the first of these. He recognizes that Thābit wrote on the subject, but he said that he is working on it because he needs the result for work he is doing on centers of gravity and because he feels Thābit's arguments are not easily understood. One wonders if the somewhat defensive tone might be because Ibrāhīm ibn Sinān had not taken well to the suggestion that the work of his grandfather, Thābit, on the parabola was less than clear.

Like his first treatise on the heptagon, al-Kūhī's treatise on the measurement of the paraboloid also attracted serious study by the following generation of mathematicians. Thus, Ibn al-Haytham says in his *Discourse on the Measurement of the Paraboloid* that he knew al-Kūhī's work on the subject.[14]

Finally, to close this discussion on centers of gravity, we remark that it appears from al-Khāzinī's 12th-century treatise, *The Balance of Wisdom*, that Ibn al-Haytham also wrote on centers of gravity, for al-Khāzinī summarizes both his and al-Kūhī's treatises on this matter in Chapter 1 of the First Discourse of that work.[15] However, al-Kūhī's treatise has not survived and is known only because al-Khāzinī summarized it together with that of Ibn al-Haytham, which has also not survived. This suggests that the topic was not regarded as one of major interest at the time, despite its impeccable credentials of being among the topics that interested Archimedes, and it did not create a lasting tradition. It seems that this was one aspect of the Archimedean tradition that did not flourish in the tenth century.

Al-Kūhī and Methods of Hellenistic Geometry

It was, however, not just the problems of Hellenistic mathematics that attracted the geometers of the tenth century but its methods as well. Thus, the method of analysis and synthesis was standard in the tenth century and one that al-Kūhī and his contemporaries used freely. As a result of this activity at least one treatise on the method was written during that century, namely that of Ibrāhīm ibn Sinān, and one in the following century, namely that of Ibn al-Haytham.[16]

One subject that appears to have occasioned some dispute during the tenth century was that of the relation of analysis to synthesis. According to

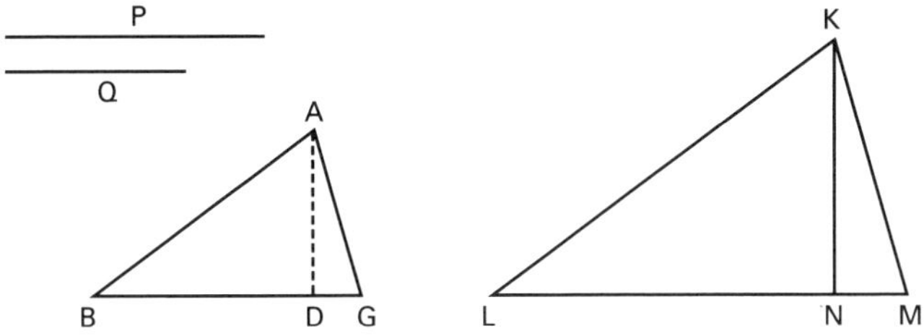

Figure 6.2

the standard, simple account of the relationship between the two, as given, for example, by al-Sijzī in his *Treatise on Geometrical Problem Solving* (Hogendijk 1996, p. 4), the synthesis was just the analysis done backward, so the steps of the two should be in virtual one-to-one correspondence. However that is not what one finds, and Ibrāhīm ibn Sinān recorded the dismay of some geometers on seeing geometrical objects appear in the synthesis that did not occur in the analysis.

Ibn Sīnān referred to this phenomenon in the following passage[17]

> I have found that modern geometers have neglected the method of Apollonius in analysis and syntheses, as they have in most of the things I have brought forward, and that they have limited themselves to analysis alone in so restrictive a manner that they have led people to believe that this analysis did not correspond to the synthesis effected.

Ibrāhīm's answer was that the lack of correspondence was only superficial, and that the reason it seemed so was that the analysis stated the steps in an abbreviated version but one which was perfectly clear to the experts.

As an example of this consider al-Kūhī's analysis of a problem from *Drawing two lines from a known point.*[18] The problem is the following (figure 6.2 illustrates the analysis and subsequent synthesis for this problem):

> A point A and a line BG are given. Draw two line segments from A to the line, containing a given angle, so that the two segments AB and AG have to each other a given ratio.

Analysis: Assume the line segments AB and AG are drawn. (i) Since $AB{:}AG$ is known and $\angle BAG$ is known, $\Delta(ABG)$ is known in form. (ii) Therefore $\angle ABG$ is known. (iii) But point A is known, hence line AB is known in position. (iv) And since $\angle BAG$ is known, line AG is also known in position. (v) Since line BG is known in position, then, points B and G are both known.

In fact no synthesis is given for this analysis, but it is not hard to supply one, as follows:

Synthesis: (i) Let the given ratio be the ratio of the lines P and Q. Draw P and Q (as LK, KM) so that the angle they contain (∠LKM) is the given angle. Draw LM. (ii) Drop a perpendicular from K to LM, defining N. Also drop a perpendicular from A to the given line, defining D. (iii) Draw DB so that AD:DB= KN:LN. Draw AB. (iv) Draw AG so that ∠BAG is the given angle. Claim: AB:AG=P:Q. (v) Δ(ABD) ~ Δ(KLN), because both have a right angle and the same ratio of sides containing that angle. (vi) Hence ∠ABG = ∠KLM; but ∠BAG = ∠LKM = the given angle, so Δ(ABG) is similar to Δ(KLM). (vii) Therefore AB:AG = KL:KM = P:Q.

Such a synthesis, using what we have elsewhere[19] called a scale model of the solution, al-Kūhī has also given for another problem, and it is exactly the kind of synthesis Ibrāhīm was talking about when he wrote:

> As for the fact that they draw lines which were not drawn in analysis, it does not establish in any way any difference between analysis and synthesis; for example, when their analysis ends up with the fact that some triangle is known in form, because its angles are known, though in the analysis it is not drawn on a line known in magnitude, they do nevertheless deduce from the ratios of its sides, one to another, a thing by which the problem gets solved. Can they avoid then, in their synthesis, setting up a triangle whose angles are equal to these known angles, in order to come to know the ratios of its sides, and make from it what gets the problem solved?

Thus auxiliary diagrams such as *KLM* above are entirely appropriate in syntheses, and do not impair the similarity between analysis and synthesis, the former being, in Ibrāhīm's word, an abridgement of the latter.

Al-Kūhī's treatise that we referred to above is unusual in that it considers about a dozen problems (variations on a single theme, admittedly) entirely by the method of analysis, where the analyses form a series of successive reductions, in the sense that there are analyses for the initial problems in the two cases when the given line *BG* is a straight line or a circular arc, and, after that, the analyses are reductions of the problem under discussion to the problem just analyzed. By the time one has about half a dozen analyses, each resting on the previous one, it becomes something of a puzzle to imagine what the synthesis of the last problem would look like.

On the other hand, it is possible to see in another treatise by al-Kūhī, his *Ratio from a Single Line Falling between Three Lines*, a more direct response to the criticism that Ibrāhīm ascribes to unknown parties, namely that the structure of and diagrams for the synthesis should be as nearly identical as possible with those of the analysis. Indeed, in that treatise al-Kūhī ensured that the analyses and syntheses contain virtually the same steps in every proposition. Perhaps he had read the critique Ibrāhīm refers to after he had written his *On Drawing Two*

Lines from a Single Point, and took the criticism seriously. (Perhaps, in fact, that treatise was one of the writings that provoked the criticism.)

It is apparent from al-Kūhī's analysis that a fundamental concept in an argument by analysis is the idea of a mathematical entity (such as position, magnitude or shape) being "known." The word is used almost exclusively in Arabic treatises where the Greeks would have used the word "given," and the basic text was Euclid's work *Givens*, best known by its Latin title *Data*. Al-Kūhī studied the work extensively and had not only thoroughly mastered its use but had written a treatise adding several propositions to it. (Since al-Kūhī's use for those propositions is not known the treatise is one of his least interesting, but that may not have been how he viewed the matter.)

However, whatever sense (or senses) the word 'given' may have had to the Hellenistic Greeks, the corresponding Arabic term, "known," had acquired a variety of different senses to the tenth-century Arabic geometers. A particularly vexing concept was the concept of a known *ratio*. According to Euclid's *Data* a ratio was given when it was equal to the ratio of two given magnitudes (a magnitude being given when it was possible to find its equal). (A classical analysis problem that depends on this latter notion is the famous problem in the *Meno* that asks for a triangle equal to a given triangle to be inscribed in a given circle.)

By al-Kūhī's time the idea of a known ratio had been divided into two separate notions: a ratio known from the point of view of quantity (*nisbat al-kamm*), where the numerical measure of the antecedent relative to the consequent is known, and a ratio simply "known," in the sense of Euclid's *Data*. Al-Kūhī called this latter an "existent ratio," and he had a long discussion with al-Ṣābī about these two different senses, the first of which he ascribes to the algebraists and astronomers, and with which he wants nothing to do!

Al-Kūhī writes as follows to al-Ṣābī:

> For we do not mean by this [geometric] aspect of "known" the amount of a thing, nor by "known ratio" the measure of one of them as compared with the other . . . . As for the existent ratio, according to the sense in which we use it, how could it not be known between the circle and the square when each of them is known. If two magnitudes are known then indeed the ratio of one of them to the other is, in our opinion, known—as Euclid proved in the first theorem of the *Data*. I am astonished at people who claim that the area of a circle is not permitted to be equal to the area of a square, and that there is no ratio between the two of them, and who say (so) since the circumference of the circle is curved and not of the same kind as the circumference of the square. . . .

Likewise, Ibn al-Haytham, referring to the same problem and, arguing for the existence of a square with the same area as the circle, states:

> The essence of the knowable notions does not require that one perceive them or that they be actually produced. Rather, if the proof of the possibility of the notion has been provided, the notion is sound, whether one has actually produced it or not. (Quoted from Hogendijk 1985a, 96.)

This debate touches directly on the famous problems of trisecting the angle and squaring the circle, for consider the following argument for squaring the circle:[20] A circle is given in position and magnitude, and, according to analysis, a square with the same magnitude is postulated. The circle and square have the known ratio of 1:1 and the circle is known in magnitude, hence by *Data*, 2 the square is known in magnitude. Then, by *Data*, 55 the side of the square is known in magnitude. A similar argument, also relying on *Data*, 2, just as easily trisects the angle.

The apparent flaw in the above analyses occurs in the application of *Data*, 2, which states that "If a known magnitude has a known ratio to another magnitude, the latter is also known in magnitude." The proof of this proposition involves a rearrangement of the ratios, which in the case of the quadrature of the circle would imply forming a ratio of a circle to a square. Pursuing a resolution of these anomalies, then, forces us to restate *Data*, 2 (and some other propositions) to forbid ratios between angles, and between circles and squares! This clearly contradicts how *Data*, 2 and other *Data* propositions were used, and puts us in the difficult position of having to rewrite the statement of a basic, heavily utilized proposition with no apparent evidence of ancient recognition of the anomaly in mathematical texts.

We believe that this conundrum provides a context for the remarks both of al-Kūhī and the later remarks of Ibn al-Haytham.

All of these issues, of course, relate to the question of when a problem has been solved or, to put it differently, what constitutes a valid solution to a problem. We have seen some of this discussion in tenth-century reaction to Archimedes' construction of the heptagon, and Hogendijk pointed out the discussion that occurred about the validity of "moving geometry," with its verging constructions, in comparison with that of the fixed geometry, that is, constructions in the Euclidean manner.[21] It appears that the adherents of fixed geometry won the day, but it was not a geometry limited to Euclidean methods. For example, Abū Sahl writes that the chord of 1° is known according to him who trisects the angle,[22] and al-Kūhī's trisection of the angle uses an elegant intersection of a hyperbola and a circle.[23]

This having been said, however, it must be added that the Hellenistic tradition expressed by the condemnation by anonymous "geometers" as "no small error,"[24] of using conic sections when circles and straight lines will solve

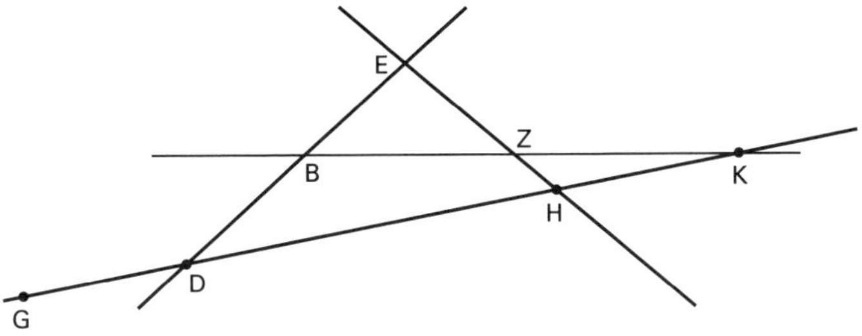

Figure 6.3

the problem, still has a lingering effect on al-Kūhī. This is apparent from his discussion in his aforementioned treatise *Ratio from a Single Line*, where he considers the following problem (figure 6.3):

> To draw a straight line from a known point G to cut three lines ED, EH, and BZ known in position so that the ratio DH:HK of one of the two segments which are formed between two of the three lines to another segment so formed is known.

At the end of the treatise, after he has dealt with the case shown above when the three given lines are straight lines in general position, al-Kūhī extends the problem by considering the case when two of the given lines are straight (and non-parallel) and the third is replaced by a curve (which al-Kūhī refers to as a 'non-straight line'). In this case he again, as in the trisection problem, effects the solution by means of a hyperbola, this time one passing through the given point and whose asymptotes are two lines parallel to the two given straight lines. The hyperbola is implicity assumed to intersect the given curve but, since that curve has no description, any consideration of when this might in fact happen is impossible.

There are two other treatises that we know of in which al-Kūhī speaks of curves other than straight lines or conic sections, one of them being his treatise on the astrolabe (Berggren 1994). But he does not do anything with them in that treatise, other than to refer to them, since, as he says, "this is not our aim in this treatise."[25] A true use of an arbitrary "non-straight" line is in his *On Centers of Tangent Circles*, in which (as in his *Ratio from a Single Line Falling between Three Lines*) al-Kūhī uses conics to solve, for the case in which one of the givens is an arbitrary curve, a problem he has already solved without conics when that given is a straight line. The above evidence would seem to indicate that he still regarded a problem solved by conics as a problem solved, but a solution without conics was to be preferred to one with conics, the very view we found expressed by Pappus's anonymous "geometers."

We noted above the amount of geometric work in the tenth century that involved analysis, and the question arises of why one so often finds analyses in the Arabic texts of the tenth century. On the one hand it could be argued that this was done because of the style set by Apollonius (and perhaps Archimedes in the first few propositions of SC II), and al-Kūhī certainly followed their lead in several of his treatises, for example, in his brilliant solution to an extension of a problem of Archimedes from SC II (discussed on p. 179 above) and in his construction of an equilateral pentagon in a given square. And the example of Apollonius is cited by Ibrāhīm ibn Sinān, when he wrote, "I have found that modern geometers have neglected the method of Apollonius in analysis and syntheses," but in fact Ibrāhīm is complaining about the modern practice of not giving the syntheses. And, indeed, al-Kūhī writes, at the end of his treatise on two lines from a known point:

> And if we were to go into partition (into cases), *diorismos*, synthesis and the positions of the points by Apollonius' method in some of his books a very big book would be produced, but we hope to have the leisure to do it later, God willing.

And in the second book, chapter six, of his treatise on the astrolabe he omits all syntheses. Thus, it seems to us that at least with some authors analysis alone was an accepted method of giving proofs and one is perhaps seeing creative mathematicians, familiar with how to develop a synthesis from an analysis, saying that the synthesis can wait until they have some time. In the meantime, the excitement of new discoveries and solutions to new problems must receive their attention.

For the tenth century was a creative time in geometry, and al-Kūhī emphasizes in a treatise on the heptagon[26] that he has solved in many ways that which Archimedes was unable to complete in even one way. According to al-Samawʾal "the construction of the regular heptagon in the circle (was done) by Wayjan ibn Rustam al-Kūhī,"[27] and al-Samawʾal wrote this to argue for his belief in the continued progress of the sciences, a belief that was shared not only by ʿUmar al-Khayyāmi and Abu'l-Qāsim al-Asṭurlābī, his predecessors in the twelth century, but by Abū Sahl who wrote in his preface to the treatise on the regular heptagon, "They (earlier geometers who failed to find the side of the regular heptagon) were sure, nevertheless, according to what their conviction has promised them, that the science of geometry will endure and it will continue to grow in contrast to man's life-span, which comes to an end."

Other Works of al-Kūhī

We leave the domain where the great Hellenistic geometers wrote when we come to Abū Sahl's treatise *Constructing the Astrolabe with Proofs*. Here Abū Sahl is working in what was virtually an Islamic tradition, and his work was the subject of a commentary by his contemporary, Abū Saʿd, and was also known to al-Bīrūnī.

It is instructive to compare this work with other tenth-century works on the astrolabe, such as that of al-Sijzī, whose exhaustive treatise on astrolabes was a sort of encyclopedia of astrolabes, and those of Ḥabash on the melon-form astrolabe and al-Ṣaghānī on planispheric astrolabes in which the projection of the sphere onto the plane through its equator is not from a pole of the sphere but from some other point on the axis. Abū Sahl's treatise is distinguished from much of the surviving Islamic literature on the astrolabe written up to his time in all of the following three features: a general treatment of stereographic projection, a discussion of a considerable number of variants of stereographic and non-stereographic projection, and being principally concerned with an extensive body of problems of slight interest to the astrolabe maker but of considerable interest to a geometer. As regards the treatment of stereographic projection the reader finds in this treatise a systematic exposition of the general theory of stereographic projection developed in the context of Book I of Apollonius's *Conics*. Al-Kūhī proves the circle-preserving property of the projection in its most general form,[28] and, in using the theory to produce the curves on the plate of the astrolabe, he shows that only one technique is necessary to project any circle symmetric to a meridian provided its inclination to the equator is known. (See Berggren 1991 and 1994.)

Thus, it was all branches of geometry that attracted al-Kūhī, not only those represented in the *Elements*, *Conics* and *Sphere and Cylinder* but also problems related to the construction of such astronomical instruments as the astrolabe and the sundial. Thus, in addition to the fact that his treatise on the astrolabe is the longest of his known works, he closes his *Complete Compass* as follows. After he has shown how to use this instrument that he designed for drawing lines, circles and conic sections satisfying certain given conditions he says, "It is now evident how to draw by means of this compass, the intersections of conic surfaces with any one of a variety of surfaces, according to a given position.[29] Consequently from this it will be easy for us to construct astrolabes on plane or axial surfaces, and to construct sundials on any surfaces, and in the same way all the instruments on which are found lines of intersection of conic surfaces with an arbitrary surface. But God knows best."

As regards al-Kūhī's treatment of projections other than that of the planispheric astrolabe it suffices to say here that he recognized that either orthographic projection or any projection from a point on the axis of the sphere onto a surface symmetric about the axis of the sphere[30] produces a possible astrolabe. He also states that the projection, which al-Bīrūnī informs us was invented by al-Ṣaghānī,[31] of circles on the sphere from a point on the axis other than a pole produces on the plane of projection conic sections (other than circles), and he cites the relevant proposition from the *Conics*. Finally, he points out the cases in which these mappings are, as we would say, not one-to-one. It is these features of the treatise that provoked the most extensive comment by Abū Sa'd.[32] Abū Sahl's treatise resembles those of Ḥabash and al-Ṣaghānī in being about mathematical curiosities, but instead of developing new mathematical projections Abū Sahl regards the ordinary planispheric astrolabe as providing an opportunity for using Euclid's *Data* to analyze, and his *Elements* to solve, problems in geometrical construction.

Finally, the following three works deal with problems not found in the Greek literature, but with techniques that are continuations of Hellenistic techniques.[33] I know of no other treatises like them in the tenth-century geometrical literature. They are:

1. On the distance from the center of the earth to the shooting stars.
2. On the area of the earth's surface visible from a given height.
3. On the possibility of an infinite motion in a finite time.[34]

Al-Kūhī seems to have been utterly uninterested in the number-theoretic facets of the Greek tradition, exemplified by *Elements* VII–IX, and the writings of Nicomachus and Diophantus, and in this he separates himself from Abu'l-Wafā', Abū Ja'far al-Khāzin and the later Ibn al-Haytham. Neither did he contribute to the development of algebra, as did al-Karajī, nor to any of the lively traditions of arithmetic,[35] which was a popular topic with many tenth-century writers, and his works show no interest in the rapidly developing trigonometry or in the numerical methods of which there was a strong tradition going back to antiquity. Indeed, he specifically denies any expertise in numerical methods in a passage in a work without any title.[36] The work begins "Some of our friends . . ." and the passage comes just after he has shown how by applying the Transversal Theorem one may calculate successively, for a given degree of the ecliptic, its declination, where it rises on a given horizon, its rising time both at the given locality and at the equator, and the equation of daylight on the day when the sun is in that degree. He says:

> Notwithstanding . . . I am not sure if there is not among the solutions of this problem an easier, shorter or more expeditious [route] than the one we followed, used by the skilled mathematicians and the expert zīj-makers, especially if they give priority to multiplication over division, or delay the one until after the other, or substitute the one for the other. . . .

In fact, in this treatise he spells out his interests fairly explicitly. He states that although he has a sufficient knowledge of Ptolemaic astronomy he has not spent the time in detailed work and calculations that, alone, can lead to real expertise. In fact he says,

> Despite the sublimity of these astronomical investigations we do not confine ourselves to their investigation and ignore the other sciences which scholars usually investigate, for example, the science of the centers of gravity, the science of optics, and the science of the characteristics of the forms of the conic sections, which is the most astonishing of them all. . . . Add to that the theories of Archimedes and the derivation of the geometrical theories . . . a single science, which stands by itself. The obsession of scholars with the derivation of such things (i.e., the sciences mentioned above) is greater than their obsession with the derivation of the other sciences beside these. *We, likewise, [follow in the same path] without any boasting or insinuation.*

Thus al-Kūhī obviously thinks of himself as being in the mainstream of mathematical investigations of his time, and one notes, again, no mention of the algebraic or numerical side of mathematics. It is possible, I think, to explain why he does not think of the numerical side of mathematics as being in the mainstream, a reason suggested by the characterization he gives of his mathematics as "demonstrative," and his denouncing the mathematics found in Archimedes' *Measurement of a Circle* as approximative and of a type that does Archimedes no credit—so much so that he says that he believes the treatise is spurious. Although he mentions trigonometric tables specifically when he compares the value of π implicit in Ptolemy's *Almagest* with the much cruder value in *Measurement of a Circle*, and although he says Ptolemy uses a finer division, it is clear that he feels is is only comparing various degrees of bad mathematics—not the bad with the good.

Apart from the reference to optics (an interesting allusion in view of other known connections between Abū Saʿd al-ʿAlāʾ ibn Sahl and al-Kūhī as well as the latter's influence on Ibn al-Haytham), Abū Sahl's account of his scientific interests accords very well with what we would have deduced from his existing treatises. This suggests that our list, if not complete, is at least representative.

Concluding Remarks

I would also like to add something to this discussion about Abū Sahl's character. Although afflicted by no false humility—he knew when he had done a good piece of work and was not reluctant to say so—he appears to have been a man who was, from some points of view, dedicated to the search for truth in the sciences he investigated and who was quick to point out when he was speaking without really expert knowledge. For example in his discussion of the complete compass he admits that he might be using terminology that is not completely standard because there might be a treatise on the subject which he has not seen. In his work on calculating the rising times of arcs of the ecliptic he is quick to remind the reader of his own lack of expertise in the "tricks of the trade" of the expert table-makers, and says at the end of the treatise, "Now if there is anything in this discourse which is redundant to the argument, and not in its right place, please accept my apology. This was mentioned only because one statement led to another."

He was also acquainted with failure. For example, ʿUmar al-Khayyāmī mentions him as a one of the mathematicians around ʿAḍud al-Dawlah who were unable to solve the problem of dividing 10 into two parts a and b ($a > b$) so that, as we would state it, $a^2 + b^2 + a/b = 72$. And there was also the incident in which he misled himself on the value of π because he convinced himself, on the basis of a numerical pattern, of an incorrect result for the location of the center of gravity of a semicircle. (This latter incident was somewhat discreditable to him because of what can only be described as his stubborn refusal to acknowledge that he had made a mistake that contradicted the estimate for π in Archimedes' *Measurement of a Circle*.)

To close this discussion, I would like to raise a question, one which is phrased specifically in terms of al-Kūhī's work but which equally could be asked of that of several of his contemporaries in the tenth century, and that is: What, if anything, about al-Kūhī's work reflects its origins in Islamic civilization? Or, to put it differently, if al-Kūhī's work had been translated anonymously into Latin and stripped, as such works sometimes were, of the dedications and the occasional "By Allah," what is there that might cause us to suspect that we were dealing with a translation from Arabic and not from Greek? Such a question could not be asked of any zīj, with its references to the coordinates of Mecca or interest in azimuths. It could not be asked of the geometrical solutions to the problem of finding the direction of Mecca given by Ibn al-Haytham. Neither would one ask it of Ibrāhīm ibn Sinān's work on sundials, nor of Abu'l-Wafāʾ's work on the geometry of craftsmen.

There are, of course, some approaches to problems that definitely are not found in the classical literature, and his use of conic sections for constructing a regular heptagon, his consideration of arbitrary curves, and his approach to analysis seem to be products of his own age. One even finds a hint of a specifically Islamic context in his interest in azimuths in his treatise on astrolabes. But it would appear, on the whole, from the problems he addressed and the methods used to address them that his works could have been written by any sufficiently able Greek geometer after the times of Archimedes and Apollonius. And I confess that I have a feeling on reading his works that I am reading the works of one who might have been considered at his time as one of the "old guard." Certainly, one can hardly avoid such a feeling in looking at Kūhī's work on rising times, where he explicitly states that he wrote it to show that one can do very nicely with Menelaus's theorem what was beginning to be done in his time by means of the newly developing trigonometric theory. Perhaps, to adapt Keynes's description of Newton, al-Kūhī was the last mathematician to look on mathematics with the eyes of the great Hellenistic geometers.

Notes

1. This is simply one possible version of his name, based on a number of variants in the manuscripts. The most common reading is, in fact, Abū Sahl Wījan ibn Wustam al-Qūhī. Some standard bio/bibliographical sources on al-Kūhī are Sezgin 1974, 314–321; Dold-Samplonius *Dictionary of Scientific Biography*, vol. XI (1975), pp 239–241.

2. According to Bayard Dodge, trans., *The Fihrist of al-Nadim*, New York 1970, 669.

3. Quoted from Sesiano 1979, 281.

4. According to Sezgin 1974, 339, Abū Naṣr ibn ʿIrāq cites Kūhī's *Points on Lines in the Ratio of Areas* in his *Al-Masā'il al-Handasiyya*.

5. Quoted in Hogendijk 1985b, 101, n. 5.

6. The requirement that a circle be "tangent" to a point means that it passes through the point.

7. This is No. 9 in Sezgin 1974, 319. It has been published by Abgrall 1995.

8. Hogendijk 1985b.

9. Abu'l-Jūd Muḥammad ibn al-Layth, al-Sijzī, al-Ṣaghānī, al-Shannī, al-ʿAlāʾ ibn Sahl, and Abū Sahl al-Kūhī.

10. For details on the medieval and modern literature on this treatise and its history see Hogendijk 1984.

11. Hogendijk's translation in Hogendijk 1984, 214.

12. Anbouba pointed this out in Anbouba 1977.

13. The fact that both al-Kūhī and al-Ṣaghānī's treatises were dedicated to the same patron supports, in view of the other evidence, the supposition that al-Kūhī developed al-Ṣaghānī's solution, rather than discovering the whole approach by conic sections independently.

14. Rashed 1981, 258.

15. Khanikoff 1857, pp. 26–33.

16. Recently published by Rashed in *MIDEO* 20 (1991), 21–231 along with Ibn al-Haytham's treatise on knowns in *MIDEO* 21 (1993), 87–275.

17. Quoted from Rashed's article on Ibrāhīm ibn Sinān in the *DSB* VII, 2–3.

18. The author, jointly with Glen Van Brummelen, has published this in *Suhayl* 2, 2001 pp. 161–198.

19. Berggren 1983.

20. I thank Glen Van Brummelen for this argument.

21. See Hogendijk 1984, esp. p. 200 note 3.

22. The passage from al-Kūhī's correspondence is in Berggren 1983, 54.

23. See the accounts in Sayili 1962 and in Knorr 1989, 301–309.

24. In Book iv of Pappus's *Math. Coll.* (Vol. I, pp. 270–272). I am not, of course, suggesting that Pappus was a source for al-Kūhī.

25. Berggren 1994, 150.

26. Preserved in BN 4821 and dedicated to Abu'l-Fawāris, the son of ʿAḍud al-Dawlah.

27. This ascription of the construction of the regular heptagon to Abū Sahl al-Kūhī supports (and perhaps reflects) al-Kūhī's own words in his treatise on the regular heptagon, "Now the easiest among those pursuits is the science of the side of the regular heptagon in a circle to which the intellects of the renowned among the famous geometricians have been applied . . . . Nevertheless, no one among them was able to attain an iota of it. Nevertheless, when the servant of our lord . . . ʿAḍud al-Dawlah . . . looked into it he was able to find it."

28. That is, in the case that any circle on the sphere not containing the pole is projected from either pole onto any plane perpendicular to the axis.

29. Woepcke 1874, 11 notes in a footnote here: "To give to the preceding constructions this extension it will suffice to determine the position and the two openings of the compass relative to the plane tangent to the given surface at the point where the axis of the compass rests on the surface."

30. Abū Sahl's choice of words ("surfaces having an axis which coincides with that of the sphere") shows he intended to be completely general, but the only specific examples he mentions are cones, cylinders or planes.

31. See Berggren 1982.

32. We shall publish this commentary separately.

33. One would normally add here his treatise *On the azimuth of the qibla*, but, despite Sezgin's attribution of it to al-Kūhī, it seems unlikely that it is by him. The treatise is not attributed to him in the manuscript and the fact that the author of the treatise gives the working-out of an example in numerical terms seems utterly unlike al-Kūhī.

34. Discussed in Sayili 1956, and recently in Rashed 1999.

35. Hindu-Arabic, sexagesimal, and the so-called "finger arithemetic."

36. Number 13 in Sezgin's list of his works. (Sezgin 1974, 319.)

Bibliography

Anbouba, ʿAdel. 1977. "Tasbīʿ al-dāʾira (Construction of the Regular Heptagon)," *Journal for the History of Arabic Science* 1, no. 1, 73–105 (Arabic pagination).

Abgrall, Philippe. 1995. "Les cercles tangents d'al-Quhi." *Arabic Sciences and Philosophy* 5, no. 2 (September), 263–295.

Bellosta, Hélène. 1991. "Ibrāhīm ibn Sinān: On Analysis and Synthesis." *Arabic Sciences and Philosophy* 1, 211–232.

Berggren, J. L. 1982. "Al-Bīrūnī on Plane Maps of the Sphere." *Journal for the History of Arabic Science* 6, 47–112.

Berggren, J. L. 1983. "The Correspondence of Abū Sahl al-Kūhī and Abū Isḥāq al-Ṣābī: A translation with commentaries." *Journal for the History of Arabic Science* 7, 39–124.

Berggren, J. L. 1991. "Medieval Islamic Methods for Drawing Azimuth Circles on the Astrolabe." *Centaurus* 34, 309–344.

Berggren, J. L. 1994. "Abū Sahl al-Kūhī's Treatise on the Construction of the Astrolabe with Proof: Text, Translation and Commentary." *Physis* 31, 141–252.

Hogendijk, J. P. 1984. "Greek and Arabic Constructions of the Regular Heptagon." *Archive for History of Exact Sciences* 30, no. 3/4, 197–330.

Hogendijk, J. P. 1985a. *Ibn al-Haytham's Completion of the Conics*. New York: Springer-Verlag.

Hogendijk, J. P. 1985b. "Al-Kūhī's Construction of an Equilateral Pentagon in a Given Square." *Zeitschrift für Geschichte der Arabisch-Islamischen Wissenschaften* 1 (1985), 100–144.

Hogendijk, J. P. 1996. *Al-Sijzī's Geometrical Treatise on Problem Solving*. Teheran: Fatemi Publishing Co.

Jones, A. 1986. *Pappus of Alexandria Book 7 of the Collection*. Springer-Verlag: New York.

Khanikoff, N. 1857. “Analysis and extracts of Book of the Balance of Wisdom . . . by al-Khāzinī in the Twelfth Century.” *Journal of the American Oriental Society* 6, 1–128.

Knorr, W. 1989. *Textual Studies in Ancient and Medieval Geometry.* Boston: Birkhaüser.

Pappi Alexandrini Collectionis quae supersunt. (3 vols.; F. Hultsch, ed.) Berlin, 1876–1878.

Rashed, R. 1979. “Ibn al-Haytham wa ʿamal al-musabbaʿi fī al-dāʾirati” and “La Construction de l’heptagone régulier par Ibn al-Haytham.” *Journal for the History of Arabic Science* 3, no. 2, 309–357 (French) and 358–376 (Arabic).

Rashed, R. 1981. “Ibn al-Haytham et la mesure du paraboloïde.” *Journal for History of Arabic Science* 5, 91–262

Rashed, R. 1999. “Al-Qūhī *vs.* Aristotle on Motion,” *Arabic Sciences and Philosophy* 9, 7–24.

Sayili, A. 1956. “A Short Article of Abū Sahl Waijan ibn Rustam al-Qūhī on the Possibility of Infinite Motion in a Finite Time.” *Actes VIII Congrès International d’Histoire des Sciences.* Florence-Milan, 248–249

Sayili, A. 1962. “The trisection of the angle by Abū Sahl Wayjan ibn Rustam al Kūhī.” *Proceedings of the 10th International Congress on the History of Science.* Ithaca, 546–547.

Sesiano, J. 1979. “Note sur trois théorèmes de Mécanique d’al-Qūhī et leur conséquence.” *Centaurus* 22, no. 4, 281–297.

Sezgin, F. 1974. *Geschichte des Arabischen Schrifttums*, Vol. V: Mathematik. Leiden: E. J. Brill.

Woepcke, F. 1851. *L’algébre d’Omar al-Khayyami.* Paris.

Woepcke, F. 1874. Trois traités Arabes sur le compas parfait, *Notices et Extraits des Manuscrits de la Bibliothèque Impèriale et autres bibliothèques* 22, 1–175.

IV

Numbers, Geometry, and Architecture

7

Quadratus Mirabilis

Jacques Sesiano

§1 Preliminaries

1.1 Introduction

One of the most impressive achievements in Islamic mathematics is the development of general methods for constructing magic squares. A magic square of order n is a square divided into n^2 cells in which different natural numbers must be arranged in such a way that the same sum appears in each of the rows, columns, and two main diagonals (figures 7.1 and 7.2); mostly, the n^2 first natural numbers are placed, and then the constant sum amounts to $\frac{1}{2}n(n^2 + 1)$, the n-th part of the sum of the natural numbers from 1 to n^2. If, in addition to this basic property of *simple* magic squares, the square remains magic when the borders are successively removed, it is called a *bordered square* (figure 7.3). If the sum in any pair of complementary diagonals (i.e., pairs of parallel diagonals lying on each side of a main diagonal and having together n cells) shows the constant sum, the square will be called *pandiagonal* (figure 7.4).

Squares are usually divided into three categories: squares of odd order—also called *odd* squares—($n = 2k + 1$, k natural); *evenly-even* squares ($n = 4k$); *oddly-even* squares ($n = 4k + 2$). There are general methods that make it possible to construct squares of any order from one of these three categories. Except for the smallest possible order, $n = 3$, there are numerous possibilities of forming magic squares of any given order. There may be, however, some limitations concerning additional magical properties; for instance, bordered squares cannot be constructed if $n = 4$, and there are no pandiagonal squares of oddly-even order.

Information about the beginning of Islamic research on magic squares is lacking; it may have been connected with the introduction of chess into Persia. Initially, the problem was a purely mathematical one; thus, the Arabic ancient designation for magic squares is *wafq al-aʿdād*, that is, "harmonious disposition of the numbers." We know that treatises on magic squares were written in the ninth century, but the two earliest extant texts date back to the tenth

1	31	22	15	30	12
2	32	14	23	29	11
3	33	24	13	28	10
34	4	16	21	9	27
35	5	17	20	8	26
36	6	18	19	7	25

Figure 7.1

57	72	62	97	92
73	91	85	52	79
74	71	76	81	78
82	66	67	95	70
94	80	90	55	61

Figure 7.2

92	17	4	95	8	91	12	87	16	83
99	76	31	22	77	26	73	30	69	2
1	20	64	41	36	63	40	59	81	100
3	19	67	58	47	51	46	34	82	98
96	80	33	52	45	57	48	68	21	5
7	78	35	49	56	44	53	66	23	94
90	27	62	43	54	50	55	39	74	11
13	72	42	60	65	38	61	37	29	88
86	32	70	79	24	75	28	71	25	15
18	84	97	6	93	10	89	14	85	9

Figure 7.3

century: the *Treatise on the magic disposition of numbers in squares* by Abu 'l-Wafā' al-Būzjānī (940–997 or 998) and a chapter in Book III of ʿAlī b. Aḥmad al-Anṭākī's (d. 987) *Commentary on Nicomachos's Arithmetic*.[1] It appears that, by that time, the science of magic squares was already established; bordered squares of any order $n \geq 5$ could be constructed, while simple magic squares could be obtained for small orders.[2] Various general methods for the construction of simple magic squares of odd and evenly-even orders, and also for pandiagonal squares of evenly-even orders, were devised in the early eleventh century. The science of magic squares can be said to be at its apogee around 1100; by that time, the remaining problem of constructing simple magic squares of oddly-even order had been solved. From the thirteenth century onwards, magic squares become increasingly associated with magic purposes. Consequently, some texts merely picture squares and mention their attributes. Some others do keep the general theory alive, though often only to enable the reader to construct amulets for himself.

In Europe, interest in magic squares was first aroused toward the end of the Middle Ages, when two sets of squares associated with the seven planets were learned of through Arabic magic texts (whence the name), but without any indication as to their construction. Methods of construction spread eastward around the twelfth century toward India and China; and, to a lesser extent, toward Byzantium (c. 1300).

It is interesting to note that, from the very beginning of the Islamic studies on magic squares, attempts were made to consider squares with additional conditions, constructed by adapting or modifying the general methods. The present study will show how elementary procedures led, already in the tenth century, to a highly intricate form of construction.

A magic square of order three filled with the first nine natural numbers can only take one form, where the even numbers occupy the corner cells and thus surround the area containing the odd numbers (figure 7.5). Since the square of order three was the smallest possible and the first to be obtained, this property might have given rise to the idea of extending such a separation to squares of higher odd order. In this case the odd numbers are to occupy a central rhombus with corners meeting the middle of the square's sides and the even numbers will be in the four remaining triangles.

It is easy to obtain a simple magic square in which the even and odd numbers are separated thus. Principles of construction were devised at about the same time as those for common simple magic squares, probably in the early eleventh century, and simplified rules appeared in the centuries that followed.[3] Concern for distinction by parity is also seen in some eleventh-century constructions of squares of even orders, and in one case the odd numbers form a kind of hexagon in each quadrant.[4] The idea of constructing magic squares with separation by parity was therefore common in the eleventh century. The

1	62	59	8	9	54	51	16
60	7	2	61	52	15	10	53
6	57	64	3	14	49	56	11
63	4	5	58	55	12	13	50
17	46	43	24	25	38	35	32
44	23	18	45	36	31	26	37
22	41	48	19	30	33	40	27
47	20	21	42	39	28	29	34

Figure 7.4

4	9	2
3	5	7
8	1	6

Figure 7.5

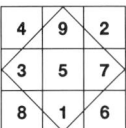

same idea already occurs in the tenth century, at a time when general methods were only known for bordered magic squares. Now the construction of a *bordered* square of odd order showing in addition the above-mentioned *separation* between odd numbers and even numbers is much more difficult. A first attempt seems to survive in Abu 'l-Wafā''s treatise. A general method of construction is explained by al-Anṭākī, but the unequal value of his treatise suggests that this very elaborate construction did not originate with him. Whoever the author, it is one of the gems of the Islamic science of magic squares (or, indeed, of Islamic science in general).

There is no specific Arabic denomination for these squares or for their construction. I have chosen to call a square thus constructed *quadratus mirabilis* owing to a wonder comparable to that of my countryman Jakob Bernoulli three hundred years ago when he discovered the properties of his *spira mirabilis*.

After some preliminaries (§1, 2–4), we shall analyze this construction (§2), and this will be followed by the translation of al-Anṭākī's relevant section (§3).

1.2 Construction of Bordered Squares of Odd Orders in the Tenth Century

Abu 'l-Wafā' and al-Anṭākī start by showing how to construct bordered squares of any odd order n ($n \geq 5$) as follows (figure 7.6, with $n = 11$).[5] Beginning next to a corner cell in the outer border, write the first odd and even numbers alternately along the column and row meeting at this corner, and continue as far as

10	120	118	116	114	113	14	16	18	20	12
103	28	100	98	96	95	32	34	36	30	19
105	87	42	84	82	81	46	48	44	35	17
107	89	75	52	72	71	56	54	47	33	15
109	91	77	67	60	65	58	55	45	31	13
111	93	79	69	59	61	63	53	43	29	11
7	25	39	49	64	57	62	73	83	97	115
5	23	37	68	50	51	66	70	85	99	117
3	21	78	38	40	41	76	74	80	101	119
1	92	22	24	26	27	90	88	86	94	121
110	2	4	6	8	9	108	106	104	102	112

Figure 7.6

their middle cells. Write the next (odd) number in the middle cell last reached, then put the following number in the corner cell at the other end of the column where 1 was placed, the next one (= n) in the middle cell on the other side, the following number in the other upper corner cell; and finally arrange the subsequent numbers along the remaining column and row, alternately as before, but this time choosing to start next to the middle cell of the column. At this point, half of the border cells are occupied. Fill each of the remaining blank cells with $n^2 + 1$ minus the number in the opposite cell.[6] Repeat this procedure until the central square of order 3 is reached, and fill it with the remaining numbers arranged as in figure 7.5. This completes the construction of the square. Note that, with this arrangement, in each row of the successive squares all numbers are even except the number in the middle and in each column they are all odd except the two numbers at the corners.

1.3 Geometrical Structure of a Bordered Square with Separation

Let the main square have the order $n = 2k + 1$. The rhombus which is to contain the odd numbers consists then of $2k + 1$ "rows" and "columns" in which the number of cells is alternately $k + 1$ and k starting with the outer rows and columns. Hence, within the rhombus there is a total of $(k + 1)^2 + k^2 = 2k^2 + 2k + 1$ cells, which is the quantity of odd numbers among the first n^2 natural numbers; the remaining $2k^2 + 2k$ even numbers will then occupy the four corner triangles, each of which contains $\frac{1}{2}k(k + 1)$ cells. The rhombus itself includes inner squares, the largest of which has the order k if k is odd and $k + 1$ if k is even. Consequently, the largest inner square for both $n = 4t + 1$ and $n = 4t + 3$ has the order $2t + 1$, the only difference being that its corner cells are not fully included in the first case but are included in the second case (figure 7.7a and 7.7b, for $t = 2$). Consider finally the borders surrounding this largest inner square. The quantity of cells they contain which do not lie within the rhombus depends upon the form of n. For the p-th border, counted from the one surrounding the inner square, it will be

- for order $n = 4t + 1$: $16p - 4$ altogether, and $4p$ in each row and column (with a common corner cell), where $p = 1, \ldots, t$;
- for order $n = 4t + 3$: $16p - 12$ altogether, and $4p - 2$ in each row and column, where $p = 1, \ldots, t + 1$.

The inner square is easy to fill in, following the above-mentioned method for bordered squares (figure 7.8): after putting $\frac{n^2+1}{2}$ (odd) in the central cell, write next to a corner cell the number $4t^2 + 1$ when $n(\geq 9)$ has the form $4t \pm 1$, and continue as explained above but using odd numbers only. With the complements placed in the remaining blank cells, the inner square becomes a bordered magic square, occupied by a continuous sequence of odd numbers. Thus far our two tenth-century authors proceed in the same way. From now on, however, they take a very different approach.

1.4 Abu 'l-Wafā''s Construction

After completing the inner square as above, Abu 'l-Wafā' suggests using for the remaining part the arrangement obtained by the method for bordered squares but omitting those numbers which do not satisfy the requirements of parity and those which have already been used for the inner square (figure 7.9; cf.

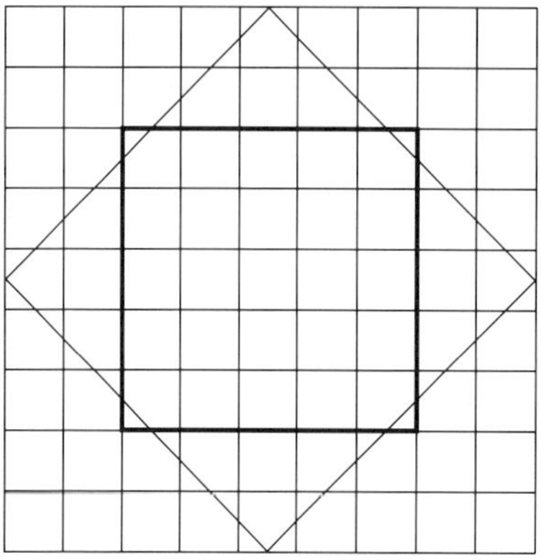

Figure 7.7a

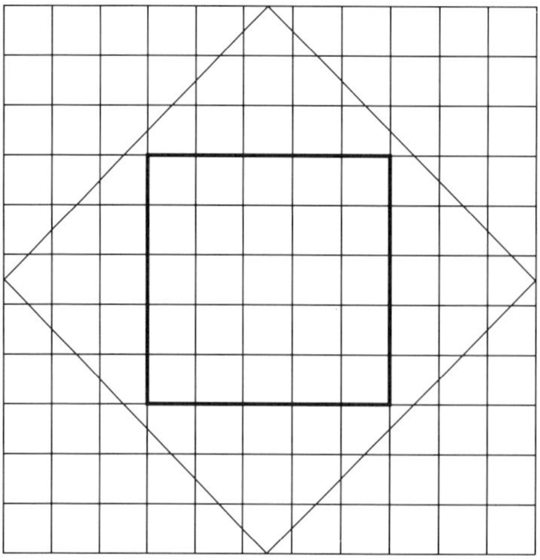

Figure 7.7b

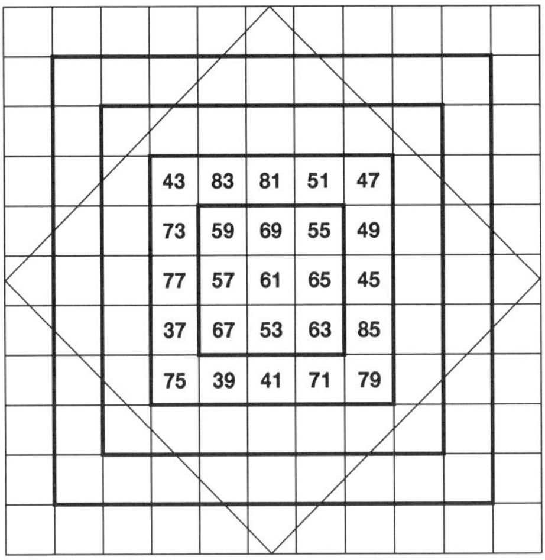

Figure 7.8

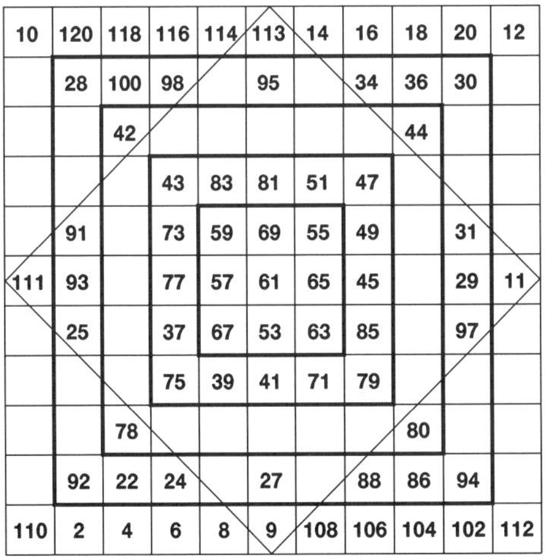

Figure 7.9

figure 7.6). The blank cells are then filled with the remaining numbers, with due consideration to parity and the amount needed to complete the magic sum. (One possibility is seen in figure 7.10.)

But this procedure does not resolve the main problem, which resides in the increasing complexity of the computations as the order increases: in the columns outside the rhombus and in the rows inside it most of the cells will remain empty because of the unequal distribution by parity mentioned at the end of §1.2 (and seen in figure 7.6). To sum up, this (apparently) early approach

10	120	118	116	114	113	14	16	18	20	12
26	28	100	98	13	95	115	34	36	30	96
90	40	42	105	101	99	35	1	44	82	32
38	50	3	43	83	81	51	47	119	72	84
46	91	5	73	59	69	55	49	117	31	76
111	93	89	77	57	61	65	45	33	29	11
48	25	103	37	67	53	63	85	19	97	74
68	70	107	75	39	41	71	79	15	52	54
66	60	78	17	21	23	87	121	80	62	56
58	92	22	24	109	27	7	88	86	94	64
110	2	4	6	8	9	108	106	104	102	112

Figure 7.10

does not meet the fundamental requirement of methods for constructing magic squares, which is to leave no room for uncertainty in placing the numbers once the order is known.

§2 Direct Construction

2.1 Placing the Remaining Odd Numbers in the Rhombus

The method reported by al-Anṭākī is reminiscent of, and no doubt inspired by, the method for common bordered squares described in §1.2, the main difference being the points of departure in the border of the rhombus. Al-Anṭākī's method is valid for any rhombus, irrespective of the parity of k in the order $n = 2k + 1$ of the main square. It should also be noted that for any two consecutive orders $n = 4t \pm 1$ the arrangement of the remaining smaller odd numbers (1 to $4t^2 - 1$) in the corner triangles of the rhombus will be the same.

To write these numbers in the rhombus, put 1 next to (say) the lower left corner cell of the inner square, 3 next to its upper left corner cell and then alternate the next odd numbers along the border row and column of the rhombus as far as the corner cells. Leave these blank and fill in the other corner cells, first the lower cell and then the right cell. The subsequent numbers are placed alternately along the other border row and column, taking the corner cell just filled as point of departure and stopping at the sides of the inner square.[7] The process, which is repeated in the subsequent borders, ends when only two cells are left (one in each of the main diagonals of the rhombus); the lower cell should be filled in first, then the right cell. After the complements have been written in, the rhombus is complete, and no odd number is left (figures 7.11 and 7.12, with orders of the form $4t \pm 1$).

2.2 Placing the Even Numbers: Preliminaries

a) Required sum in each Border

The magic sum being $\frac{1}{2}n(n^2 + 1)$, we may compare the sum in m cells already filled in with the sum they should contain on the average, namely $\frac{1}{2}m(n^2 + 1)$. (We shall refer to this hereafter as the "sum due" for m cells.) If the odd numbers are arranged in the rhombus as above, each upper border row of the main square shows an excess over the sum due, and so does each of its left columns, while the corresponding opposite rows and columns, being filled with complements, show a deficit of the same amount. For the p-th upper border row, counted from the border surrounding the inner square, the excess Γ_p is as follows:

• For order $n = 4t + 1$:

$\Gamma_p = 8pt + 4$, thus

$$\Gamma_1 = 8t + 4$$

							213							
						11	*189*	*207*						
					7	**35**	*173*	*183*	*203*					
				3	**31**	**51**	*165*	*167*	*179*	*199*				
			201	75	159	155	153	83	87	79	**25**			
		205	*181*	141	95	135	133	103	99	85	**45**	**21**		
	209	*185*	*169*	145	125	111	121	107	101	81	**57**	**41**	**17**	
211	*187*	*171*	*163*	149	129	109	113	117	97	77	**63**	**55**	**39**	**15**
	9	**33**	**49**	69	89	119	105	115	137	157	*177*	*193*	*217*	
		5	**29**	65	127	91	93	123	131	161	*197*	*221*		
			1	147	67	71	73	143	139	151	*225*			
				223	*195*	*175*	**61**	**59**	**47**	**27**				
					219	*191*	**53**	**43**	**23**					
						215	**37**	**19**						
							13							

Figure 7.11

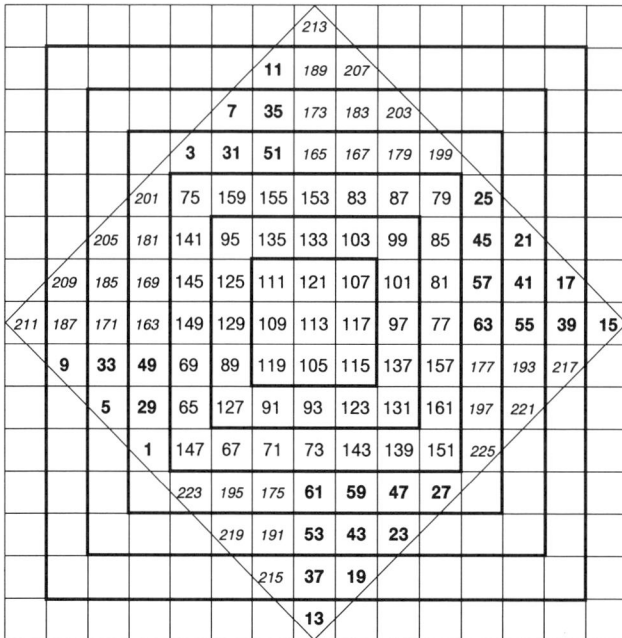

								277								
							11	*253*	*271*							
						7	**35**	*237*	*247*	*267*						
					3	**31**	**51**	*229*	*231*	*243*	*263*					
				79	223	219	215	213	87	91	95	83				
			265	197	107	191	187	185	115	119	111	93	**25**			
		269	*245*	201	173	127	167	165	135	131	117	89	**45**	**21**		
	273	*249*	*233*	205	177	157	143	153	139	133	113	85	**57**	**41**	**17**	
275	*251*	*235*	*227*	209	181	161	141	145	149	129	109	81	**63**	**55**	**39**	**15**
	9	**33**	**49**	73	101	121	151	137	147	169	189	217	*241*	*257*	*281*	
		5	**29**	69	97	159	123	125	155	163	193	221	*261*	*285*		
			1	65	179	99	103	105	175	171	183	225	*289*			
				207	67	71	75	77	203	199	195	211				
					287	*259*	*239*	**61**	**59**	**47**	**27**					
						283	*255*	**53**	**43**	**23**						
							279	**37**	**19**							
								13								

Figure 7.12

$\Gamma_2 = 16t + 4$
$\Gamma_3 = 24t + 4$
$\Gamma_4 = 32t + 4$
...

For the left columns, the excesses are each 2 less than the corresponding excesses for the rows; they are therefore

$\Gamma'_p = 8pt + 2$, thus

$\Gamma'_1 = 8t + 2$
$\Gamma'_2 = 16t + 2$
...

• For order $n = 4t + 3$, the excess Δ_p is:

$\Delta_p = 8(p - 1)(t + 1) + 4$ for the p-th upper row, thus

$\Delta_1 = 4$
$\Delta_2 = 8(t + 1) + 4$
$\Delta_3 = 16(t + 1) + 4$
$\Delta_4 = 24(t + 1) + 4$

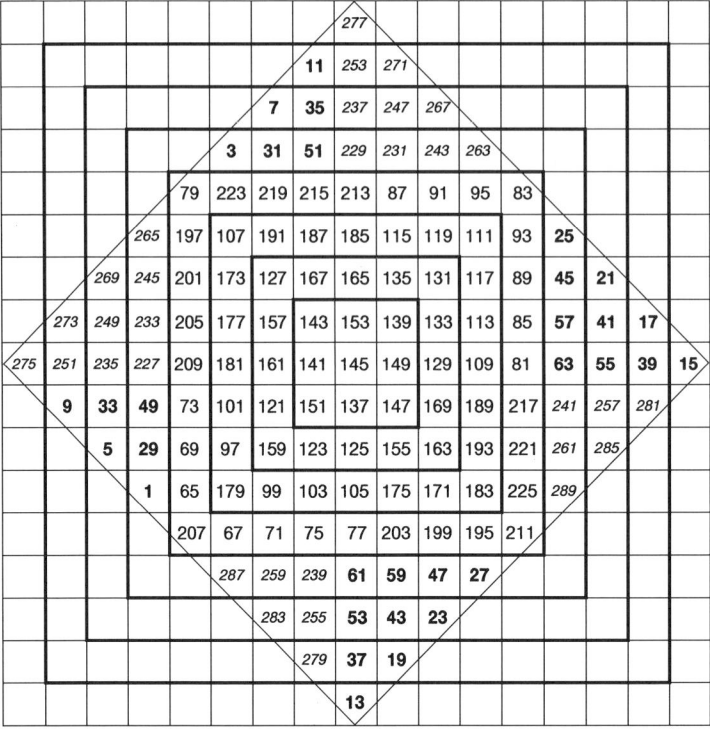

...

and for the p-th left column, the excess is:

$\Delta'_p = 8(p-1)(t+1)+2$, thus

$$\Delta'_1 = 2$$
$$\Delta'_2 = 8(t+1)+2$$

...

We have already determined the quantity of even numbers needed for each border row and column (§1.3) and have now found the excess or deficit in each incomplete border row and column after placing the odd numbers. Thus we may calculate the sum of the even numbers to be placed in each border row or column: this is the sum due ($\frac{n^2+1}{2}$ multiplied by the number of empty cells left in each incomplete border row or column) reduced or increased by the excess or deficit Γ or Δ.

b) Placing the even numbers: basic rules

Let us consider the set of even numbers to be placed. For any order $n = 2k+1$, there is an even quantity $2k^2+2k$ of even numbers: $k(k+1)$ smaller ones (less than $\frac{n^2+1}{2}$) and their $k(k+1)$ complements. Since $k(k+1)$ is even, all these numbers can not only be aligned vertically in pairs of complements but also grouped horizontally by pairs, as follows:

$$\begin{array}{cccccccccc} 2 & 4 & ; & \ldots & ; & \frac{n^2-8j+3}{2} & \frac{n^2-8j+7}{2} & ; & \ldots & ; & \frac{n^2-5}{2} & \frac{n^2-1}{2} & ; \\ n^2-1 & n^2-3 & ; & \ldots & ; & \frac{n^2+8j-1}{2} & \frac{n^2+8j-5}{2} & ; & \ldots & ; & \frac{n^2+7}{2} & \frac{n^2+3}{2} & . \end{array}$$

We shall henceforth call “dyad” any such pair of even numbers and characterize it by the value of j ($j = 1, 2, \ldots, \frac{1}{2}k(k+1)$ for those of the upper line, from right to left).

The method for the initial elimination of the excesses and deficits and the subsequent filling of the remaining empty cells rests on the appropriate choice and arrangement of these dyads.

• The initial elimination of the excesses and deficits (referred to hereafter as *equalization*) relies for $n = 4t+1$ and $n = 4t+3$ on four rules:

I. Writing the number α in one side (row or column) and $\alpha + 2s$ in the other and then filling the opposite cells with the complements will produce in the side of α the sum

$$\alpha + [n^2 + 1 - (\alpha + 2s)] = n^2 + 1 - 2s$$

and in the other side

	α	$n^2+1-(\alpha+2s)$		$\Rightarrow n^2+1-2s$
	$n^2+1-\alpha$	$\alpha+2s$		$\Rightarrow n^2+1+2s$

Figure 7.13

	α	$n^2+1-(\alpha+2)$		$\Rightarrow n^2+1-2$
	$n^2+1-\alpha$	$\alpha+2$		$\Rightarrow n^2+1+2$

Figure 7.14

		α	$n^2+1-(\alpha+4)$			$\Rightarrow n^2+1-4$
$\alpha+2$					$n^2+1-(\alpha+2)$	
$n^2+1-(\alpha+6)$					$\alpha+6$	
		$n^2+1-\alpha$	$\alpha+4$			$\Rightarrow n^2+1+4$
$\Downarrow$					$\Downarrow$	
n^2+1-4					n^2+1+4	

Figure 7.15

	$\frac{n^2-8j+3}{2}$	$\frac{n^2-8j+7}{2}$		$\Rightarrow n^2+1-(8j-4)$
	$\frac{n^2+8j-1}{2}$	$\frac{n^2+8j-5}{2}$		$\Rightarrow n^2+1+(8j-4)$

Figure 7.16

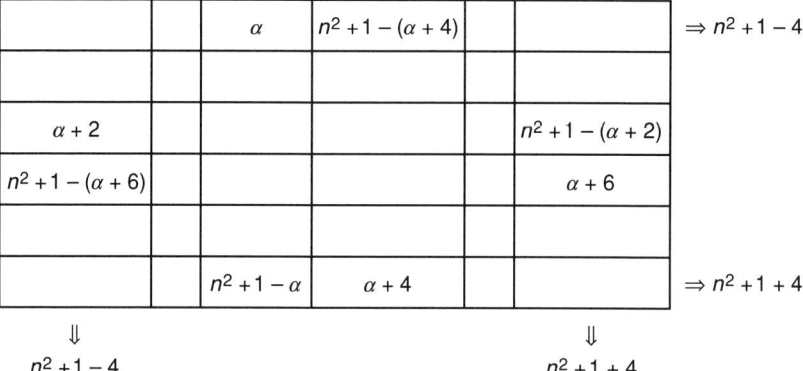

$\frac{n^2-8j+3}{2}$		$\frac{n^2-8j+7}{2}$	$\Rightarrow n^2+1-(8j-4)$
$\frac{n^2+8j-5}{2}$		$\frac{n^2+8j-1}{2}$	$\Rightarrow n^2+1+(8j-4)$
$\Downarrow$		$\Downarrow$	
n^2+1-2		n^2+1+2	

Figure 7.17

$[n^2 + 1 - \alpha] + \alpha + 2s = n^2 + 1 + 2s$ (figure 7.13).

This rule enables us to eliminate any difference $2s$ by using two even numbers which differ by $2s$. It should be noted, however, that taking $s \neq 1$ "breaks" two dyads, since this uses only one element of each, which is a situation to be avoided if possible, as we shall see below. The following particular cases, which only involve dyads that have been left complete, are mainly used by al-Anṭākī for the equalization:

II. Taking any dyad of smaller numbers and writing its elements in opposite sides, then filling the opposite cells with the complements, will produce the smallest difference possible, 2, in the form of a deficit in the side containing the smallest element (figure 7.14).

III. Placing two consecutive dyads of smaller numbers around the border, then filling the opposite cells with the complements, will produce a difference of 4, in the form of a deficit in the row and column containing the first dyad (figure 7.15).

IV. Writing the j-th dyad of smaller numbers in one side and their complements in the opposite side will produce a deficit of $8j - 4$ in the side of the j-th dyad and a corresponding excess in the other (figure 7.16). If this dyad is written in the corner cells, the same will hold, but in addition there will be a deficit of 2 in the perpendicular side containing the smallest element, and a corresponding excess in the opposite side (figure 7.17).

The application of these four rules allows us to eliminate any difference of the form $8u + v$ with $v = 0, 2, 4, 6$, thus all the possible differences, and without breaking dyads unless we apply the first rule.

• From rule I we further deduce a "neutral" arrangement, which produces the

α	$n^2+1-(\alpha+2s)$	$n^2+1-\beta$	$\beta+2s$	$\Rightarrow 2(n^2+1)$
$n^2+1-\alpha$	$\alpha+2s$	β	$n^2+1-(\beta+2s)$	$\Rightarrow 2(n^2+1)$

Figure 7.18

sum due, without any excess or deficit:

> V. Consider two pairs of numbers having an equal difference: α, $\alpha + 2s$; β, $\beta + 2s$. Writing the two extreme terms in one side and the two middle terms in the other, then filling the opposite cells with the complements, will produce in the two sides the sum $2(n^2 + 1)$, which is the sum due for four cells (figure 7.18). Here again, we should choose $s = 1$, so that the two pairs of numbers are two (separate or consecutive) dyads.

Thus, if by applying rules I–IV we are able to fill a certain (even) number of cells in each border row or column, including the cells in the corners, so as to arrive at the sum due for the number of cells already filled; and if the number of remaining (empty) cells in this row or column is divisible by 4 while the set of still available even numbers consists of dyads, these cells can be easily filled using rule V. If the method is to be generally applicable, the first step (the equalization process) must be uniform and involve the smallest possible number of cells so as to apply to all orders, from the smallest one possible.

This is the method followed by al-Anṭākī. In order to show that this method is indeed generally applicable, it will be described below in modern symbolism.

3.3 Placing the Even Numbers for the Order $n = 4t + 1$ ($t \geq 2$)

a) Rows

We shall consider only the upper rows, since the corresponding lower rows will be occupied by the complements of the numbers in the upper rows.

As we have seen (§1.3), in each border row and column there is a number of cells remaining empty which is divisible by 4. Since the differences from the sum due must be eliminated by filling the smallest possible number of cells, including those at the corners, and there must remain a number of empty cells divisible by 4, we shall equalize each border row and column by means of four or at most eight numbers. Al-Anṭākī's first step is thus to fill four cells in each row, including the two cells at the corners, so as to leave some excess for $p \geq 2$ but none in the row closest to the inner square since it has only four empty cells; the next step will be to eliminate the remaining excess by means of four other numbers, thus leaving for $p \geq 3$ a number of empty cells divisible by 4 and no excess or deficit. The four numbers to be initially placed are chosen as follows (see figure 7.19 — $n = 17$, thus $t = 4$ —, numbers in Roman type).

• First (lowest) upper row ($p = 1$): Put

$\frac{n^2-5}{2}$ in the left corner

$\frac{n^2-1}{2}$ in the right corner.

These numbers form the largest dyad of smaller numbers (corresponding to $j = 1$). Since their sum is $n^2 - 3$ instead of the sum due $n^2 + 1$, while the initial excess was $\Gamma_1 = 8t + 4$, the remaining excess is $8t$. By rule I we can eliminate this excess by means of any two even numbers with a difference of $8t$. We may, for instance, put the smallest even number 2 in the upper row and $2 + 8t$ in the opposite side. The first upper row will then contain as even numbers, in addition to the ones in the corner cells, 2 and $n^2 + 1 - (2 + 8t)$, and will thus no longer show an excess.

• For the other rows ($p \geq 2$), the initial excess is $\Gamma_p = 8pt + 4$. This may be, at least partly, compensated using an appropriate dyad. Rule IV will enable us to determine which j will suit. Let us thus put

$$8pt + 4 = 8j - 4.$$

Since we have to eliminate the excess by means of either four or eight numbers (but in any case not two), we take $j = pt$ (and not $j = pt + 1$) and put the pt-th dyad of smaller numbers

$$\frac{n^2-8pt+3}{2}, \quad \frac{n^2-8pt+7}{2}$$

in the p-th upper row. Then the remaining excess in each upper row is 8.

Above, we used the first dyad ($j = 1$) for the first border. Let us accordingly take the p-th dyad and put its smaller element in the upper left corner of the p-th border and the larger element in the lower corner in the same *column*. Since the sum of this smaller element, namely

$$\frac{n^2-8p+3}{2},$$

and the complement of the larger element, namely

$$\frac{n^2+8p-5}{2},$$

is $(n^2 + 1) - 2$, the excess left in each upper p-th row ($p \geq 2$) will be equal to 6, and $4(p - 1)$ cells in this row remain empty.

b) Columns

We shall consider only the left columns, since the right columns are to be filled with the complements of the numbers in the left column. Now the corner cells are already occupied; we shall therefore eliminate the differences in the p-th column by means of six numbers for $p \geq 2$ but only two for the case $p = 1$, where just two empty cells are left. This is done as follows (see figure 7.19, numbers in italics).

130	82	84						277								158
94	134	98	100				11	253	271						154	196
96	106	138	114	116		7	35	237	247	267				150	184	194
	108	118	142	2	3	31	51	229	231	243	263	256	144	172	182	
		120	4	79	223	219	215	213	87	91	95	83	286	170		
			265	197	107	191	187	185	115	119	111	93	25			
		269	245	201	173	127	167	165	135	131	117	89	45	21		
	273	249	233	205	177	157	143	153	139	133	113	85	57	41	17	
275	251	235	227	209	181	161	141	145	149	129	109	81	63	55	39	15
	9	33	49	73	101	121	151	137	147	169	189	217	241	257	281	
		5	29	69	97	159	123	125	155	163	193	221	261	285		
			1	65	179	99	103	105	175	171	183	225	289			
			254	207	67	71	75	77	203	199	195	211	36			
			146	288	287	259	239	61	59	47	27	34	148			
		140	176	174		283	255	53	43	23				152		
	136	192	190				279	37	19						156	
132	208	206						13								160

Figure 7.19

• First column (p = 1): Since the corner cells have been occupied by

$$\frac{n^2-5}{2} \text{ and } \frac{n^2+3}{2},$$

with sum $n^2 - 1$, the initial excess $\Gamma'_1 = 8t + 2$ has been reduced to $8t$. Proceeding as for the first row, we put 4 in the left column and $4 + 8t$ in the right column. Thus the first left column, now completed by 4 and $n^2 + 1 - (4 + 8t)$, has no excess, and with the choice of 4 there are no more broken dyads: the dyad consisting of the smallest numbers 2 and 4 (corresponding to $j = 2t^2 + t$) and the associated dyad formed by $2 + 8t$ and $4 + 8t$ (corresponding to $j = 2t^2 - t$) have both been used.

• Other columns ($p \geq 2$): As we know, the initial excess is $\Gamma'_p = 8pt + 2$. Now the two numbers in the corners are

$$\frac{n^2-8p+3}{2} \text{ and } \frac{n^2-8p+7}{2},$$

with sum $n^2 - 8p + 5$, and so the excess remaining in each left column has become $8pt + 2 - 8p + 4 = 8p(t - 1) + 6$. By rule IV, we can reduce each of these excesses to a constant: putting

$$8p(t-1) + 6 = 8j - 4,$$

we choose $j = p(t - 1) + 1$ and write the dyad

$$\frac{n^2-8(pt-p+1)+3}{2}, \qquad \frac{n^2-8(pt-p+1)+7}{2}$$

in the p-th left column ($p \geq 2$). Thus the excesses are now reduced to the constant amount of 2.

Remark: After placing the odd numbers in a square with order $n = 4t + 1$, $t \geq 2$, we were left with t unfinished borders, numbered $p = 1, \ldots, t$. Now we have seen that all the dyads of smaller even numbers initially at our disposal can be written as

$$\frac{n^2-8j+3}{2}, \quad \frac{n^2-8j+7}{2} \qquad \text{with } j = 1, \ldots, 2t^2 + t.$$

Among these dyads, the following ones have been used:

(i) $j = 1$ (in the corner cells of the first upper row, thus for $p = 1$)
(ii) $j = 2t^2 \pm t$ (elsewhere in the first border, thus for $p = 1$)
(iii) $j = p$, that is, $j = 2, 3, \ldots, t$ (in the left corner cells for $p \geq 2$)
(iv) $j = pt$, that is, $j = 2t, 3t, \ldots, t^2$ (in the upper rows for $p \geq 2$)
(v) $j = p(t - 1) + 1$, that is, $j = 2t - 1, 3t - 2, \ldots, t^2 - t + 1$ (in the left columns for $p \geq 2$).

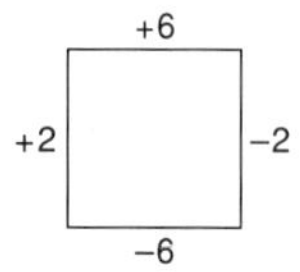

Figure 7.20

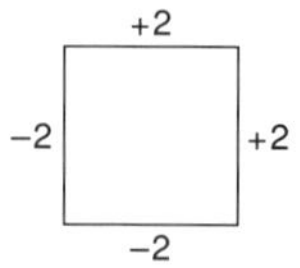

Figure 7.21

Each of the dyads used corresponds to exactly one j among these $3t$ values, none of which is identical to another. Thus none of the dyads used can occur twice.

c) Completing the square

There is now left in each row and column (for $p \geq 2$) a number of empty cells divisible by 4. Furthermore, the excess is 6 for each incomplete upper row and 2 for each incomplete left column (figure 7.20). The set of remaining even numbers consists solely of dyads or also tetrads of consecutive terms; for the numbers placed previously always formed dyads, which were either grouped or separated by one or two (or more) unplaced dyads. The remainder of each border may now be filled in as follows.

(1) Take, according to rule III, one tetrad of consecutive even numbers (α, $\alpha + 2$, $\alpha + 4$, $\alpha + 6$) and place them (anti-clockwise) around the border, starting with the top row, and then write the respective complements. This will increase the sums in the border rows and columns as in figure 7.15. Thus, each of the former excesses and deficits is replaced by ± 2 (figure 7.21).

(2) Take any two (consecutive or separate) dyads (β, $\beta + 2$, γ, $\gamma + 2$) and, following rule II, write each dyad in a pair of opposite sides, with the smaller elements on the side of the excesses, and write the complements. This will eliminate the previous differences, since the sums in the border rows and columns are increased as seen in figure 7.14. Thus, the rows and columns of the borders now show the sum due for the quantity of cells already filled in: that is, as many times $\frac{1}{2}(n^2 + 1)$ as there are occupied cells; and, for $p \geq 3$, in each row and column the number of remaining empty cells is divisible by 4.

(3) Take any two (consecutive or separate) dyads (δ, $\delta + 2$, ε, $\varepsilon + 2$) and, following rule V, place the extreme terms in one row or column and the middle terms in the opposite row or column. After the complements have been written in, this will produce the sum due for four cells. This enables us to fill the empty cells with the numbers which have not yet been used.

The above steps are seen in figure 7.22 (where the borders, starting from the outer one, have been equalized successively, using the first two steps

130	82	84	6	280	14	274	58	277	230	228	64	66	222	220	72	158
94	134	98	100	22	264	30	11	253	271	258	74	214	212	80	154	196
96	106	138	114	116	42	7	35	237	247	267	244	50	238	150	184	194
8	108	118	142	2	3	31	51	229	231	243	263	256	144	172	182	282
278	24	120	4	79	223	219	215	213	87	91	95	83	286	170	266	12
272	262	44	265	197	107	191	187	185	115	119	111	93	25	246	28	18
20	252	269	245	201	173	127	167	165	135	131	117	89	45	21	38	270
86	273	249	233	205	177	157	143	153	139	133	113	85	57	41	17	204
275	251	235	227	209	181	161	141	145	149	129	109	81	63	55	39	15
202	9	33	49	73	101	121	151	137	147	169	189	217	241	257	281	88
200	40	5	29	69	97	159	123	125	155	163	193	221	261	285	250	90
92	122	242	1	65	179	99	103	105	175	171	183	225	289	48	168	198
102	166	236	254	207	67	71	75	77	203	199	195	211	36	54	124	188
186	164	56	146	288	287	259	239	61	59	47	27	34	148	234	126	104
180	128	140	176	174	248	283	255	53	43	23	46	240	52	152	162	110
112	136	192	190	268	26	260	279	37	19	32	216	76	78	210	156	178
132	208	206	284	10	276	16	232	13	60	62	226	224	68	70	218	160

Figure 7.22

described above, by means of the available numbers ranging discontinuously from 6 to 56; and the third step was applied first to the rows and then to the columns). The completed squares of the smaller orders 9 and 13 are also shown in figures 7.23 and 7.24. Al-Anṭākī restricts himself to constructing the square for order 9.

Remark: These three final steps appeared in the tenth century for the construction of bordered squares of even order, which implies that they were not specifically devised for the *quadratus mirabilis*.

2.4 Placing the Even Numbers for the Order $n = 4t + 3$ $(t \geq 1)$

a) Rows

We shall again consider only the upper rows, since the lower rows are to be filled with the complements of the numbers in the upper rows.

As we have seen (§1.3), the number of empty cells is of the form $4p - 2$. In order to eliminate the difference from the sum due, we wish to fill the smallest

34	26	28	6	77	72	14	66	46
30	38	2	3	69	71	64	40	52
32	4	23	63	61	31	27	78	50
8	73	53	39	49	35	29	9	74
75	67	57	37	41	45	25	15	7
70	1	17	47	33	43	65	81	12
60	62	55	19	21	51	59	20	22
24	42	80	79	13	11	18	44	58
36	56	54	76	5	10	68	16	48

Figure 7.23

74	50	52	6	160	14	161	154	42	126	124	48	94
58	78	62	64	30	7	145	155	136	22	146	90	112
60	66	82	2	3	23	137	139	151	144	84	104	110
8	68	4	47	131	127	125	55	59	51	166	102	162
158	32	153	113	67	107	105	75	71	57	17	138	12
152	157	141	117	97	83	93	79	73	53	29	13	18
159	143	135	121	101	81	85	89	69	49	35	27	11
20	5	21	41	61	91	77	87	109	129	149	165	150
54	134	1	37	99	63	65	95	103	133	169	36	116
114	132	142	119	39	43	45	115	111	123	28	38	56
100	40	86	168	167	147	33	31	19	26	88	130	70
72	80	108	106	140	163	25	15	34	148	24	92	98
76	120	118	164	10	156	9	16	128	44	46	122	96

Figure 7.24

possible number of cells, including those at the corners, and leave a number of empty cells divisible by 4. Thus we may ask whether a single dyad placed in the upper corner cells can directly produce the sum due. Since the initial excess is $\Delta_p = 8(p-1)(t+1)+4$, we apply rule IV and put $8(p-1)(t+1)+4 = 8j-4$. Thus, Δ_p will be eliminated at once if we take $j = (p-1)(t+1)+1 = pt+p-t$. We therefore write in the corner cells of the p-th upper row ($p \geq 1$) the dyad

$$\frac{n^2-8(pt+p-t)+3}{2}, \qquad \frac{n^2-8(pt+p-t)+7}{2},$$

62							213							64
126	78					11	189	207					80	100
128	122	94			7	35	173	183	203			96	104	98
50	124	118	110	3	31	51	165	167	179	199	112	108	102	176
52	70	120	201	75	159	155	153	83	87	79	25	106	156	174
	72	205	181	141	95	135	133	103	99	85	45	21	154	
	209	185	169	145	125	111	121	107	101	81	57	41	17	
211	187	171	163	149	129	109	113	117	97	77	63	55	39	15
	9	33	49	69	89	119	105	115	137	157	177	193	217	
		5	29	65	127	91	93	123	131	161	197	221		
		90	1	147	67	71	73	143	139	151	225	136		
		92	114	223	195	175	61	59	47	27	116	134		
		130			219	191	53	43	23			132		
	146					215	37	19					148	
162							13							164

Figure 7.25

with the smaller term in the column with the excess (see figure 7.25, numbers in Roman type). Thus we shall choose

$$\text{for } p = 1: \quad \frac{n^2-5}{2}, \quad \frac{n^2-1}{2};$$

$$\text{for } p = 2: \quad \frac{n^2-2n-7}{2}, \quad \frac{n^2-2n-3}{2};$$

$$\text{for } p = 3: \quad \frac{n^2-4n-9}{2}, \quad \frac{n^2-4n-5}{2};$$

and so on.

b) Columns

We shall again consider only the left columns, for the right columns are to be filled with the complements of the numbers in the left columns.

Since the corner cells are already occupied, the number of remaining empty cells is a multiple of 4 (for $p \geq 2$). We shall therefore eliminate the excess left in the p-th column by means of four numbers (see figure 7.25, numbers in italics). Since the initial excess before the corner cells were filled was $\Delta'_p = 8(p-1)(t+1) + 2$ and they are now occupied by

$$\frac{n^2-8(pt+p-t)+3}{2}, \quad \frac{n^2+8(pt+p-t)-5}{2},$$

with sum $(n^2+1)-2$, the remaining excess is $8(p-1)(t+1) = 8(pt+p-t-1)$.

As we must place four numbers in each column, it might be possible to start by choosing a dyad which leaves an excess of the form $8u \pm 4$ and then —by rule IV— another dyad to eliminate the remaining difference. Now we have already used the first dyad for the first border; if we use accordingly the p-th dyad for the p-th border ($p \geq 2$) and write it in the *right* column, this column will contain the smaller numbers

$$\frac{n^2-8p+3}{2}, \quad \frac{n^2-8p+7}{2}$$

and the left p-th column the corresponding p-th dyad of larger numbers

$$\frac{n^2+8p-1}{2}, \quad \frac{n^2+8p-5}{2}.$$

As a result, the previous excess in the p-th left column ($p \geq 2$) has now increased to $8(pt+p-t-1) + 8p - 4 = 8(pt+2p-t-1) - 4$. Following rule IV, we put $8(pt+2p-t-1)-4 = 8j-4$ and take accordingly $j = pt+2p-t-1$; we therefore write in the p-th left column the dyad

$$\frac{n^2-8(pt+2p-t-1)+3}{2}, \quad \frac{n^2-8(pt+2p-t-1)+7}{2}.$$

Thus, we have reached a point where for each column the cells filled in show the sum due and the number of empty cells left (for $p \geq 3$) is divisible by 4.

Remark: After placing the odd numbers in a square with order $n = 4t+3$, we are left with $t+1$ incomplete borders, numbered $p = 1, \ldots, t+1$. Of all the dyads of smaller even numbers initially at our disposal, that is,

$$\frac{n^2-8j+3}{2}, \quad \frac{n^2-8j+7}{2} \quad \text{with } j = 1, \ldots, 2t^2+3t+1,$$

the following ones have been used:

(i) $j = pt+p-t$, that is, $j = 1, t+2, 2t+3, \ldots, t^2-t-1, t^2, t^2+t+1$ (in the corner cells of the upper rows, thus for $p \geq 1$)

(ii) $j = p$, that is, $j = 2, 3, \ldots, t+1$ (in the right columns for $p \geq 2$)

(iii) $j = pt+2p-t-1$, that is, $j = t+3, 2t+5, 3t+7, \ldots, t^2+t-1, t^2+2t+1$ (in the left columns for $p \geq 2$).

Each of the dyads used corresponds to exactly one j among these $3t+1$ values, none of which is identical to another. Thus none of the dyads used can occur twice.

62	2	222	220	8	10	214	213	212	16	18	206	204	24	64
126	78	26	198	196	32	11	189	207	34	190	188	40	80	100
128	122	94	42	182	7	35	173	183	203	180	48	96	104	98
50	124	118	110	3	31	51	165	167	179	199	112	108	102	176
52	70	120	201	75	159	155	153	83	87	79	25	106	156	174
54	72	205	181	141	95	135	133	103	99	85	45	21	154	172
170	209	185	169	145	125	111	121	107	101	81	57	41	17	56
211	187	171	163	149	129	109	113	117	97	77	63	55	39	15
168	9	33	49	69	89	119	105	115	137	157	177	193	217	58
60	82	5	29	65	127	91	93	123	131	161	197	221	144	166
66	142	90	1	147	67	71	73	143	139	151	225	136	84	160
158	140	92	114	223	195	175	61	59	47	27	116	134	86	68
152	88	130	184	44	219	191	53	43	23	46	178	132	138	74
76	146	200	28	30	194	215	37	19	192	36	38	186	148	154
162	224	4	6	218	216	12	13	14	210	208	20	22	202	164

Figure 7.26

c) Completing the square

Since all the rows and columns contain the sum due for the quantity of occupied cells and the number of remaining empty cells is divisible by 4, we shall conclude by applying rule V with the remaining dyads. The result is shown in figure 7.26 (where the tetrads of available even numbers have been used first for the rows, starting with the top row, then for the columns). For the smaller orders $n = 7$ and $n = 11$ see figures 7.27 and 7.28. Both squares are constructed by al-Anṭākī.

2.5 Placing the Even Numbers for the Order $n = 5$

As al-Anṭākī mentioned, the above construction for the case $n = 4t + 1$ is invalid for $t = 1$. (The numbers to be used in the corner cells appear elsewhere in the border.) Thus the square he presents (figure 7.29) is not constructed according to these rules. Abu'l-Wafā''s simplification is of no help either. The square of order five he refers to is presumably the same as al-Anṭākī's (the only extant manuscript omits the figures altogether). Abu'l-Wafā' does, however, point out that if we put 2 in one of the upper corner cells, the other one can be

14	2	46	45	44	8	16
30	22	3	37	39	24	20
32	41	23	33	19	9	18
43	35	21	25	29	15	7
10	1	31	17	27	49	40
12	26	47	13	11	28	38
34	48	4	5	6	42	36

Figure 7.27

34	2	118	116	8	113	10	110	108	16	36
70	46	18	102	7	97	107	100	24	48	52
72	66	58	3	23	89	91	103	60	56	50
26	68	105	43	83	81	51	47	17	54	96
28	109	93	73	59	69	55	49	29	13	94
111	95	87	77	57	61	65	45	35	27	11
30	5	21	37	67	53	63	85	101	117	92
90	42	1	75	39	41	71	79	121	80	32
84	44	62	119	99	33	31	19	64	78	38
40	74	104	20	115	25	15	22	98	76	82
86	120	4	6	114	9	112	12	14	106	88

Figure 7.28

occupied by 4, 6, 8 or 12. There are in fact twenty-one possibilities, which are listed below (see figure 7.30).

• Take $\alpha = 2$ and

$\beta = 4$ with 14, 20 in the upper row and 8, 10 in the left column, or with 16, 18 in the upper row and 6, 12 in the left column (as in figure 7.29);

$\beta = 6$ with 10, 22 and 8, 12 or with 14, 18 and 4, 16;

$\beta = 14$ with 4, 20 and 10, 18 or with 8, 16 and 6, 22;

$\beta = 18$ with 4, 16 and 12, 20 or with 6, 14 and 10, 22;

$\beta = 20$ with 4, 14 and 16, 18 or with 8, 10 and 12, 22.[8]

• Take $\alpha = 4$ and

$\beta = 6$ with 12, 18 (upper row) and 2, 16 (left column).

2	16	25	18	4
6	11	21	7	20
23	9	13	17	3
12	19	5	15	14
22	10	1	8	24

Figure 7.29

α		25		β
	11	21	7	
23	9	13	17	3
	19	5	15	
$26-\beta$		1		$26-\alpha$

Figure 7.30

- Take $\alpha = 6$ and
  - $\beta = 10$ with 2, 22 and 8, 12;
  - $\beta = 22$ with 2, 10 and 14, 18.
- Take $\alpha = 8$ and
  - $\beta = 2$ with 14, 16 and 4, 6;
  - $\beta = 10$ with 2, 20 and 4, 14;
  - $\beta = 16$ with 2, 14 and 4, 20.
- Take $\alpha = 12$ and
  - $\beta = 4$ with 8, 16 and 2, 6.
- Take $\alpha = 14$ and
  - $\beta = 8$ with 2, 16 and 4, 6;
  - $\beta = 18$ with 2, 6 and 4, 16.
- Take $\alpha = 16$ and
  - $\beta = 8$ with 4, 12 and 2, 6;
  - $\beta = 18$ with 2, 4 and 6, 12.

§3 Description of the Construction by al-Anṭākī

Al-Anṭākī's description of the construction leading to the *quadratus mirabilis* is very concise, but sufficient to confirm our mathematical reconstruction. We think it in any case appropriate to add a translation of the relevant text. This is the second part of the section in which al-Anṭākī deals with squares of odd orders. (In the first part, he explains the construction of bordered squares summarized in §1.2.)

> Consider the odd numbers from 1 to the last of those which will be in the square. Arrange them inside the square so as to form the shape of a rhombus within the large square, thus leaving (empty) cells forming triangles with a same number of cells on each side. Write there the even numbers from 2 to the last to be found in the square in such a way that the sums be equal everywhere. Then the odd numbers will be inside the larger square in a rhomboid figure and the even numbers will surround them on the four sides, as is shown in this figure (figure 7.31).

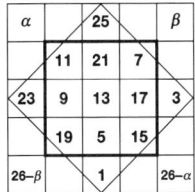

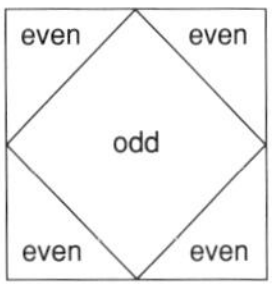

Figure 7.31

4	3	8
9	5	1
2	7	6

Figure 7.32

The way to place the odd numbers is the following.[9] Take 1 and the last term belonging to this square, namely its largest number, then 3 and the number preceding the largest one, and so forth until you reach its middle term. You put this middle term in the central cell of the square. You put its adjacent odd terms in the cells where you placed the two of the nine terms which were adjacent to the middle term in the square of 3 by 3 (figure 7.32). You do the same for the remaining numbers until the square of 3 by 3 is completed, if this is the object of your treatment, or you do the same for the squares of 5 by 5, 7 by 7, 9 by 9, if this is the object of your treatment. You will always do this until the whole square contained by the rhombus is completed.

This being done,[10] take the two odd terms reached, and put the smaller one in the middle cell of the first left-hand line and its complement facing it in the first right-hand line, the next small number in the middle cell of the lower line, and facing it in the upper line its complement. Take then the two terms reached and put the smaller one below, next to the middle cell on its left, and, facing it above, its complementary term. Put then the next small term on the left, just above the middle cell, and facing it on the right its complement. Then put the next small number in the middle cell of the second left-hand line, and facing it in the second right-hand line its complement. Then put: the small number following this term in the middle cell of the second lower line, and facing it its complement; then the following small term next to the middle cell in the first upper line, and facing it in the first lower line its complement; the following small number in the second right-hand line, below the middle cell, and facing it on the left its complement. When this is done for the two squares (of order 7 and 9), 1 and the last term are reached and the placing has been performed in the desired way. In the case of the square of 9, the treatment is then completed.

In the case of the squares of 11 and 13,[11] do the same until the above arrangement is attained. Then you write the small number following the small number placed lastly in the third cell from the middle cell,[12] on the left of it, in the first lower line, and facing it its complement. Then you put the following small number in the third cell above the middle cell in the first left-hand line, and its complement facing it on the right. Then you put the subsequent small term in the second lower line, next to the middle cell, and its complement facing it above. Then you put the subsequent small term next to the middle cell in the second left-hand line, and its complement facing it on the right. Then you put the small term reached in the middle cell of the third left-hand line, and facing it its complement. Then you put the subsequent small term in the middle cell of the third lower line, and facing it above its complement. Then you put the subsequent small term next to the middle cell of the second upper line, and facing it below its complement. Then you put the subsequent small number in the second right-hand line, next to the middle cell, and facing it on the left its complement. Then you put the subsequent small number in the first upper line, in the third cell from the middle cell, and facing it below its complement. Then you put the subsequent small number in the third cell from the middle cell in the first right-hand line, and facing it on the left its complement. When this is done, the odd numbers are placed in these two squares, starting from the center.

In the case of the squares of 15 and 17, you do the same as for the square of 13 until you reach the third border. Then you put the subsequent small number in the fourth cell from the middle one in the first lower line, and facing it its complement; then the subsequent number in the fourth cell from the middle cell in the first left-hand line, and facing it its complement; then the subsequent number in the third cell from the middle cell in the second lower line, and facing it its complement; then the subsequent number in the third cell from the middle cell in the second left-hand line, and facing it its complement; then you put the subsequent number next to the middle cell in the third lower line, and facing it its complement; then the subsequent number next to the middle cell in the third left-hand line, and facing it its complement; then the subsequent number in the middle cell of the fourth left-hand line, and facing it its complement; then the subsequent number in the middle cell in the fourth lower line, and facing it its complement; then the subsequent number next to the middle cell in the third upper line, and facing it its complement; then the subsequent number next to the middle cell in the third right-hand line, and facing it its complement; then the subsequent number in the third cell from the middle cell in the second upper line, and facing it its complement; then the subsequent number in the third cell from the middle cell in the second right-hand line, and facing it its complement; then the subsequent number in the fourth cell from the middle cell in the first upper line, and facing it its complement; then the subsequent number in the fourth cell from the middle cell in the first right-hand line, and facing it its complement. When this is done, you have finished with the odd numbers for these two squares in the desired way. If you wish to proceed with larger squares, continue placing step by step in the same way.

At this point we find that the squares are divided into classes requiring each its own treatment for the arrangement of the even numbers in the corners. There are those of (order) 5, 9, 13, 17, and so forth by steps of 4; and those of (order) 7, 11, 15, 19, and so forth by steps of 4.

For constructing that of 5 and those of the same kind (observe this).[13] The remaining empty cells in the first border, which surrounds the inner square, are the four at the corners and eight adjacent to them, two on each side; in the second border, the empty cells are the four at the corners and twenty-four cells adjacent to them. It will always be like that: each border has 16 angular cells more than the border below.

For the square of 5 only, the treatment fails: it requires a different method. For the squares belonging to the same class, the situation is the following.

Square of 9.[14] After placing the odd numbers as indicated, the first upper line is in excess over the sum due by 20, the lower line is in deficit by 20; the first right-hand line is in excess over the sum due by 18, and the left-hand line is in deficit by 18. The second upper line is in excess by 36, and the lower line is in deficit by the same amount; the second right-hand line is in excess by 34, and the left-hand line is in deficit by the same amount.

Square of 13. The first upper line is in excess by 28, and the lower line is in deficit by the same amount; the first right-hand line is in excess by 26, and the left-hand line is in deficit by the same amount. The second upper line is in excess by 52, and the lower line is in deficit by the same amount; the second right-hand line is in excess by 50, and the left-hand line is in deficit by the same amount. The third upper line is in excess by 76, and the lower line is in deficit by the same amount; the third right-hand line is in excess by 74, and the left-hand line is in deficit by the same amount. Similarly for the others.

In the square of 9, a second line has 16 more than the line before. In the square of 13, a first line has 8 more than the first of the square of 9, then each line has 24 more than the line before. In the square of 17, a first line has 8 more than the first of the square of 13, then each line 32 more than the line before. Each line will always have more than the line before in the same manner.

"Excess over the sum due" means the following. The required amount for the central cell equals the number in the middle. Thus each cell of the square of 5 has a sum due of 13. Therefore you will add the odd numbers in each line and divide the result by the number of cells filled; if the quotient is less than the middle term, the sum in the cells will be less than the sum due by the product of this deficit and the number of cells; analogously if the quotient is in excess. So the subsequent placing of the even numbers in the empty cells of each line must be such that it compensates the deficit, if any, or falls short by the amount of the excess, if any. We shall show this in the appropriate place, when explaining how to deal with the even numbers.

For constructing the square of 7 and the others of this class (observe this).[15] The remaining empty cells are as follows. In the first border, which surrounds the inner square, there are the four cells at the corners; in the second border, there are the cells at the corners and 16 cells adjacent to them, 4 on each side; in the third border, there are the cells at the corners and 32 cells adjacent to them. And so forth always: each border has 16 angular cells more than the border below.

In this class of squares, the first upper line has an excess of 4 over the sum due, and the lower line has a deficit of the same amount;[16] the first right-hand line has an excess of 2 over the sum due, and the left-hand line has a deficit of the same amount.

The second upper line of the square of 7 has an excess of 20 over the sum due, and the second right-hand line an excess of 18.

In the square of 11, the second upper line has an excess of 28 over the sum due, the second right-hand line an excess of 26, the third upper line an excess of 52, the third right-hand line an excess of 50, and all the opposite lines have a deficit equal to the excess.

Square of 15. The second upper line exceeds the first upper line by 32, the third the second by 32.

And so forth for the others: the excess increases each time by 8.[17]

Thus, writing the even numbers in the lines must be done so as to equalize them, and we must then search for numbers the sum of which will produce the required excess or the required deficit.

To determine this, you associate 2 and the last even term, then 4 and the corresponding opposite term, and so forth until you reach the two middle terms of these even numbers.[18] This being done, you see that if the last small number and the preceding one are placed on one side, and facing them their complements, the side containing the two small numbers will be less than the sum due by 4, and the other more by 4; placing the next two numbers, the differences will be 12 and 12, with the next two numbers 20 and 20, then 28, 36, 44, 52, 60 and so forth to the numbers 4 and 2.[19] Now you see that the borders which require the placing of even numbers have indeed, on two of their sides, this succession of excesses and deficits, namely 4, [12,] 20, 28, 36 and so forth. Thus, bringing the excess of these even numbers to where the deficit is will equalize the borders on two sides, and the other two sides will need to be equalized with the remaining even numbers (by means of what follows).

Taking four (consecutive) small numbers, the sum of the first and the second has a certain deficit, and the sum of the third and the fourth a deficit smaller by 8. Then adding the first and the fourth you find that they have a deficit equal to half the sum of the two deficits; adding similarly the second and the third gives the same result. For instance, the sum of the last two small numbers[20] is less than their sum due by 4, and the sum of the two preceding numbers less by 12; thus, adding the last and the first, and the second and the number before, will produce a deficit of 8, that is, half the sum of 12 and 4. The knowledge of this is necessary, for you will use it constantly.[21]

When the first small number, or any arbitrary small number, is written on one side and the subsequent number is written on the other side, and the two large numbers which are their complements are written on the opposite, the side where the first small number is written will be less than the sum due by 2, whereas the other side will be in excess by 2.[22] If some small number is written on one side and on the other side the third small number counted from this one is written, the side containing the first small number will be less than the sum due by 4. And so forth: whenever the distance between them is increased by one number the deficit is increased by 2.[23] For instance, putting 4 on one side and 6 on the other side, and facing them their complements, the side containing 4 will be less than the sum due by 2 and the side containing 6 will be more by 2. Writing 8 instead of 6 and doing the same, the side containing 4 will be less by 4 and the side of 8 more by 4, and the amount will increase together with the interval between the two numbers.

Placing four numbers in the corners, two in consecutive corners and their complements in the corners diagonally opposite, if the two consecutive small numbers are on the upper side this side will be less than the sum due by 4 if they are the last two, by 12 if they are the two previous numbers, and so forth with a regular increment of 8 until 2 and 4 are reached. The right-hand sides will have a uniform difference, excess or deficit, of 2 from the sum due, without any increment or diminution.[24]

All this must be understood: it will be necessary for the writing of the even numbers in this class of squares.

Examples of treatments for all that we have explained.

Treatment for the square of 5 by 5 (figure 7.33).[25] You put the odd numbers in the inner square of 3 as explained. Those remaining are 1, 25, 3, 23. You put 1 in the lower middle cell and 25 facing it above, 3 in the middle left-hand cell and 23 on the opposite side, on the right. Next, you put 2 in the upper right-hand corner and facing it diagonally, in the lower left-hand corner, its complement, namely 24. You put 4 in the upper left-hand corner and facing it diagonally, in the lower right-hand corner, its complement, namely the even number 22. You put 6 on the right side and its complement, 20, facing it on the left. You put 10 and 8 below, 12 on the right, and you put facing each its complement.

Passing from 5 to 9, 13, 17 and those of this kind, put the last small even term in the upper left-hand corner of the first border, that is, the border following the inner square filled with odd numbers, and facing it diagonally, in the lower right-hand corner of the first border, its complement.[26] Put the preceding small term in the upper right-hand corner of the first border, and facing it diagonally, in the lower left-hand corner, the term which is its complement. When this is done, you find that the excess of the upper line is 16,[27] the excess of the right-hand line 16, thus the excesses of the upper line

4	18	25	16	2
20	11	9	19	6
3	21	13	5	23
14	7	17	15	12
24	8	1	10	22

Figure 7.33

and the right-hand line are the same. Put then 2 in the upper line; consider the excess of the upper line, take its half, and count after 2 as many small even numbers as this half, and put the number reached in the lower line. Put facing each of these two numbers its complementary term. Put 4 on the right side;[28] then count after 4 as many small numbers as half of the excess, and put the number reached in the left-hand line. Put facing each of these two numbers its complementary term. After doing this all the sides of the first border will be equalized for this kind of square.

Put then the two small terms the sum of which is less than their sum due by 12 in the right-hand corners of the second border, with the lesser one above;[29] put in the diagonally opposite corners, on the left, their complements. Then look for the pair of small numbers such that their sum is less than their sum due by an amount equal to the (initial) excess of the second right-hand line less 14; put them in the second right-hand line and, facing them on the left, their complements. Then look for the pair of small numbers such that their sum is less than their sum due by an amount equal to the (initial) excess of the upper line less 8; put them in this line and, facing them below, their complements.

Put then the pair of small numbers the sum of which is less than their sum due by 20 in the right-hand corners of the third border, with the lesser one in the upper corner; put facing them diagonally, in the corners of the third left-hand line, their complements. Then look for the pair of small numbers which have a sum less than their sum due by an amount equal to the (initial) excess of the third right-hand line less 22; put them in this line, and their complements facing them, in the third left-hand line. Then look for the pair of small numbers such that their sum is less than their sum due by an amount equal to the (initial) excess of the third upper line less 8; you put them in this line and, facing them in the third lower line, their complements.

Put then the pair of small numbers such that their sum is less than their sum due by 28 in the right-hand corners of the fourth border, with the lesser one above, and facing them diagonally on the left their complements. Then look for the pair of small numbers such that their sum is less than their sum due by an amount equal to the (initial) excess of the fourth right-hand line less 30; put them in this line and, facing them in the fourth left-hand line,

46	66	14	72	77	6	28	26	34
52	40	64	71	69	3	2	38	30
50	78	27	31	61	63	23	4	32
74	9	29	39	37	47	53	73	8
7	15	25	49	41	33	57	67	75
12	81	65	35	45	43	17	1	70
22	20	59	51	21	19	55	62	60
58	44	18	11	13	79	80	42	24
48	16	68	10	5	76	54	56	36

Figure 7.34

their complements. Then look for the pair of small numbers such that their sum is less than their sum due by an amount equal to the (initial) excess of the fourth upper line less 8; put them in this line and, facing them below, their complements.

Proceed always likewise, with increments of 8 for the deficits in a right-hand line and uniform deficits of 8 in an upper line. When this has been done, the excess of each upper line over its sum due will be 6, the excess of each right-hand line over its sum due will be 2, and the number of remaining empty cells will be four cells in each second line, eight in each third line, twelve in each fourth line, and so forth by successive additions of four.

Let us now turn our attention to the remaining numbers.[30] You take a tetrad of small numbers and put the first number in an upper line, the third in the opposite line below, the second on the right and the fourth on the left, and you put facing each of these four numbers its complement. This placing is performed for each border. When this is done, you take a pair of small numbers; put the first above, the second below, and facing them their complements. Then take another pair; put the first on the left side of the same border, the second on the right, and facing them their complements. Do the same for all borders. When this is done, each line and its opposite will be equalized and none will be in excess.

If there are remaining empty cells, it can be only four facing four, eight facing eight, twelve facing twelve (and so forth); they will be equalized by groups of four with available sequences of four numbers, in the way explained at the beginning of the section for four numbers of which each pair is in progression: you put the first of the first pair and the second of the second pair on one side, the second of the first pair and the first of the second pair on the facing side, and opposite to each its complement (figure 7.34).

Treatment for 7, 11, 15, 19 and the like.

You put the last of the small even terms in the still empty upper left-hand corner of the first border, which surrounds the inner square filled with odd numbers, and facing it diagonally, in the lower right-hand corner of the same border, its complement. Put the preceding small term in the upper right-hand corner of the same border, and, facing it diagonally, in the lower left-hand corner, its complement. When this is done the first border is equalized for all squares of this kind.[31] After that, consider the excess of each upper line and look for the pair of small numbers such that their sum is less than their sum due by the same amount. Put them in the corners of this line, the lesser number on the right, and facing them diagonally below their complements. Complete in this way all the remaining corners. When this is done all the upper and lower lines will be equalized, and their remaining empty cells will be four facing four, and so forth by additions of 4. Each group of four is then equalized with four numbers in the way we have explained previously.

We are left with the right-hand lines, which exceed the sum due by 16, 24, 32, 40, 48, and so forth by increments of 8, and the remaining empty cells are four facing four, eight facing eight, and so forth by additions of 4.[32] You then look for a pair of large numbers such that their sum exceeds their sum due by an amount which, when added to the excess of the right-hand line, equals the deficit of the sum of two small numbers; put then the two small numbers on the same side (as the large numbers), and put on the left the complements of the four numbers, each pair facing its complements. When this is done, each pair of corresponding sides will be equalized, and the remaining empty cells will be four facing four, and so forth by successive additions of 4. Each group of four is then equalized with four numbers as we have explained previously.

Example of the treatment of the right-hand line for the square of 7.

You find that the excess of the upper line is 20 and the excess of the right-hand line, 18. Putting the two numbers having a sum less than their sum due by 20 in the two upper corners, with the lesser one on the right, leaves 16 as the excess in the right-hand line. Consider then the two large numbers with an excess of 12; for adding 12 to 16, which is the right-hand excess, gives an excess of 28, equal to the deficit of the sum of two small numbers. You put then the two large numbers on the right, as also the two small numbers having a sum less than their sum due by 28, and you put on the left, facing each one, its complement (figure 7.35). The treatment is the same for the other squares (of this kind) (figure 7.36).

16	8	44	45	46	2	14
18	24	39	37	3	22	32
20	9	23	21	31	41	30
7	15	33	25	17	35	43
40	49	19	29	27	1	10
38	28	11	13	47	26	12
36	42	6	5	4	48	34

Figure 7.35

36	16	108	110	10	113	8	116	118	2	34
50	48	24	100	107	97	7	102	18	46	72
52	56	60	103	91	89	23	3	58	66	70
96	54	17	47	51	81	83	43	105	68	26
94	13	29	49	59	57	67	73	93	109	28
11	27	35	45	69	61	53	77	87	95	111
92	117	101	85	55	65	63	37	21	5	30
32	80	121	79	71	41	39	75	1	42	90
38	78	64	19	31	33	99	119	62	44	84
82	76	98	22	15	25	115	20	104	74	40
88	106	14	12	112	9	114	6	4	120	86

Figure 7.36

NOTES

1. For the first text, see J. Sesiano, "Le traité d'Abu 'l-Wafā' sur les carrés magiques," *Zeitschrift für Geschichte der arabisch-islamischen Wissenschaften,* 12 (1998), pp.121–244; an edition of the second treatise, in MS (Ankara) Saib 5311, fol. 1^r–36^r will be published in the same journal.

2. Although methods for simple magic squares are easier to apply than methods for bordered ones, the latter are easier to discover; this explains why they appeared first.

3. See J. Sesiano, *Un traité médiéval sur les carrés magiques,* Lausanne 1996, pp. 35–40.

4. Ibid., p. 68.

5. Here, and whenever the original (Arabic) order is from right to left, we have chosen to reverse it; the rules for construction have been changed accordingly.

6. Throughout this study, we call "opposite cells" those at either end of any border row, column or main diagonal of a bordered magic square. Such pairs of cells will always contain a "smaller" number α ($\alpha < \frac{n^2+1}{2}$) and its "complement" $n^2 + 1 - \alpha$.

7. The need to reserve opposite cells in the main square for complements makes it impossible to proceed here in exactly the same manner as for a bordered square.

8. Thus, Abu 'l-Wafā' has indeed covered all possibilities for two consecutive corner cells when one of these numbers is taken to be 2, as can be seen by considering the complements of the last three values of β.

9. This explains the arrangement in the inner square. See §1.3, *in fine.*

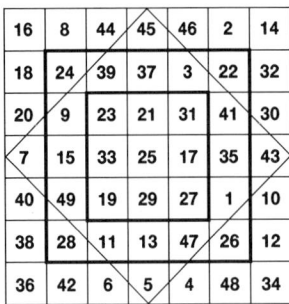

10. Here begins the arrangement of the remaining odd numbers in the rhombus. The explanation can be followed on figures 7.11–7.12 (provided the terms "left" and "right" are exchanged), starting with 63.

11. Squares are taken by pairs since the arrangement for orders $4t \pm 1$ is the same. See §2.1.

12. In the "third cell from the middle," the middle cell is also counted.

13. See §1.3 ($n = 4t + 1$).

14. See §2.2.*a* (Γ_p, Γ'_p).

15. See §1.3 ($n = 4t + 3$).

16. See §2.2.*a* (Δ_p, Δ'_p).

17. That is: the excess of one row or column over the row or column underneath is constant for the same order, but increases by 8 for each order.

18. See §2.2.*b* (pairs of complements and dyads).

19. See §2.2.*b*, Rule IV, first part.

20. That is, $\frac{n^2-5}{2}$ and $\frac{n^2-1}{2}$.

21. See Rule V.

22. See Rule II.

23. See Rule I.

24. See Rule IV, second part.

25. See §2.5.

26. See §2.3.*a* ($p = 1$).

27. Case $n = 9$ only; the excess increases by 8 for the other orders.

28. See §2.3.*b* ($p = 1$).

29. See §2.3.*a-b* ($p \geq 2$).

30. See §2.3.*c* (and Rule III).

31. See §2.4.*a* & *c*.

32. See §2.4.*a* & *c*. The values of the excesses are 16, 24, 32, 40, . . . if $p = 2$; 48, 64, 80, . . . if $p = 3$; and so forth.

8

Calculating Surface Areas and Volumes in Islamic Architecture

Yvonne Dold-Samplonius

Introduction

As long as man has been constructing his dwellings, he has wanted to know how many bricks are needed or how much earth has to be removed for the foundations of his house. Problems of this kind are already found in the oldest texts on arithmetic, such as the Chinese *Nine Chapters on Arithmetical Techniques*. In chapter V,[1] entitled *Shang kung* (≈ Evaluation of Work), there are calculations of volumes of regular solids, and often also of the number of people required for excavating and transport. The same kind of problems are treated in Arabic manuals, such as al-Karajī's *Sufficient Arithmetic* [*Kāfī fī 'l-ḥisāb*]. This work explains how to determine the amount of sun-dried or fired bricks needed for a building or how to level the ground. In this paper I do not want to talk about these general construction problems but rather about calculations in connection with other specific elements of Islamic architecture.

Islamic Architecture

What does "Islamic architecture" mean? As we can define "Christian architecture" mainly by churches, chapels, monasteries, etc., so we can define "Islamic architecture" by the principal forms of Islamic buildings, which are not only, but mainly religious. The oldest of these buildings are mosques which date from the first beginnings of Islam. According to Creswell's account,[2] Arabia, at the rise of Islam, did not possess anything worthy of the name of architecture. Only a small proportion of the population was settled, and these lived in dwellings which were scarcely more than hovels. When Muḥammad migrated to Medina he built a house for himself and his family in AD 623. Like many mud-brick houses in the Middle East, Muḥammad's house consisted of a square courtyard with two rooms (later increased to nine to accommodate his wives) on the south-east side. The first communal prayers were held in this courtyard. For the comfort of the worshippers a portico made of palm-trunks and branches was built on the north side of the courtyard, together with a smaller one which gave shelter to visitors who sometimes spent the night there. The portico also served as a place for deliberations on community affairs; hence to this day the mosque has retained its multivalent role as a place of prayer, social activities and political debate. Practical needs thus contributed to the house of the Prophet becoming the first mosque of Islam.

The men who formed the Arab armies of conquest were mainly Bedouin, but even those who came from permanent settlements, such as Mecca and Medina, knew nothing of art or architecture. In these early days, the Muslims, when they conquered a town in Syria, usually took one of the churches and used it as a mosque, or merely shared one of the churches if the town had surrendered without resistance. At Jerusalem they made use of the remains of the basilical hall of Herod, which ran along the south side of the Temple enclosure, but had been destroyed by the army of Titus. In Persia, at Persepolis and Qazwīn, they appear to have taken hypostyle audience-halls of the Persian kings, which had flat roofs resting on columns with double bull-headed capitals. But the situation was different in Iraq, for there the Arabs founded new towns, so pre-existing buildings could not be employed, and they had to construct some sort of place for themselves. What kind of buildings were the first mosques of the earliest towns in Islam?

At Baṣra, founded about 635, the first mosque was simply an area marked out on the ground and the people prayed there without any building. At Kufa, founded in 638, the first mosque was equally primitive. The first mosques to be worthy of the name of architecture were the second Great Mosques at Basra (665) and Kufa (670). According to Creswell, it is apparent that the roofing system resembled that of a hall of columns of the Achaemenian kings, exactly as was the case in the first Great Mosque at Baghdad. The oldest extant monument of Muslim architecture is the Dome of the Rock at Jerusalem, completed in 691, and restored and embellished during the following centuries. The harmony of its proportions and the richness of its decorations make the Dome of the Rock one of the most beautiful buildings in the world.

Style and methods of construction changed from generation to generation, especially in respect to the material, the gates, the façades and minarets, the profile of the arches in the interior, and the ornamentation. But the ground-plan of the mosque remained largely the same. In figure 8.1 two basic categories of mosque design in distinctive regional style are shown:[3] on the left the hypostyle hall and open courtyard, as found in the Arabian heartland (Syria, Iraq, Saudi Arabia, Egypt, and Yemen), Spain and North Africa, and on the right, the bi-axial four-iwan type, as found in Iran and Central Asia, with four *iwans*. Specific elements are the minarets, cupolas, and arches.

Both for the role they play in respect of the mosque and in their own right, there are three fields in which Islam has made a unique contribution to archi-

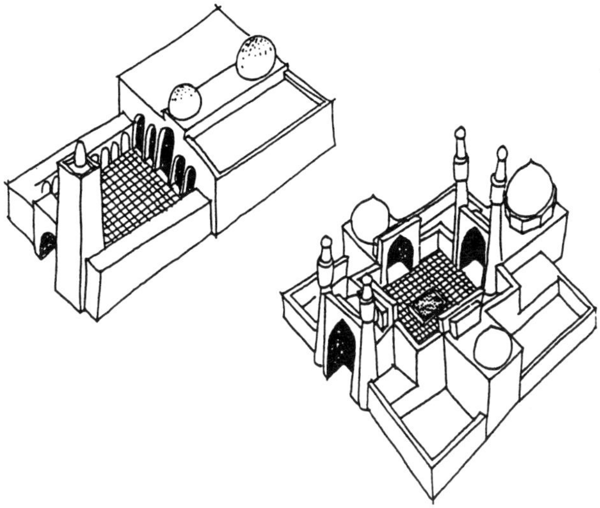

Figure 8.1
Two categories of mosques, as found in the Arabian heartland (left) and in Iran and Central Asia (right) (Frishman).

tecture and the architectural arts: calligraphy, garden design, and geometry. The use of geometry in decoration is ubiquitous and serves to cover flat, curved and convoluted surfaces in two- or three-dimensional forms. Thus it can enrich and beautify an interior through the use of uniquely Islamic devices such as the *muqarnas* (stalactite vault). The *muqarnas* was an invention of the Islamic world. Although it can be of structural value, associated with the transitional zone of a dome, more often it is a purely decorative feature connected to, or suspended from, a structural member.

Thus the specific elements characterizing Islamic architecture are the *muqarnas*, and the *qubba* (cupola), as well as *arches* and *vaults*. These we want to calculate. Another specific aspect of Islamic architecture, *ornament*, as found in mosaics, tilings, geometrical writing, will not be treated in this paper. In earlier papers I explained the calculation of *muqarnas* and *qubba*, hence in the present chapter the emphasis is on the calculation of arches and vaults. These calculations were *not* the basis for constructions but an appraisal of the necessary manpower and building materials.

The Calculators

Much is still unknown about the tradition of practical mathematics in Islamic civilization. In 1992 Rebstock published an important survey on calculation in the Islamic Orient, based on a study of more than 100 Arabic and Persian arithmetic texts, most of which remain to be edited. This study has considerably improved our insight into the practical use of mathematics in the medieval Islamic tradition.

Under the reign of Caliph al-Maʾmūn (813–833) the mathematician and astronomer Abū Jaʿfar Muḥammad b. Mūsā al-Khwārizmī[4] wrote in Baghdad his well-known *Algebra*, a work of elementary practical mathematics. Its purpose is, according to the author, to provide "what is easiest and most useful in arithmetic, such as men constantly require in cases of inheritance, legacies, partition, lawsuits, and trade, and in all their dealings with one another, or where the measuring of lands, the digging of canals, geometrical computations, and other objects of various sorts and kinds are concerned."[5] Only the small second part of al-Khwārizmī's treatise concerns practical mensuration.[6] He gives rules for finding the areas of various plane figures, including the circle, and for finding the volume of a number of solids, including the cone, the pillar with a circular base, the pyramid, and the truncated pyramid. This section is really concerned with the practical application of mensuration, as the first few lines already demonstrate: "Know that the meaning of the expression *one by one* is mensuration: one cubit (in length) by one cubit (in width) understood."

The earliest manual on practical arithmetic is a textbook by Abu 'l-Wafāʾ al-Būzjānī,[7] written between 961 and 976. This *Book on Settling What Is Necessary from the Science of Arithmetic for Secretaries and Businessmen* [*K. al-manazil fī-mā yaḥtāju ilayhi al-kuttāb wa al-ʿummāl min ʿilm al-ḥisāb*][8] consists of seven parts. The first three are purely mathematical: on ratio, multiplication and division, and mensuration. The other four contain the solutions of practical problems: on taxes, exchange and shares related to the harvest, problems concerning payment for work, and construction estimates.[9] The other manual by Abu 'l-Wafāʾ, *The Book on What the Artisan Needs to Know of Geometric Constructions* [*K. fī-mā yaḥtāju ilayhi al-ṣāniʿ min aʿmāl al-handasa*] seems to have been compiled from class notes by one of his pupils. This practical geometry outlines basic mechanical methods for constructing, proportionally subdividing, and symmetrically extending geometric figures, further simplified by the use of only a single opening of the compass.[10]

Interesting problems on mensuration are found in Abū Bakr Muḥammad al-Karajī's[11] *Sufficient Arithmetic* [*Kāfī fī 'l-ḥisāb*], written between 1010 and 1016. The composition of this treatise was part of his duties as a mathematician holding an official position: to write a simple textbook in a way accessible to civil servants. In the introduction al-Karajī states that "his work presents what people of different classes need for their various activities." Al-Karajī's aim was to write a practical manual, like al-Khwārizmī and Abu 'l-Wafāʾ before him, but he does not deal with the important practical problem of taxes and heritage. The treatise contains the elements of arithmetic with integers and fractions (common and sexagesimal), the extraction of square roots, the determination of areas and volumes, and elementary algebra. Chapters 44 to 49 deal with measuring plane figures. Chapter 50 is on the mensuration of solids, in chapter 52 the amount of sun-dried or baked bricks needed for a construction is determined, and chapter 53 describes three instruments for levelling the ground.[12] Al-Karajī explains the same three instruments in more detail in his treatise *Locating Hidden Waters* [*K. inbāṭ al-miyāh al-khafiyya*].[13]

Two recently studied texts are the *Book of the Levels in the Explanation of the Measurements* [*K. al-ṭabaqāt fī sharḥ al-misāḥāt*] by Qāḍī Abū Bakr,[14] written before the end of the twelfth century, and the *Calculators' Riches* [*Ghunyat al-ḥussāb*] by Aḥmad b. Thabāt who died 1234.[15] Their calculations of domes concern hemispheric domes and are not faultless; in the case of vaults their calculations are simple and correct.

Ghiyāth al-Dīn Jamshīd Masʿūd al-Kāshī ranks among the greatest mathematicians and astronomers in the Islamic world. He was a master calculator/ mathematician of extraordinary ability, he applied iterative functions widely, he laid out his calculations in such a careful way that errors could not creep in undetected, and he did a running check at every stage. In short, his talent to optimize a problem let him appear as the first modern mathematician. When Ulūgh Beg decided to construct the Samarkand observatory, he invited al-Kāshī to his court—some time after 1416. Al-Kāshī died in June 1429 outside the observatory, probably murdered on the command of Ulūgh Beg.[16] Two years earlier he had finished the *Key of Arithmetic* [*Miftāḥ al-ḥisāb*], one of his major works. The work is intended for everyday use, as is clear from al-Kāshī's remark: "I redacted this book and collected in it all that is needed for the one who calculates carefully, avoiding tedious length and annoying brevity." By far the most extensive part is Book IV, *On Measurements*.

The main aim of the last chapter, *Measuring Structures and Buildings*, is practical: "The specialists merely spoke about this way of measuring for the arch and the vault, and did not think that anything else was necessary. But I present this application among the necessities together with the rest, because it is more often required in measuring buildings than in the rest." Al-Kāshī uses geometry as a tool for his calculations, *not* for constructions. Besides arches/ vaults and *qubbas*, al-Kāshī calculates here the surface area of a *muqarnas*, that is to say, he establishes approximate values for such a surface. He is able to do so because a *muqarnas*, although it is a complex architectural structure, is based on relatively simple geometrical elements. Only elementary geometrical rules are used in the calculation.

To summarize: In the extant Arabic literature we find several calculations of arches and vaults as well as treatises on the measurement of the *qubba*. The most accomplished explanations and calculations, however, are those of al-Kāshī, who also considers more elaborate forms of arches and domes than the others. His skill in finding practical approximations is equally shown in measuring the *muqarnas*. No other treatise on the calculation of a *muqarnas* surface has yet been found.

CALCULATING ARCHES AND VAULTS

From about 200 BC to AD 1100 the main structural components of all large buildings in the Roman and Byzantine Empires were the semicircular arch (the barrel vault is an elongated arch and the dome is a rotated arch), columns and walls.[17] From the Byzantine Empire the Umayyads inherited the system of round arcading which rarely showed a tendency towards becoming slightly pointed.[18] The innovation of the pointed, or ogival, arch came from the East.

The ogival arch is first found in Buddhist India in the second century AD and had reached Syria, possibly by way of Sassanid Iran, by AD 561. A number of such arches appear in Syrian buildings of the eighth century and it became common in Egypt in the ninth century. A beautiful early example is the mosque of Ibn Ṭūlūn in Cairo, built in 876–879. Under Umayyad rule the round arch persisted, but developed into the two-centered form showing an increasing tendency towards pointedness. A round arch is struck from a single center; a pointed arch has more than one center and can be thought of in its simplest form as being struck from two centers with overlapping arcs which produce an increasingly pointed arch as they are moved apart horizontally (figure 8.2). In the great mosque at Damascus, built in 715, the arches are very slightly pointed with the two centers being only one-eleventh of the span apart.

In the succeeding two centuries the same trend is still apparent, but is complicated by the three- and four-centered arch. Figure 8.4 shows a three-centered arch, with point E as a (double) center and the other two centers situated in the two lower points Z and H. These two points are obtained by intersecting the extensions of the radii through E with the perpendiculars dropped from the extremities of the span of the arch. When the two lower centers move, the arch will change its acuteness. The four-centered arch is similar to the three-centered one except that the (double) center splits into two points, which are displaced from the center of the first circle toward the extremes. The smaller the displacement of the two centers, the closer the profile will be to the profile of the three centered arch. The greater the displacement, the shallower the profile.

Since ogival arches are the exception in the first centuries of Islamic architecture, only the calculation of semicircular arches is found in the early arithmetic books. This practice continued well into al-Kāshī's time, probably because most manuals were based on earlier manuals. Al-Kāshī remarks that,

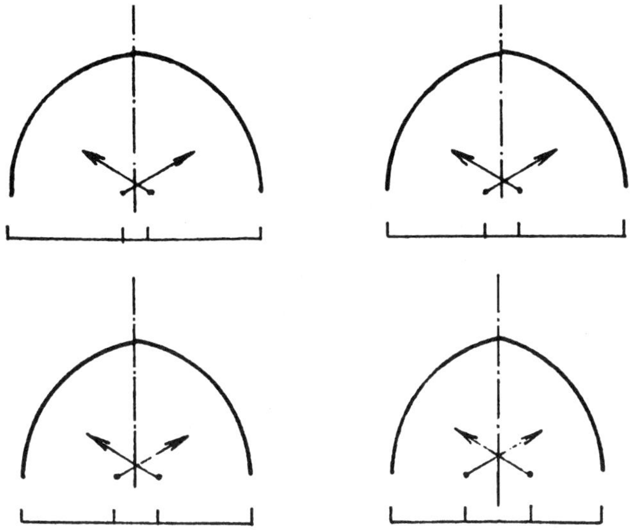

Figure 8.2
Diagram showing pointed arches formed with constant radii on centers with successive separation of one tenth, one seventh, one fifth, and one third of a span (Warren).

"Our predecessors determined those (i.e., arch and vault) as half a circular hollow cylinder, but we did not see anything like it, either in old or in new buildings. We have mostly seen ones that are pointed in the middle, and in a few cases they are smaller than half a hollow cylinder."

Among these predecessors we find Aḥmad b. Thabāt (d. 1234), who calculates arches as follows[19] in figure 8.3; all given and calculated quantities are written out in words [MS Ayasofya 2728, fol. 117b]:

The authors of these computations use "arch" and "vault" interchangeably. The difference between an arch and a vault is that the depth of an arch is not greater than its span, but in the case of the vault it is greater. The distance between the two façades is called the depth of the arch. What we call in the arch its depth, is called in the vault its length.

Aḥmad b. Thabāt discusses the following calculation (figure 8.3): When a vault is said to have an exterior curve of [$u_1 =$] 20*d* [where *d* stands for one *dhirā*ʿ, i. e., one *l*], and an interior curve of [$u_2 =$] 12*d*, and the distance between the two curves amounts to [$b =$] 2*d*, and its length to [$l =$] 50*d*, then how much is its surface area and its volume?

Solution: Aḥmad b. Thabāt considers the surface area of the vault as the sum of "plane" surfaces. These consist of the visible surfaces inside and outside

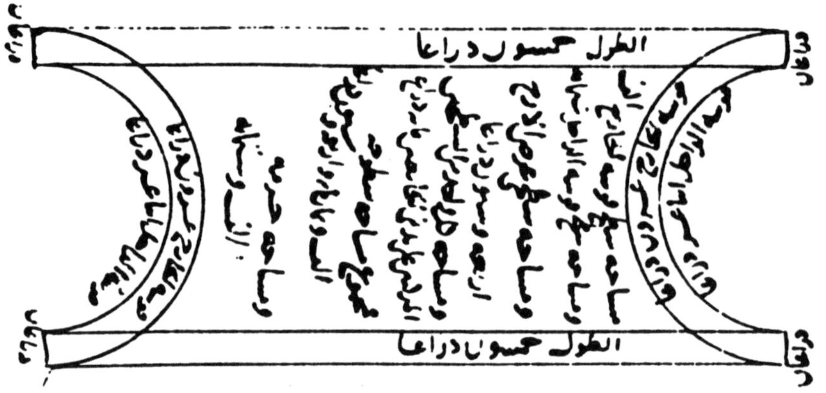

Figure 8.3
Calculation of a barrel vault.

the vault together with the surfaces at both ends, plus the invisible ones, namely, the two surfaces which support the vault. Thus we have:

Outside surface area: $l \times u_1 = 50d \times 20d = 1000d^2$,

Inside surface area: $l \times u_2 = 50d \times 12d = 600d^2$,

Two ends: $2 \times \dfrac{u_1 + u_2}{2} \times b = (20d + 12d) \times 2d = 64d^2$,

Two supports: $2 \times l \times b = 2 \times 50d \times 2d = 200d^2$, therefore the total surface area equals $1864d^2$.[20]

The volume of the vault is obtained by multiplying the surface area of one end of the vault by the length of the vault:

$$\frac{u_1 + u_2}{2} \times b \times l = 16d \times 2d \times 50d = 1600d^3.$$

The same method is applied by the later mathematician Ibn al-Ḥanbalī (d. 1564) in his *Book on Measurements* [*Kitāb al-misāḥa*], sometimes considered as a comprehensive commentary to Aḥmad b. Thabāt's work.[21]

The calculation of the volume of a wreath or discus is already formulated by al-Karajī as follows [chapter 50, section 4]: "Multiply half the sum of the outer and inner circumference by the width of the solid to be measured and multiply this product by the length."

This formula is identical to the one applied above. In a barrel vault with radii $r_1 > r_2$, the circumferences of the outer half circle and the inner half circle are πr_1 and πr_2, and the surface areas of the outer and inner circle are πr_1^2 and πr_2^2.

Hence the surface area of the half ring is

$$\left[\frac{\pi r_1^2 - \pi r_2^2}{2}\right] = \frac{\pi r_1 + \pi r_2}{2} \times \left(r_1 - r_2\right) = \frac{u_1 + u_2}{2} \times b.$$

Multiplying this product by the length (width) yields the formula

$$\frac{u_1 + u_2}{2} \times b \times l.$$

The vault or arch are not mentioned by al-Karajī in this context.

How does al-Kāshī deal with these matters? At first he explains at length the different elements of an arch and how these are connected and which part could disappear into a wall. This is followed by five methods for drawing the façade of an arch.[22] The first two are three-centered arches, the third is a four-centered arch, the fourth and fifth are two-centered. The fourth method is the one shown in figure 8.2 on the lower right, its two centers being separated by one third of the span. According to al-Kāshī, the second façade was the most common in his time; he therefore uses it to illustrate his calculation method. This kind of façade is convenient, as al-Kāshī remarks, when you need a span of five to ten, or up to fifteen cubits.

Construction of the second façade (figure 8.4 is taken from the oldest extant ms. with Roman letters added):

- draw a semicircle on diameter *AD* , the span of the arch
- extend *AD* at both sides by the thickness of the arch to the points *I* and *M*. *E* is the center of the semicircle;
- divide this semicircle into four equal parts through the points *A*, *B*, *C*, *G*, *D*
- extend *BE* and *GE* by *EZ* and *HE*, equal to *AC*, and by *BK* and *GL*, equal to *DM*, the thickness of the arch
- on center *E* draw the arcs *ML* and *KI*, on center *H* the arc *GT*, and on center *Z* the arc *BT*
- connect *HT* and *ZT* and extend them by the thickness of the arch to the points *O* and *S*
- draw arc *LO* on center *H* and arc *KS* on center *Z*
- erect the perpendiculars *SN* and *ON* on the lines *TS* and *TO*

The sections *AK*, *KT*, *TN*, *TL*, and *LD* form together the façade of the arch with *TN* as the keystone.

When we construct area *AFQD* with parallel sides and right angles, we obtain the spandrels of the arch, the sections *tNQ* and *JNF*. Section *tMD* as well as section *JIA* could disappear into a wall.

After al-Kāshī has explained and carried out all five methods for constructing the façade of an arch and has completed the characterization of the arch and the vault, he continues by surveying them. He explains that he has

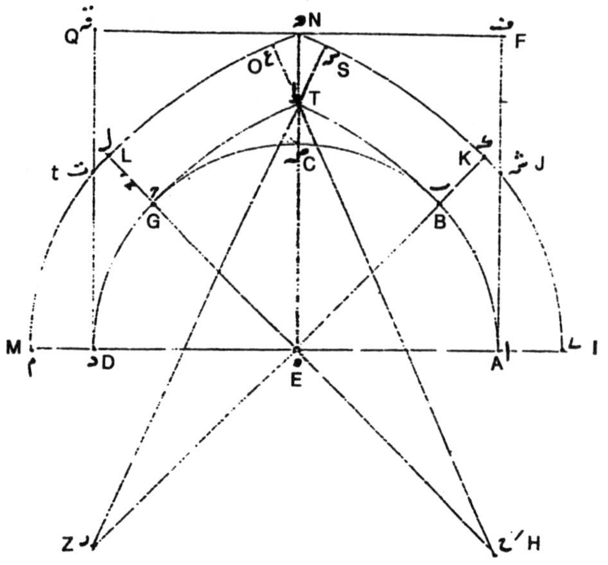

Figure 8.4
Construction of the second façade in a ms. of al-Kāshī's treatise.

calculated factors relating some measurements of an arch to its span and to its thickness. He has put these factors down in a table together with an explanation of the method. These factors have also been transformed into Indian numerals and been added to the table. He then gives an example of how to calculate with these factors. This is followed by a detailed account of how these factors were accomplished.[23]

Example, How to use the table:

Al-Kāshī assumes (figure 8.4) the span *AD* of the second façade to be equal to 20 and the width *DM* of the arch to be equal to 5. I call the exterior curve u_1 the line of convexity and the interior curve u_2 the line of concavity, and b is the width of the arch at its base. Al-Kāshī does not render the calculations but gives only the rounded results. I now list the approximations given by al-Kāshī:

• column 1: With this factor al-Kāshī calculates the interior curve :

$$ABTGD = 1.651 \times \mathrm{AD} = 1.651 \times 20 = 33.02 \approx 33.$$

• column 2: Multiplying this factor with the width al-Kāshī obtains half the difference between the exterior and the interior curve. Adding this amount to the interior curve, found by means of column 1, he obtains half the sum of the exterior and interior curve. Multiplying this quantity with the width of the arch he obtains the surface area (A) of the façade:

$$1.599 \times b = \frac{u_1 - u_2}{2};\ \left(\frac{u_1 - u_2}{2} + u_2\right) \times b = \frac{u_1 + u_2}{2} \times b = A$$

In our practical example: $u_1 = IJNtM$; $u_2 = ABTGD$;

$$\frac{u_1 - u_2}{2} = 1.599 \times 5 \approx 8;\ \frac{u_1 - u_2}{2} + u_2 = 8 + 33 = 41;\ A = 41 \times 5\ .$$

- column 3: Inner height of the arch:
  $ET = AD \times 0.598 = 20 \times 0.598 = 11.96 \approx 12.$
- column 4: Upper width of the arch:
  $TN = AI \times 1.099 = 5 \times 1.099 = 5.495 \approx 5.5.$

	When we multiply the span of the arch with this number, then we obtain the line of concavity of the arch.				When we multiply the width of the arch with this number, add the product to the line of concavity of the arch, and multiply the sum with the width, then we obtain the surface of the façade.				When we multiply the span of the arch with this number, then we obtain the lower height of the curvature.				We multiply the span of the arch with this number to obtain the width of the curvature, add this to the lower height of the curvature and thus obtain the upper height of the curvature.				We multiply the square of the span of the arch with this number, then we obtain the surface area of the opening, which the constructors call the concavity.			
These quantities are in Jumal-numbers	units	minutes	seconds	tertia	units	minutes	seconds	tertia	units	minutes	seconds	tertia	units	minutes	seconds	tertia	units	minutes	seconds	tertia
first façade	1	37	26	6	1	35	37	28	0	34	7	38	1	1	58	4	0	24	28	42
2nd façade	1	39	2	19	1	35	55	42	0	35	55	16	1	5	55	12	0	25	9	13
3rd façade	1	42	44	3	1	36	21	47	0	38	17	30	1	6	55	38	0	27	2	34
4th façade	1	45	26	17	1	34	34	47	0	38	43	47	1	5	55	12	0	28	41	41
These quantities are in Indian numbers	units	decimals	decimal seconds	decimal tertia	units	decimals	decimal seconds	decimal tertia	units	decimals	decimal seconds	decimal tertia	units	decimals	decimal seconds	decimal tertia	units	decimals	decimal seconds	decimal tertia
first façade	1	6	2	4	1	5	9	4		5	6	9	1	0	3	3		4	0	8
2nd façade	1	6	5	1	1	5	9	9		5	9	8	1	0	9	9		4	1	9
3rd façade	1	7	1	2	1	6	0	6		6	3	8	1	1	1	5		4	5	1
4th façade	1	7	5	7	1	5	7	6		6	4	5	1	0	9	9		4	7	8

Table to calculate arches and their parts.

• column 5: To obtain the surface area of the concavity, *ABTGDE*, al-Kāshī multiplies the square of the span by 5 and divides by 12. This calculation is equivalent to multiplying with the value found in the table, 0.419, or 0;25,9,13 ≈ 5/12.

With these values we can now calculate many different parts of the arch:

To calculate the *volume* of the arch we proceed in the same way as Aḥmad b. Thabāt above: After the surface area of the arch has been found, by means of the second column of the table, we multipy this number with the depth of the arch and obtain its volume.

Sometimes the arch disappears partly inside a wall and we want to know how much is visible and how extensive the *segments inside the wall* are, namely section *tDM* and section *JAI*: These segments are calculated by taking the difference of the circle segment *MtE* and the triangle *tDE*, which can be found from *MD* and *ED*.

When we subtract this amount from the total surface area of the arch we obtain the surface area of the *visible part of the arch*.

It might be necessary to calculate the *spandrels*, section *NQt* and section *NFJ*: In this case we calculate the area *AFQD* and subtract from this amount the area of the visible part of the arch, calculated above, and the area of the opening of the arch, area *ABTGDE*, found by means of the fifth column. The difference gives the surfacc arca of the spandrels. When we multiply this amount with the depth of the spandrels, we obtain the volume of the two spandrels.

As explained above, al-Kāshī's book is for practical use. Hence he rightly shows how to make life easier by working with rounded values. When arches other than the five models given by al-Kāshī are involved, approximation again is used, and one takes the model closest to the required arch. Golombek and Wilber[24] have considered existing examples of Timurid arches in the order outlined by al-Kāshī. Examples have been recorded for all but the fifth model, which was, however, common in small windows. In comparing the models described by al-Kāshī with actual examples of Timurid arches we have to bear in mind that al-Kāshī's purpose was to *calculate* volumes and surfaces, not to *construct* them. This means that an elegant approximation, which leads us to an easy calculation, is the ultimate goal.

Bulatow[25] has analyzed arches from the twelfth to fifteenth centuries in Central Asia and he suggests that some pointed arches were constructed as intersections of ellipses. For he notes that for spans exceeding 10 m. these were easier to construct than four-centered arches. The architects were, in addition, familiar with the stability of the ellipse, because its construction was known from Sasanian examples. According to his analysis, this kind of arch is found in

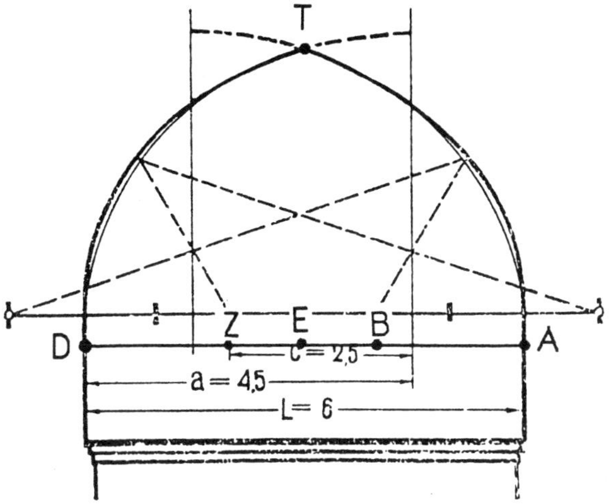

Figure 8.5
Dome of the Gur-i Amir (Bulatow's analysis with additions).

some of the most important Timurid buildings of the period, such as the Gur-i Amir in Samarkand, the mausoleum of Tamerlane and Ulūgh Beg, namely in its dome, interior niches, arches in zones of transition, and entrance portal. The same arches have elsewhere been identified as three- and four-centered arches and can be considered as such for all practical purposes.

Al-Kāshī does not mention an elliptical profile either for arches or for domes. There are a number of domes for which the profile may be interpreted as the intersection of symmetrical elliptical arcs. Bulatow has demonstrated that the dome of the Gur-i Amir was probably designed by using a pair of foci and a string. However, an analysis of the Gur-i Amir (figure 8.5) suggests that this dome could also have been originated by the fourth method (figure 8.2, far right): With line AD as the span and the points *B* and *Z* dividing the span into three equal parts we obtain the circle segments just inside the curve drawn by Bulatow. The difference between the two curves lies within the error range accepted by modern architects. The ellipse may be easier to construct but circular segments are easier to calculate. For calculating an elliptical dome, like the dome of the Gur-i Amir, al-Kāshī's factors are excellent.

The section on calculating arches ends in al-Kāshī's *Key of Arithmetic* with the following remark (my translation) "I have talked a lot about the subject of this section, as this section is very important, and my predecessors did not treat it as they should have done."

The Qubba

Besides mosques and *madrasas*—the second important religious building—monasteries, drinking fountains and elementary schools, the mausoleum, or *qubba*, is an important Islamic monument. From the earliest times most *qubbas* were of the same style: a cubic room on a square base with a vaulted roof. The problem of erecting a cupola on a square base finds in Muslim architecture a large variety of solutions. The space of transition from the square, or polygon, to the circle is occupied either by planks covered with stucco, or by corner-trompes made at first of bricks and later of stone, or by stalactite pendentives, called *muqarnas*. The Arabic word for dome or cupola is *qubba* pl. *qibāb* or *qubab*. By extension *qubba* also means: cupolaed structure, dome-shaped edifice; domed shrine, memorial shrine (esp. of a saint). At the end of this section we shall see that the term *parabolic qubba* [*al-qubba al-mukāfiya*] is used in a purely mathematical sense to denote a paraboloid.

Calculation of the *qubba* follows the same pattern as that of vaults: In the common arithmetic texts only hemispheric domes are calculated. Al-Kāshī however, gives an elegant and precise method to calculate a dome originated by turning the façade of an arch around its axis. His exposition is based on his fourth model, that is, on a two-centered arch with the two centers one-third of the span apart (figure 8.2, below right). Applying al-Kāshī's factors, excellent results can be obtained with an error range of less than 3 percent. "To simplify the procedure" as al-Kāshī says, he does not explain how he arrives at these factors. For a practical application the rules are enough on their own. The following question arises: Why is al-Kāshī so concise in the case of domes, when he uses such elaborate calculations for surveying arches? Could these factors be used for several kinds of domes, which deviated to varying extents from the model discussed by al-Kāshī?

In all extant Arabic treatises the *qubba* is assumed to consist of a solid shell between two concentric surfaces. In practice, however, the inner and outer surfaces of the shell are never really parallel, because pressure occurs in the lower part (up to 61°) and pull in the upper part. Al-Kāshī does not carry out the calculation of the hemispheric *qubba* but refers to his calculation of the sphere. There he uses, as expected, the right formulas for area and volume, expressing π as the ratio between the circumference and the diameter of a circle. Abu 'l-Wafā', who is the first to mention the measurement of the *qubba*, also gives the right definitions and calculation method, mentioning Archimedes.[26] From then on the calculation of volume and surface areas of hollow hemispheric *qubbas* form a regular part of practical arithmetical textbooks, although a wrong "formula" for the calculation of the volume frequently occurs.

When the inner and outer diameter of a hemispheric *qubba* are known, its volume and the inner and outer surface areas can be calculated from the volume and the surface area of a sphere.

The correct formulas for the area (A) and the volume (V) of a sphere with diameter $2r$ are:

$$A = (2r)^2 \times \pi;\ V = r \times \frac{A}{3} = \frac{4}{3}\pi r^3.$$

In Islamic mathematics π is usually expressed as: discard one seventh and half of one seventh and multiply with four:

$$\pi = \left(1 - \frac{1}{7} - \frac{1}{2} \times \frac{1}{7}\right) \times 4 = \frac{11}{14} \times 4 = 3\frac{1}{7}.$$

Mathematically, it is more difficult to prove a formula for calculating the surface area of a sphere than for the volume. However, in the common arithmetic books the computation of the inner and outer surface areas of the *qubba* is based on the correct formula:

$$A = (2r)^2 \times \left(1 - \frac{1}{7} - \frac{1}{2} \times \frac{1}{7}\right) \times 4$$

The volume of the *qubba*, on the contrary, is usually based on the following wrong formula:

$$V = (2r)^3 \times \left(1 - \frac{1}{7} - \frac{1}{2} \times \frac{1}{7}\right) \times \left(1 - \frac{1}{7} - \frac{1}{2} \times \frac{1}{7}\right)$$

Comparing the two results calculated by means of the correct and the erroneous formula, the difference (D) is:

$$D = (2r)^3, \left\{\left(\frac{11}{14}\right)^2 - \frac{\pi}{6}\right\}, \text{ and with } \pi = 3\frac{1}{7},$$

$$D = (2r)^3\left(\frac{121}{196} - \frac{22}{42}\right) = 0.0935(2r)^3.$$

This means that the calculated volume is 17.86 percent(!) more than the correct volume, independent of the size of the diameter. This is especially significant when the architect is paid according to the measurements of the building (see conclusion).

Even al-Karajī in *Sufficient Arithmetic* gives this wrong formula, but he adds a second method, in which the volume of a sphere is calculated by making a waxen model:[27] "You take a rectangular solid, made out of wax and with its three dimensions of equal length, and weigh it. Let its weight be 30 dirham. Now you make from this (material) a sphere, as perfect as possible, with the diameter equal to the dimension of the solid. The weight of this sphere is found to be a little less than $18\frac{2}{3}$.

Hence, the diameter is raised to the third power and then

$\frac{1}{3}$ and $\frac{1}{9} \times \frac{2}{5}$ is subtracted.

The difference between the two results is very small."

This implies the formula:

$$V = (2r)^3 \times \left(1 - \frac{1}{3} - \frac{1}{9} \times \frac{2}{5}\right)$$

Although this sounds like practical mathematics, it gives a result which is even worse than that of the first method—being off by almost 20 percent! Thus one wonders whether the experiment was made at all.

Muḥammad Bahāʾ al-Dīn al-ʿĀmulī (1547–1622) says in his *Essence of Arithmetic* (chapter 6, section 3, *Mensuration of Solids*, ca. 1600),[28] "To measure the sphere, multiply half its diameter with one third of its surface area; or, subtract three fourteenths from the cube of its diameter, and again from the remainder, and again from the remainder."

Hence two formulas are given:

$V = r \times \frac{A}{3}$, the correct mathematical formula, or,

$$V = (2r)^3 \times \left(1 - \frac{3}{14}\right) \times \left(1 - \frac{3}{14}\right) \left[\times \left(1 - \frac{3}{14}\right)\right],$$

the formula for practical use.

Here we find the two formulas side by side, each in its own right, the mathematically correct formula and the common practical formula. The brackets have been added, because I have assumed, with only Nesselmann's edition at my disposal, that the second "and again from the remainder" is due to the scribe.[29] Whether Bahāʾ al-Dīn or the scribe made this change, this formula gives a result that is out of the question, the calculated formula being 7.4 percent less than the correct volume. But who wants this? Throughout the centuries constructors, architects, and artisans, probably profited by using the wrong formula.

In her commentary to the Topkapi scroll Necipoğlu reasons:[30]

> It is unfortunate that Ibn al-Haytham's *Book of Buildings and Constructions* [*K. al-Abniya wa 'l-ʿuqūd*] and al-Karajī's *Book of Architectural Constructions* [*K. ʿUqūd al-abniya*] have not survived. The latter, which dealt with the construction of buildings, bridges, and engineering, was described by Ibn al-Akfānī as a work exclusively based on practical geometry. Ibn al-Haytham's lost work, on the other hand, was referred to by the physician and biographer Ibn Abī Uṣaybiʿa (1203–1270) as a "treatise on the construction of ditches and buildings, with all the figures of geometry joined together, a work con-

cluding with the three conic sections, namely, the parabola, hyperbola, and ellipse." These lost treatises relying on practical geometry, therefore, appear to have been illustrated with geometric constructions that included conic sections, constructions probably adapted to simplified mechanical procedures.

Ibn Khaldūn testified to the use of conic sections in architecture and those crafts dealing with bodies, "Conic sections are a branch of geometry. This discipline is concerned with the study of figures and sections occurring in connection with cones. It proves the property of cones by means of geometrical proofs based upon elementary geometry. Its usefulness is apparent in practical crafts that have to do with bodies, such as carpentry and architecture. It is also useful for making remarkable statues [monuments] and large objects [*hayākil*, "effigies" or "edifices"] and for moving loads and transporting large objects [*hayākil*] with the help of mechanical contrivances, engineering [techniques], pulleys and similar things.". . .

Treatises on the mensuration of parabolas and paraboloids were written by such mathematicians as Thābit b. Qurra (836–901), Ibrāhīm b. Sinān (909–946), al-Sijzī (2nd half 10th c.), Abū Sahl al-Qūhī (fl. 980–100), and Ibn al-Haytham (died c. 1040), some of them including sections on arches, vaults and domes. Al-Sijzī, for example, wrote a work exclusively dealing with the mensuration of domes, entitled *Epistle about the Characteristics of Hyperbolic and Parabolic Domes* [*R. fī khawāṣṣ al-qubba al-zāʾida wa 'l-mukāfiya*].

The many Arabic treatises on conic sections, written after Apollonius of Perga's *Conica* was translated, often deal with practical application. Their contribution to architectural practice (particularly designing pointed arches, vaults and domes) and to the decorative arts awaits assessment by historians of science so that the ways in which theory and praxis interacted can be understood more clearly.

So far, the inspection of the above-mentioned treatises has not confirmed that they were for practical use. Jan Hogendijk has looked at al-Sijzī's treatise on the hyperbolic and parabolic domes. According to him, the work is on pure mathematics only, with no connection to architecture. The other treatises have been studied by Suter[31] and recently edited by Rashed.

Thābit b. Qurra's[32] long treatise *On Measuring Paraboloids* [*Fī misāḥat al-mujassamāt al-mukāfiya*][33] deals with solids created by rotating different sections of a parabola. In proposition 36, the last proposition of the treatise, he proves the following property of the "parabolic *qubba*" (figure 8.6, top), that is:

> The volume of every parabolic *qubba* is equal to half the volume of the cylinder, with as base the circular base of the *qubba*, if the *qubba* has a regular vertex, or of the lower basic circle, if the *qubba* does not have a regular vertex, and with the height equal to the axis of the *qubba*.

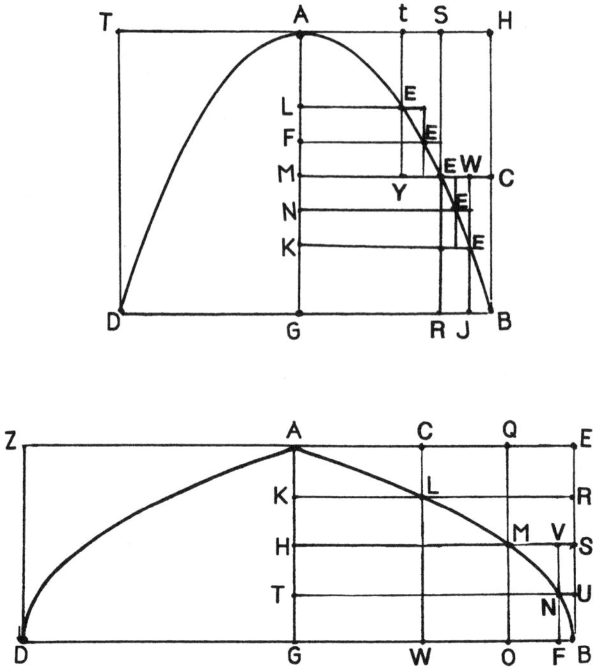

Figure 8.6
Parabolic domes after Ibn al-Haytham (Suter).

Abū Sahl al-Qūhī[34] wrote a much shorter treatise on the subject *Determining the Volume of the Paraboloid* [*Fī istikhrāj misāḥat al-mujassam al-mukāfī*].[35] Here he criticizes the "famous and well-known" treatise by Thābit b. Qurra, "the only one on the subject," saying that it is "too voluminous and long, there are about forty propositions, numerical, geometrical and others, there are all these lemmata for just one proposition: how to know the volume of a paraboloid."[36] Al-Qūhī expounds that examining Thābit's treatise, he found it very difficult to understand, whereas Archimedes' treatise *On the Sphere and the Cylinder* seemed much easier to him. Thinking that many people might have gained the same impression, he feels compelled to determine the volume of the paraboloid afresh. He is able to do this by using a method that is easily understandable and does not need any lemmata. Al-Qūhī proves in only three propositions that: "Every paraboloid is equal to half its cylinder."

Al-Qūhī's treatise was criticized by Ibn al-Haytham[37] in his *Treatise on Measuring the Paraboloids* [*Maqāla fī misāḥat al-mujassamāt al-mukāfiya*].[38] In an almost philosophical introduction he writes: "The person who pronounces or composes any discourse or essay has a motive which leads him to say what he says or write what he writes. We examined with great care Thābit b. Qurra's treatise *On Measuring Paraboloids*. We found that he followed a road devoid of any plan, and that his way of explaining was long and involved painful difficulties. We then got hold of a treatise on *Determining the Volume of the Paraboloid* by Abū Sahl al-Qūhī. We found it bare and concise and we learned that, according to the author, the reason he did his research and wrote the treatise, was the difficulties encountered in Thābit b. Qurra's treatise. However, we have established that Abū Sahl's treatise, although easier and simpler, only contains the proof of the measurement of one of the two kinds of paraboloids." (In fact, Thābit b. Qurra's treatise too discusses only the first kind of paraboloid.)

Ibn al-Haytham continues by saying that there are two kinds of paraboloids, one is comprehensible and easy, the other difficult and painful. His proof for the first kind of paraboloid needs several lemmata and is more detailed than al-Qūhī's demonstration. Then he turns to the second kind of paraboloid generated by rotating the parabola around its ordinate (figure 8.6), and proves in an elegant way that: "The paraboloid generated by rotating the parabola around its ordinate is equal to one third plus one fifth of the circumscribed cylinder."

Both formulas are easy and exact, so they could well be used for practical calculations but they do not provide the surface areas. Also, we have not yet found traces of these easy formulas in arithmetic manuals. Until we find these, we should consider the above results on paraboloids as highschool mathematics. Another serious question is, were parabolic *qubbas* ever built? The paraboloid, shown in figure 8.6, above, would certainly make a beautiful *qubba*, but where do we find an example of a parabolic *qubba*? Elliptical arches and qubbas existed in pre-Islamic times. Exact formulas for ellipsoids are more difficult to develop, but, as we have seen in the section on arches, their surface areas and volumes can be approximated. However, how these structures were calculated before al-Kāshī established his factors, remains to be looked at.

These treatises on conic sections are of great interest from the point of view of the history of mathematics but were probably never used by artisans. They do not give the impression of being meant for practical use.

One important question remains: why are the above-mentioned treatises by Ibn al-Haytham and al-Karajī lost? Were they never studied, and therefore not copied? Or, were they used so intensively, that the soiled and worn out copies were thrown away? During the Dibner conference in November 1998, David King remarked that the extant astrolabes might well be those that were never used. Could this apply similarly to treatises on construction or containing ornaments, like the Topkapi scroll?

Approximating the Muqarnas

The *muqarnas* (figure 8.7) developed around the eleventh century in Iran and North Africa. Whether or not the two developments are related, two points about these new forms must be made. One is that, from the late eleventh century onward, all Muslim lands adopted and developed the *muqarnas*, which became almost as common a feature of an elevation as the Corinthian capital was in Antiquity. The second and far more important point is that, from the moment of its first appearance, the *muqarnas* acquired four characteristic attributes whose evolution and characteristics form its history: it was three-dimensional and therefore provided volume wherever it was used, the nature and depth of the volume being left to the discretion of the maker; it could be used both as an architectonic form, because of its relationship to vaults, and as an applied ornament, because its depth could be controlled; it had no intrinsic limits, since not one of its elements is a finite unit of composition and there is no logical or mathematical limitation to the scale of any one composition; and it was a volume that could be a solid or a void, a projecting mass of complex shapes or a complex outline—a three-dimensional unit which could be resolved into a two-dimensional outline.[39]

The *muqarnas* was used in domes, in niches, on arches, and as an almost flat decorative frieze. In each instance the module as well as the depth of the composition is different and adapts to the size of the area involved or to the required purpose. In ceilings it serves a clear architectonic aim, or at the very least provides the structural illusion of ascending movement culminating in a small cupola. The *muqarnas* is at the same time a linear system and an organization of masses. Despite occasional textual references to plans, there are no known Islamic architectural working drawings from the pre-Mongol era. The earliest known example of a construction plan is a 50 cm. stucco plate showing the projection of a quarter muqarnas vault which was found at the Takht-i Suleiman in Iran. This plate from the 1270's was the basis for Ulrich Harb's reconstruction of the collapsed vault. Fourteenth-century sources frequently mention architectural drawings produced either on clay tablets or on paper. In the fifteenth-century Timurid world drawings seem to have been more widely used than before. Their extensive use had become essential because of the increasing intricacy of the geometric design. Up until Necipoğlu's discovery of the Topkapi scroll the earliest known examples of such architectural drawings were a collection of fragmentary post-Timurid design scrolls made of sixteenth-century Samarqand paper, preserved at the Uzbek Academy of Sciences in Tashkent. These scrolls almost certainly reflect the sophisticated Timurid drafting methods of the fifteenth century. In 1876 the English architect

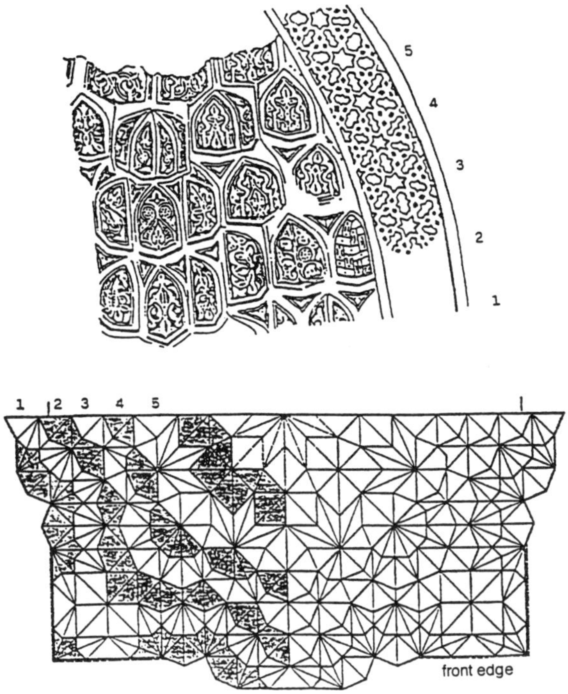

Figure 8.7
Curved Muqarnas with its Plane Projection.

C. Purdon Clarke brought back from Teheran some scrolls and working drawings from the eighteenth and the nineteenth centuries that he had collected following the death of the official state architect, Mirza Akbar; these scrolls are now preserved in the Victoria and Albert Museum. In 1981 similar material, still in the hands of the master-artisan, was examined by W. K. Chorbachi in two Arab towns. The collection of these scrolls was not only the basic reference manual but also served as a design book. A few years ago I visited a workshop at Fez/Morocco, where the artisans used a construction-plan for a *muqarnas* on a scale of 1:1. The pieces cut out for constructing the *muqarnas* could actually be put on the draft such that the cross section of the element, i.e., the cross section of the beam, matched the figure in the draft exactly. The Timurid scrolls show a decisive switch to the far more complex radial *muqarnas*, with an increasing variety of polygons and star polygons. Also the Akbar scrolls are more elaborate than the twentieth-century Fez drawing. Despite their simplicity, however, the more recent scrolls testify to a relatively unbroken tradition of architectural practice in the Islamic world from at least the Timurid period onward. A continuous tradition from the thirteenth-century Takht-i Suleiman plate to the twentieth-century Fez drawing is evident. As in the Ilkhanid period, 700 years earlier, the plane projection of the elements in the Moroccan plan consists of simple geometrical figures: squares, half-squares, rhombuses, half-rhombuses, rectangles, almonds, bipeds (figure 8.7).

The contents of the Topkapi and Tashkent scrolls support the commonly expressed view that the key to Timurid–Turkmen architecture lies not only in its fascination with complicated vaulting systems but also in its extensive surface decoration. The numerous two-dimensional geometric patterns and epigraphic compositions in these scrolls condense complex compositions into shorthand formulas meant to act as guidelines for the simpler working methods employed on the construction site. They thus provide a valuable glimpse into the processes of design and execution. The contents of the Topkapi scroll, which resemble those of a pattern book, can be seen as an index of the unprecedented Timurid-Turkmen emphasis on surface decoration, an emphasis that turned the flat façades of buildings into stage props for the display of virtuoso ornamental panels and fragmented vaults into multifaceted compartments with no structural role. The Topkapi scroll reflects a "painterly" aesthetic of architecture informed by the cultural prestige of drawings on paper.

Mohammad al-Asad describes[40] the *muqarnas* as:

> . . . a vaulting system based on the replication of units arranged in tiers, each of which supports another one corbeled on top of it. The final result is a stair-like arrangement that is sometimes referred to as honeycomb or stalactite vaulting. The units are made of wood, brick, plaster, or stone and can be painted, or, as in the case of the brick or plaster ones, covered with glazed tiles. Muqarnas compositions can be located in different parts of a building, articulating a column capital, supporting a minaret's balcony, or vaulting over an entry portal, niche, or hall. Muqarnas vaults are usually part of a double–shell arrangement and are therefore visible only from the inside of a building. In some cases, as in the mausoleums of Nūr al-Dīn in Damascus (1172) and Imam Dur in Samarra (circa 1085), the muqarnas is also reflected on the outside.

Al-Kāshī defines the *muqarnas* in his practical way as follows[41] (figure 8.7):

> The *muqarnas* is a ceiling like a staircase with facets and a flat roof. Every facet intersects the adjacent one at either a right angle, or half a right angle, or their sum, or another combination of these two. The two facets can be thought of as standing on a plane parallel to the horizon. Above them is built either a flat surface, not parallel to the horizon, or

two surfaces, either flat or curved, that constitute their roof. Both facets together with their roof are called one cell. Adjacent cells, which have their bases on one and the same surface parallel to the horizon, are called one tier.

Hence the elements of a *muqarnas* consist of cells and intermediate elements, connecting the roofs of two adjacent cells.

The plane projection of an element consists of simple geometrical forms, called by al-Kāshī squares, half-squares (cut along the diagonal), rhombuses, half-rhombuses (isosceles triangles with the shorter diagonal of the rhombus as their base), almonds (deltoids), and large and small bipeds. Rectangles also occur. To make these elements fit, they have to be constructed according to the same unit of measure. Al-Kāshī uses in his computation the so-called *module* of the *muqarnas*, defined as the base of the largest facet, that being the side of the square.

On this foundation he calculates the surface area of the *muqarnas*. In the following I shall only sketch his method—the full explanation can be found in two former papers.[42]

Definitions

- A facet of a cell is a vertical side.
- A roof of a cell is a surface, not parallel to the horizon, or two joined surfaces, either flat or curved.
- A cell consists of two facets plus their roof.
- An intermediate element is a surface, or two joint surfaces, connecting the roofs of two connecting cells.
- An element is a cell or an intermediate element.
- A tier is a row of cells, with their bases on the same surface parallel to the horizon.
- The module is defined as the base of the largest facet, that being the side of the square. It is the measure-unit of the *muqarnas* and is equated with "one."
- A rhombus is a parallelogram with all sides equal to the module and with the acute angles equal to 45°.
- A half-rhombus is an isosceles triangle with an angle of 45° at its vertex and with a base equal to the shorter diagonal of the whole rhombus.
- A rhomboid is a parallelogram with two opposite sides equal to the module and with the acute angles equal to 45°.
- An almond is a quadrilateral with two opposite right angles, an acute angle of 45°, and the two sides adjoining at the acute angle equal to the module.

- A small biped is the complement of an almond to a rhombus.
- A jug is a quadrilateral with two opposite angles of 67°30′, a right angle, and the two sides adjoining at the right angle equal to the module. It consists of two half-rhombuses connected along the sides equal to the module.
- A large biped is the complement of a jug to a square.
- A barley-kernel is a quadrilateral with two opposite equal obtuse angles, a right angle, and the two sides adjoining at the right angle equal to the module.

Al-Kāshī distinguishes four types of *muqarnas*: The simple *muqarnas*, the clay-plastered [*muṭayyan*] *muqarnas*, both have plane facets and roofs. In the curved *muqarnas* and the Shīrāzī *muqarnas* the roofs of the cells and the intermediate elements are curved.

The *simple muqarnas* is the one in which the underlying elements in the plan are simple. The surfaces of the cells' facets in the lowest tier are based on a square, a rhombus, or a rhomboid; in the other tiers they have an additional almond or half-rhombus, and their roofs are shaped like squares, half-squares, rhombuses, half-rhombuses, almonds, bipeds, or *barley-kernels*. The barley-kernels do not occur except on the upper tier. All these elements had been treated earlier in chapter two of the same book IV, which is entitled, *Measuring Quadrilaterals and What Is Connected with It*.

To measure the area of the simple *muqarnas*, called *minbar*-like by the masons, al-Kāshī tells us to proceed tier by tier. Al-Kāshī computes the area of the facets of a tier by multiplying the sum of the bases of the facets of the tier by the height, which for the simple *muqarnas* is in most cases equal to the module. The next step is to count all the surfaces of the roofs of the cells. Their sum plus the area of the facets is the area of one tier. Then he adds all the tiers to obtain the surface area of the *muqarnas*. Al-Kāshī continues: "If we measure the surface on which the *muqarnas* is constructed, we would obtain the surface area of the entire roof of the *muqarnas*." This means that for al-Kāshī the *muqarnas* is an ornament, a decoration. We already have a roof, or a vault, on which the *muqarnas* is constructed. The surface area of the entire structure consists of the *muqarnas* plus the roof or vault on which the *muqarnas* is constructed.

Note: Al-Kāshī gives in his calculations the sexagesimal as well as the decimal values in order to avoid errors in later copies of his work. He must have calculated in the sexagesimal system, as these are the only numbers found in the addendum, *Method of the Masons*, although the result, that is, the coefficient, is given in the sexagesimal as well as in the decimal form.

N.B. The sexagesimal results are more accurate than the decimal ones.

Al-Kāshī saw the *clay-plastered* or *muṭayyan muqarnas* in ancient buildings in Isfahan. It is similar to the simple *muqarnas*, except that the heights of its tiers might differ and a few tiers might have only a roof and no facets. Calculations on the *muṭayyan muqarnas* are analogous to the calculations on the simple *muqarnas*.

The *curved muqarnas* is like a simple *muqarnas* in which the roofs of the cells are curved. Curved surfaces are located between two roofs. These intermediate elements are shaped as a triangle or as two triangles, which together form a biped. Such triangles are also found on its roof as well as curved almonds and barley-kernels. The facets of the cells are only squares or rectangles. Their bases are either equal to the module of that *muqarnas* (i.e., the cell is standing on a square or a rhombus), or half the diameter of its square (i.e., standing on a half-square), or the difference between the diagonal and its side (which equals the shorter side of an almond or biped), or the side of an octagon with radius equal to the module. There are no other bases besides these four. One eighth of an octagon with radius equal to the module is an isosceles triangle with an angle of 45° at its vertex, that is, a half-rhombus, with a base equal to the shorter diagonal of the whole rhombus. That al-Kāshī finds it necessary to mention the octagon probably means that octagons occurred rather frequently, or that he had a *muqarnas* in mind where octagons occurred. Two eighths of an octagon with radius equal to the module give a *jug*, the complement to a large biped.

To measure the area of the curved *muqarnas* we add the bases of all the cells. Now we multiply the sum by the coefficient and obtain the area of all the cells, i.e., their facets plus their curved roofs. To this amount we add the areas of all the intermediate elements and thus obtain the area of the *muqarnas*. See my above-mentioned papers on how to obtain the coefficient and calculate the intermediate elements.

The *Shīrāzī muqarnas* is like a curved *muqarnas* but has a greater variety of elements. In the previous kind, only four possible measures for the bases of the facets occur, as was explained, but in the Shīrāzī *muqarnas* the possibilities are innumerable. Also many different elements are found on the roof: besides the curved roofs of the cells with intermediate triangles and bipeds one finds triangles, squares, pentagons, hexagons, star polygons, etc., flat as well as curved. Sometimes a facet without a roof is found with a *miḥrāb* (niche) drawn on it.

In Timur's time, when building activity exploded, local constructors could manage the simpler buildings. But for the special and more artistic monuments architects and artisans were imported from the conquered lands, first Khwārizm, then Tabrīz and Shīrāz, and finally India and Syria. It is known that Timur brought in architects from Shīrāz in 1388 and 1393, and that many migrated of their own free will.[43] The names of several Shīrāzī architects have been transmitted, the most famous being Qawām al-Dīn ibn Zayn al-Dīn al-Shīrāzī, the only active builder whose surviving structures display a distinctive architectural style. This might well be the reason why the type of *muqarnas* constructed with many variations, "innumerable possibilities" as al-Kāshī explains, was called Shīrāzī.

To measure the area of the Shīrāzī *muqarnas* we proceed in principle in the same way as in the case of the curved *muqarnas*. We first make a ruler corresponding to the module of that *muqarnas* and divide it into sixty sections, if we calculate in the sexagesimal system, or into ten sections, if we calculate "with Indian numbers," that is, in the decimal system. With this ruler we measure all the various elements and then compute as before.

Although the appearance of a *muqarnas* is complex, they usually consist of only a few basic elements. These might have been prefabricated, as in the case of the collapsed vault on the Takht-i Suleiman, or as in the decorations still made in Fez. Al-Kāshī's four types are more or less of the same kind, and constructed with similar elements. The elements of a *muqarnas* are standardized. Apart from the decoration, the difference in the appearance results mainly from the different ways in which these standardized elements are put together. The practice of using standardized elements, making construction faster and cheaper, is widely found in ancient China. In Islamic architectural practice many monuments, especially palaces, were built rapidly, either because insecurity of power made lengthy building programs unlikely to be completed or because they tended to be personal rather than dynastic and were not meant to or expected to survive their original patron. Hence standardized elements were a necessity. When we want to study practical mathematics, we have to take into consideration the practice.

Conclusions

In seventeenth-century Safavid Iran architects were paid a percentage on each building based on the cubit measure of the height and thickness of the walls:[44]

> The Persians determine the price for masons on the basis of the height and thickness of walls, which they measure by the cubit, like cloth. The king imposes no tax on the sale of buildings, but the Master Architect, that is Chief of Masons, takes two percent of inheritance allotments and sales. This officer also has a right to five percent on all edifices commissioned by the king. These are appraised when they are completed and the Master Architect, who has directed the construction, receives as his right and salary as much as five percent of the construction cost of each edifice.

Likewise in medieval Italy it was common practice to pay the artisans according to the surface area they had completed. The same custom seems to have existed in the Arab world. It is also useful to know in advance, more or less, how much material is needed like gold for gilding, bricks for construction or paint. Payment per cubit was common in Ottoman architectural practice where a team of architects and surveyors had to make cost estimates for projected buildings and supply preliminary drawings for various options. In addition to facilitating estimates of wages and building materials before the construction began, al-Kāshī's formulas may also have been used in appraising the price of a building after its completion. His sophisticated formulas were, like the simple formulas found in the manuals of arithmetics, useful for everyday life. This was al-Kāshī's objective: to present "all that is needed for the one who calculates carefully."

Al-Kāshī's *Key of Arithmetic* is the only known treatise in which an attempt is made to calculate the surface area of a *muqarnas*. Looking at a *muqarnas* structure (figure 8.8) we can see why. Although based on a relatively simple two-dimensional plan, the three-dimensional *muqarnas* is a complex, intricate vaulting. To calculate the curved surfaces of the different cells was a near impossibility for all but a master mathematician such as al-Kāshī. By suitably chosen approximations he worked out factors appropriate for calculation in everyday life, optimizing the problem like a modern mathematician. I have some doubts, however, whether these *muqarnas* factors were often applied in practice, as they are given in decimal fractions to the sixth position, or in sexagesimal fractions to the fourth position. For the calculation of arches, the factors contain decimal fractions to the third position only; this is sufficient for all practical purposes. Another possibility why al-Kāshī worked out factors for calculating a *muqarnas* could be that he knew how to do it. As in modern mathematical research, formulas are discovered for their own sake, only later, may practical applications be found and developed.

NOTES

1. Vogel, pp. 43–53.

2. Creswell, pp. 608–610. Compare also G. R. D. King's criticism on Creswell's appraisal of the early level architecture in the Arabian peninsula.

3. Frishman, p. 13.

4. See Toomer, DSB VII, pp. 358–365.

5. Rosen [1831] trans., p. 3.

6. Rosen l.c. trans., pp. 70–86.

7. See Youschkevitch, DSB I, pp. 39–43.

8. Edited by Saidan [1971], based on the Leiden ms. Or. 103, which contains only the first three parts of the ms., and the Cairo ms. Riyāḍa 42 M.

9. Saidan [1974] pp. 369–375.

10. Necipoğlu [1995] pp. 133–138 and p. 176, note 13.

11. See Sesiano, Enc. Hist. pp. 475–476.

12. Hochheim, Repr. pp. 204–222.

13. See Vernet and Bruin.

14. Hogendijk [1990].

15. Rebstock [1993].

16. According to Amīn Aḥmad Rāzī (d. 1010 A.H.) in *Tadhkira-ye haft iqlīm* [The account of the seven climates], Ms. Sepahsālār (Tehran) No. 2733, fol. 774. (Courtesy of M. Bagheri).

17. Hill [1996] pp. 98–101.

18. This is essentially Creswell's theory, see Warren, p. 59f.

19. Rebstock [1993] pp. 133–134.

20. This example implies $\pi = 4$!

21. Rebstock [1992] pp. 202–218.

22. All five constructions are performed in the video "Qubba for al-Kāshī."

23. This detailed account will be published in a forthcoming paper together with a discussion on the accuracy of the factors and the calculations.

24. Golombek [1988] pp. 153–157.

25. Bulatow [1978].

26. Saidan [1971] l.c., p. 268.

27. Hochheim [1877–80] part 2, p. 28.

28. Nesselmann [1843] p. 33 and note 19. Muḥammad Bahā᾽ al-Dīn al-ʿĀmulī is mentioned neither in DSB nor in Enc. Hist. ʿĀmul is a town in Syria.

29. Cf. Rebstock [1992] p. 216: In the edition Riyāḍīyāt, Aleppo 1976, p.88, the term in brackets does not occur. However, it is written in the Leningrad Ms.

30. Necipoğlu [1995] pp. 140–141,with some minor changes.

31. Reprinted in 1986.

32. See Rosenfeld/Grigorian, DSB XIII, pp. 288–295.

33. Rashed, Vol. I, pp. 319–457; Suter, Repr. Vol. II, pp. 435–476.

34. See Dold-Samplonis, DSB IX, pp. 239–241.

35. Rashed, Vol. I, pp. 850–871; Suter, Repr. Vol. II, pp. 435–476.

36. Rashed, Vol. I, pp. 850–852; Suter, Repr. Vol. II, p. 463.

37. See Sabra, DSB VI, pp. 189–210.

38. Rashed, Vol. II, pp. 207–293; Suter, Repr. Vol. II, pp. 369–412.

39. Grabar [1992] p. 147.

40. Necipoğlu [1995] p. 349.

41. Dold-Samplonius [1992/3] p. 202.

42. Dold-Samplonius 1992/93 and 1996, see also Özdural.

43. Golombek/Wilber [1988] pp. 187–194.

44. Necipoğlu [1995] pp. 44, 159.

BIBLIOGRAPHY

Bruin, Frans. 1970. Surveying and Surveying Instruments, being chapters 26–30 of the book *On Finding Hidden Waters* by Abū Bakr Muḥammad al-Karajī (AD 1029), Bīrūnī Newsletter No. 39. Beirut.

Bulatow, M. S. 1978. *Geometric Harmony in the Architecture of Central Asia, 9th–15th century* (Russian). Moscow.

Creswell, K. A. C. 1958. *A Short Account of Early Muslim Architecture*. Pelican Paperback.

Creswell, K. A. C. 1960. "Architecture," *Encyclopaedia of Islam*. 2nd ed. Vol. I, pp. 608–624.

DSB. 1970–1980. *Dictionary of Scientific Biography*, 16 vols. New York.

Diez, E. 1938/1986. "Ḳubba," *Encyclopaedia of Islam*, Supplement, pp. 139–146, repr. in the 2nd ed. Vol. V, pp. 289–296.

Dold-Samplonius, Yvonne. 1992. "The XVth Century Timurid Mathematician Ghiyāth al-Dīn Jamshīd al-Kāshī and his Computation of the Qubba," *"Amphora," Festschrift for Hans Wussing on the Occasion of his 65th Birthday*. Ed. S. S. Demidov, M. Folkerts, D. E. Rowe, and Ch. J. Scriba. Basel et al., pp. 171–181.

Dold-Samplonius, Yvonne. 1992/3. "Practical Arabic Mathematics: Measuring the Muqarnas by al-Kāshī," *Centaurus* 35, pp. 193–242.

Dold-Samplonius, Yvonne. 1993. "The Volume of Domes in Arabic Mathematics" *"Vestigia Mathematica," Studies in medieval and early modern mathematics in honour of H.L.L. Busard*. Ed. M. Folkerts and J. P. Hogendijk. Amsterdam & Atlanta, pp. 93–106.

Dold-Samplonius, Yvonne. 1996. "How al-Kāshī Measures the Muqarnas: A Second Look," *Mathematische Probleme im Mittelalter—Der lateinische und arabische Sprachbereich*, Ed. M. Folkerts (Harrassowitz Verlag in Kommission) Wiesbaden, pp. 57–90.

Enc. Hist. 1997. *Encyclopaedia of the History of Science, Technology, and Medicine in Non-Western Cultures*. Ed. H. Selin. Dordrecht.

Frishman, Martin, and Khan, Hasan-Uddin (Eds.). 1994. *The Mosque. History, Architectural Development & Regional Diversity.* London.

Golombek, Lisa, and Wilber, Donald. 1988. *The Timurid Architecture of Iran and Turan*. 2 Vols. Princeton, N.J.

Grabar, Oleg. 1978/1992. *The Alhambra*. London/Sebastopol, Calif. [2nd ed.]

Hill, Donald,1984/1996. *A History of Engineering in Classical and Medieval Times*. London/New York. [Paperback ed.]

Hochheim, Adolf, 1877–1880/1998. *Kāfī fī'l Ḥisāb* (Genügendes über Arithmetik). (German) Halle. Repr. Islamic Mathematics and Astronomy, Vol. 38. Frankfurt/ Main.

Hogendijk, Jan P. 1990. "A Medieval Arabic Treatise on Mensuration by Qāḍī Abū Bakr," *Zeitschrift für Geschichte der arabisch-islamischen Wissenschaften* 6, pp. 130–150.

Al-Kāshī, Ghiyāth al-Dīn, 1427. *Miftāḥ al-Ḥisāb* (Key of Arithmetic). Ms. Malek Library 3180/1, Tehran, dated 830 AH(!), copied by Mo'īn al-Dīn al-Kāshī, who went with al-Kāshī from Kashan to Samarqand. Ms. Or. 185, Leiden, dated AD 1558.

King, G. R. D. 1991. "Creswell's Appreciation of Arabian Architecture," *Muqarnas*, Vol. 8, pp. 94–102.

Luckey, Paul. 1951/1998. *Die Rechenkunst bei Jamshīd b. Mas'ūd al-Kāshī mit Rückblicken auf die ältere Geschichte des Rechnens*, (German). Wiesbaden. Repr. Islamic Mathematics and Astronomy, Vol. 56, pp. 75–226. Frankfurt/Main.

Nader, Nabulsi. 1977. *Al-Kāshī, Ghiyāth al-Dīn: Miftāḥ al-Ḥisāb* (Key of Arithmetic). Arabic edition, with French notes and introduction. Damascus.

Necipoğlu, Gülru. 1995. *The Topkapi Scroll—Geometry and Ornament in Islamic Architecture*. Santa Monica, Calif.

Nesselmann, Georg H. F., 1843/1998. *Essenz der Rechenkunst von Mohammed Behaeddin al-ʿĀmulī.* (Arabic edition, with German translation and notes.) Berlin. Repr. Islamic Mathematics and Astronomy, Vol. 59, pp. 29–166. Frankfurt/ Main.

Özdural, Alpay. 1991. "Analysis of the Geometry of Stalactites: Buruciye Medresse in Sivas." *METU Journal of Faculty of Architecture* 11 (Middle East Technical University, Ankara), pp. 57–71.

Rashed, Roshdi. 1993/1996. *Les mathématiques infinitésimales du IXe au XIe siècle*. Vol. I: Fondateurs et Commentateurs. Vol. II: Ibn al-Haytham [Travaux en mathématiques infinitésimales] (French and Arabic). London.

Rebstock, Ulrich. 1992. *Rechnen im islamischen Orient: Die literarischen Spuren der praktischen Rechenkunst* (German). Darmstadt.

Rebstock, Ulrich. 1993. *Die Reichtümer der Rechner von Aḥmad b. Thabāt* (German). Beiträge zur Sprach- und Kulturgeschichte des Orients, Vol. 32. Walldorf-Hessen.

Rosen, Frederic (ed. and trans.), 1831/1997. *The Algebra of Mohammed Ben Musa*. London. Repr. Islamic Mathematics and Astronomy, Vol. 1. Frankfurt/Main.

Rosenfeld, Boris A., and Youschkevitch, Adolf P. 1954. *Al-Kāshī, Ghiyāth al-Dīn: Miftāḥ al-Ḥisāb* (Key of Arithmetic). Russian Translation and Commentary. Matematitcheskije Traktaty, *Istoriko-matematitcheskije Issledovaniya* 7, pp. 13–326.

Saidan, Ahmad S. 1971. *The Arithmetic of Abū'l-Wafā' al-Būzjānī*, (Arabic) Edition, Introduction, Commentaries and Reference to *The Arithmetic of al-Karajī*. Amman.

Saidan, Ahmad S. 1974. "The Arithmetic of Abū'l-Wafā'," *Isis* 65, pp. 367–375.

Suter, Heinrich. 1986. *Beiträge zur Geschichte der Mathematik und Astronomie im Islam*, Nachdruck seiner Schriften aus den Jahren 1892–1922, 2 Vols. (German). Ed. Fuat Sezgin. Frankfurt am Main.

Vernet, Juan (in collaboration with A. Catalá). 1970/1979. "Un ingeniero árabe del siglo XI: al-Karajī," *Al-Andalus* 35, pp. 69–91. Repr. in *Estudios sobre historia de la ciencia medieval*, pp.147–169 (Spanish). Barcelona.

Vogel, Kurt (ed.). 1968. *Chiu Chang Suan Shu* (Neun Bücher arithmetischer Technik), with commentary (German). Braunschweig.

Warren, John. 1991. "Creswell's Use of the Theory of Dating by the Acuteness of the Pointed Arches in Early Muslim Architecture," *Muqarnas*, Vol. 8, pp. 59–65.

Video: "Qubba for al-Kāshī" (16 min.), Yvonne Dold-Samplonius, Technics: Christoph Kindel and Kurt Saetzler, IWR, Heidelberg 1995/6. Distributed by AMS 1997, *see* www.ams.org.

V

Seventeenth-Century Transmission of Astronomy

9

The *Sarvasiddhāntarāja* of Nityānanda

David Pingree

One of the most neglected areas in the history of Islamic astronomy is the development of that science in South Asia, the influence of Sanskrit astronomy on it, and its impact on the older Indian tradition of the siddhāntas.[1] Vast quantities of relevant manuscripts in Persian, Arabic, and Sanskrit survive in public and private collections in South Asia and, for the Persian and Arabic material, in Iran and Central Asia as well, but little attention has until very recently been paid to this important material. This chapter will deal with one aspect of this general question, namely, with the presentation of Muslim planetary models by the Brāhmaṇa scholar, Nityānanda, who wrote at Delhi during the reign of Shāh Jahān, his sources, and his influence.[2]

Farīd al-Dīn Masʿūd ibn Ibrāhīm al-Dihlawī apparently left the service of the ʿĀdil Shāh of Bījāpūr on the death of his patron, Ibrāhīm ʿĀdil Shāh II, in 1627, and seems to have joined the entourage of Shāh Jahān at Junnar in Mahārāṣṭra before the future Emperor marched toward Delhi on 2 December of that year.[3] He cast the horoscope, presumably beforehand, for Shāh Jahān's enthronement, which took place on 14 February 1628.[4] Shortly after that event, he was asked by Shāh Jahān's wazīr, Āsaf Khān, to prepare a new zīj, the *Zīj-i-Shāh-Jahānī*, that would be based on Ulugh Beg's *Zīj-i-jadīd*, but would employ a new calendar, the taʾrīkh-i-Ilāhī Shāhishānī.[5] Farīd al-Dīn was able to present the enormous *Zīj-i-Shāh-Jahānī* to the Emperor in October 1629, the very month in which he is said to have died.[6]

Āsaf Khān was impressed by the new zīj, and decided to have it translated into Sanskrit. The task of making the translation was assigned to Nityānanda, a Brāhmaṇa residing in Delhi; he completed the translation, which he entitled *Siddhāntasindhu*, in the early 1630s. Uniform copies of gigantic size (45 × 33 cm. and approximately 440 folia) were prepared,[7] and at least eleven were distributed to worthy individuals (mostly Muslim nobles) in Northern India.[8] Four copies of the *Siddhāntasindhu* are kept at the Palace Library in Jaipur. Three are perhaps from among the original production; one bears the seal of Shāh Jahān.[9] The fourth was copied, in identical style, by Gaṅgārāma of Kāśmīra for Mahārāja Jayasiṃha in 1727.[10] The four copies at Jayapura[11] are the only

Table 9.1

	Romaka	Brāhmasphuṭa	Sūrya
Sun	4,320,000,000	4,320,000,000	4,320,000,000
Moon	57,750,968,965	57,753,300,000	57,753,336,000
Mars	2,296,968,639	2,296,828,522	2,296,832,000
Mercury's śīghra	17,936,534,114	17,936,998,984	17,937,060,000
Jupiter	364,356,698	364,226,455	364,220,000
Venus' śīghra	7,022,180,538	7,022,389,492	7,022,376,000
Saturn	146,835,981	146,567,298	146,568,000
Days	1,577,847,748,101	1,577,916,450,000	1,577,917,828,000
Year in days	6,5;14,33,7,24,31,...	6,5;15,30,22,30	6,5;15,31,31,24

complete copies extant; four incomplete manuscripts are preserved in other libraries in Rājasthān (Alwar and Bikaner) and Madhya Pradeśa (Ujjain).[12]

The failure of the *Siddhāntasindhu* to find favor among the Hindu adherents of the traditional siddhāntas inspired Nityānanda to write an elaborate apology for using Muslim astronomy, the *Sarvasiddhāntarāja* that he completed in 1639.[13] The second and third chapters of this work, on the computation of the mean and true longitudes of the planets, is the part of this gigantic and unpublished work on which we will focus, using manuscripts from Benares and London.[14]

Whereas the *Siddhāntasindhu*, following Ulugh Beg, presents the mean motions of the Sun, the Moon, and the planets as motions in multiples of 30 Arab years, single Arab years, Arab months, days, and hours, and also presents motions in solar years and solar months given Persian names, Nityānanda in the *Sarvasiddhāntarāja* has converted these mean motions approximately into integer numbers of revolutions in a Kalpa of 4,320,000,000 years in order to make them comparable and comprehensible to his expected audience of siddhāntins. In the course of his presentation of the parameters and models of the *Siddhāntasindhu*, which he calls the *Romakasiddhānta* or *Roman Zīj*, he compares them with those of the *Brāhmasphuṭasiddhānta* composed by Brahmagupta in 628[15] and those of the *Sūryasiddhānta* composed by an unknown author in about 800.[16] Though he does not provide the numbers of rotations of these planets in a Kalpa according to these two representatives of the Indian tradition, I include them for comparison in table 9.1.

The shorter year-length in the *Romaka* should diminish the number of rotations each planet makes in a Kalpa. For instance, the difference in days between the *Romaka* and the *Brāhmasphuṭa* amounts to about 188,095 years, in which there are about 2,508,000 rotations of the Moon; the *Romaka*'s rotations of the Moon are 2,331,035 less than those of the *Brāhmasphuṭa*. However, the rotations of Mars, Jupiter, and Saturn according to the *Romaka* are all greater in number than they are according to the *Brāhmasphuṭa*.

How Nityānanda derived his rotations in a Kalpa from the mean motions of the *Siddhāntasindhu*—that is, of the *Zīj-i-Shāh-Jahānī*—has yet to be determined. But the yearly mean motions they imply are easily computed from $\frac{360\ R}{4{,}320{,}000{,}000}$ (the Sun obviously travels 360° in a year).

Moon	13 rotations + 2, 12; 34, 50, 39, . . .°
Mars	3, 11; 24, 50, 34, . . .°
Mercury's śīghra	4 rotations + 54; 42, 40, 12, . . .°
Jupiter	30; 21, 46, 58, . . .°
Venus' śīghra	1 rotation + 3, 45; 10, 54, 9, . . .°
Saturn	12; 14, 10, 47, . . .°

Later on, in order to "simplify" the calculations, Nityānanda, in the fashion of an Indian karaṇa, imagines a period of 10,000,000 days, which he equates with 27,379 solar years so that each year equals 6,5;14,36,20, . . . days instead of the earlier 6,5;14,33,7, . . . days. Using the ratio of the solar years in 10,000,000 days to those in a Kalpa, one can easily compute the "integer" rotations of each planet in the shorter period (Table 9.2).

The epoch that Nityānanda chooses is the beginning of the Indian month Caitra in the year in which Shāh Jahān was enthroned—that is, 25 February 1628—at noon at Laṅkā. He gives the epoch mean longitudes of the planets as numbers of lapsed days in a period of 10,000,000 days. Since these numbers all

Table 9.2

Planet	Text	Computation
Sun	27,379	(27,379)
Moon	366,011	366,010.1...
Mars	14,557	14,557.5...
Mercury's śīghra	113,677	113,676.9...
Jupiter	2,309	2,309.1...
Venus' śīghra	44,504	44,504.6...
Saturn	930	930.6...

have seven digits, on division by 10,000,000 they become decimal fractions of a rotation. In the following table the true longitudes of the planets as given in the Tuckerman tables are provided for comparison (Table 9.3).

Mars is so far off because its elongation from the Sun is 119°, which produces an equation of about –39°.

In the midst of this computation Nityānanda reports on the theory of precession according to Maya the Asura.[17] This Maya was the person to whom the Sun revealed the *Sūryasiddhānta*; he is mentioned here because Nityānanda claims that the *Romaka* was revealed to a Yavana by the Sun.[18] This theory certainly comes from a Muslim, not an Indian source. Nityānanda states that, according to Maya, precession amounted to 16;6,32° in 1628; this puts the date of coincidence of the tropical and sidereal zero-points in AD 500. He further notes that the zero-point was once in Aries 8°, at another time in Aries 10°; these are the Babylonian norms for Systems B and A of the Moon.[19] It is odd to find them surfacing in a seventeenth century Sanskrit poem.

Since the computation of the rotations of the planets in 10,000,000 days was somewhat crude, Nityānanda offers a daily bīja or correction for each (Table 9.4).

The longitudes of the apogees of the Sun and the five star-planets and the nodes of the latter in 1628 are presented in the following table as they are given by Nityānanda (Table 9.5).

The lunar apogee and node rotate respectively 488,327,103 and 232,088,311 times in a Kalpa or 3095 and 1471 times respectively in 10,000,000 days; their epoch positions are given as 7,134,658 or 256;50, . . .° and 6,765,363 or –243;33, . . .° = +116;26, . . .°.

In an attempt to persuade his readers that the differences between his three systems are small, Nityānanda gives the corrections to derive the longitudes of the apogees and nodes of the *Sūryasiddhānta* from those of the *Romakasiddhānta*, and states those of the *Brāhmasphuṭasiddhānta*. He does not inform his readers that he has added precession (16;6,32°), as he should, to the longitudes in the *Sūryasiddhānta* (Table 9.6).

The corrections to the *Romaka*'s longitudes of the nodes are, in general, far too large to convince the reader that there is little difference between that text and the *Sūryasiddhānta*.

Nityānanda also gives the bījas or corrections to go from the *Romaka*'s epoch mean longitudes to those of the *Sūryasiddhānta* corrected by precession (he does not specify this latter correction), and from the mean longitudes of the *Sūryasiddhānta* to those of the *Brāhmasphuṭasiddhānta* (Table 9.7).

Chapter 3 of the *Sarvasiddhāntarāja* deals with the computation of the true longitudes of the planets from their mean longitudes. It begins with a bald

Table 9.3

Planet	Lapsed days	Epoch mean longitudes	True longitudes	Differences
Sun	9,565,549	344;21,35°	346°	–2°
Moon	9,589,692	345;13,44°	355°	–10°
Mars	2,897,712	104;19,3°	73°	+31°
Mercury's śīghra	9,381,010	337;36,58°	(335°)	
Jupiter	7,363,135	265;4,22°	270°	–5°
Venus' śīghra	1,087,113	39;8,9°	(8°)	
Saturn	5,054,485	181;57,41°	189°	–7°

Table 9.4

Sun	$+\ \frac{243°}{10{,}000{,}000}$	=	+ 0;0,0,5,14,55,40,48°
Moon	$+\ \frac{121°}{10{,}000{,}000}$	=	+ 0;0,0,2,36,48,57,36°
Mars	$+\ \frac{2166°}{10{,}000{,}000}$	=	+ 0;0,0,46,47,8,9,36°
Mercury's śīghra	$+\ \frac{1246°}{10{,}000{,}000}$	=	+ 0;0,0,26,54,48,57,36°
Jupiter	$+\ \frac{716°}{10{,}000{,}000}$	=	+ 0;0,0,15,27,56,9,36°
Venus' śīghra	$+\ \frac{2200°}{10{,}000{,}000}$	=	+ 0;0,0,47,31,12°
Saturn	$+\ \frac{2186°}{10{,}000{,}000}$	=	+ 0;0,0,47,13,3,21,36°

Table 9.5

Planet	Apogees		Nodes	
Sun	Cancer	5;9,33,15°		
Mars	Leo	24;40,23°	Taurus	20;40,23°
Mercury	Scorpio	7;12,3°	Aquarius	7;12,3°
Jupiter	Libra	2;15,6°	Cancer	10;15,6°
Venus	Gemini	25;9,0°	Pisces	25;9,0°
Saturn	Sagittarius	19;39,6°	Cancer	19;39,6°

Table 9.6

Apogees							
Planet	Romaka		± Correction			Sūrya with precession	Brāhmasphuṭa
Sun	95;93,15°	–	1;46,5,15°	=	93;23,28°	93;23,32°	77;56,56°
Mars	144;40,23	+	1;28,34	=	146;8,57	146;8,32	128;25,8
Mercury	217;12,3	+	19;22,23	=	236;34,26	236;33,40	224;54,53
Jupiter	182;15,6	+	5;12,43	=	187;27,49	187;25,58	172;35,49
Venus	85;9,0	+	10;49,11	=	95;58,11	95;57,5	81;17,35
Saturn	259;39,6	–	6;55,3	=	252;44,3	252;43,58	238;16,9

Nodes							
Planet	Romaka		+ Correction			Sūrya with precession	Brāhmasphuṭa
Mars	50;40,23°	+	5;29,30°	=	56;9,53°	56;10,19°	21;53,27°
Mercury	307;12,3	+	89;35,45	=	36;47,48	36;48,48	21;8,34
Jupiter	100;15,6	+	355;31,33	=	95;46,39	95;47,11	82;1,9
Venus	355;9,0	+	80;37,36	=	75;46,36	75;48,50	59;43,55
Saturn	109;39,6	+	6;48,58	=	116;46,36	116;29,17	103;3,9,42

statement of the radii of the epicycles in the three siddhāntas (the eccentricities in place of the manda-epicycles for the *Romaka*); however, the pulsating epicycles of the Sun and the Moon in the *Brāhmasphuṭasiddhānta* are ignored; only their values for meridian-crossings are presented (Table 9.8).

The author makes no comment on the fact that the *Romaka*'s lunar model has a crank-mechanism. This, however, like the other elements in Muslim astronomy that are not found in Indian science, is called by a Sanskrit name, pākṣika (relating to the half-months between the syzygies) in order to camouflage its foreignness. We shall comment more on this strategy of Nityānanda later.

After a long discursus on the geometrical method of computing a Sine-table in which he is clearly indebted to a Muslim source, he turns to the method of computing and applying the equations according to the *Romaka*.

In the case of the Sun, using Ptolemy and a table of Sines ($R = 60 = 1{,}0$) instead of Chords, one should form (see figure 9.1) a triangle whose sides are

$$\mathrm{Sin}\,\kappa \times \frac{e}{R} \text{ and } \mathrm{R} \pm \mathrm{Cos}\,\kappa \times \frac{e}{R}$$

Table 9.7

Planet	Romaka	± Correction	= Precessed	− Precession	= Corrected	± Correction	= Corrected
			Sūrya		Sūrya		Brāhmasphuṭa
Sun	344;21,35°	− 1;55,11°	= 342;26,24°	− 16;6,32°	= 326;19,52°	+ 0;3,29°	= 326;23,21°
Moon	345;13,44	− 1;46,32	= 343;27,12	− 16;6,32	= 327;20,40	+ 0;25,43	= 327;46,23
Mars	104;19,3	− 2;53,57	= 101;25,6	− 16;6,32	= 85;18,34	− 1;15,13	= 84;3,21
Mercury's śīghra	337;36,58	− 5;0,19	= 332;36,39	− 16;6,32	= 316;30,7	− 1	= 315;30,7
Jupiter	265;4,22	+ 1;13,47	= 265;18,9	− 16;6,32	= 250;11,37	+ 0;8,24	= 250;20,1
Venus' śīghra	39;8,9	+ 5;35,10	= 44;43,19	− 16;6,32	= 28;36,47	− 1;53,18	= 26;43,29
Saturn	181;57,41	− 6;59,24	= 174;58,17	− 16;6,32	= 158;51,45	+ 0;7,5	= 158;58,50

Table 9.8
Eccentricities/Manda epicycles' radii

Planets	Romaka	Sūryasiddhānta	Brāhmasphuṭasiddhānta
Sun	2;1,20	2;20–2;16,40	2;16,40
Moon	5;12,24	5;20–5;16,40	5;16
Crank mechanism	10;24,48		
Mars	6;4 or 6;14	12;30–12;40	11;40
Mercury	3	5;0–4;40	6;20
Jupiter	2;47	5;30–5;20	5;30
Venus	0;52	2;0–1;50	1;50–1;30
Saturn	3;29	8;10–8;0	5

Śīghra epicycles' radii

Planets	Romaka	Sūryasiddhānta	Brāhmasphuṭasiddhānta
Mars	39;43	39;10–38;40	40;36,40–39;30
Mercury	22;30	22;10–22	22
Jupiter	11;47	11;40–12	11;20
Venus	43;10	43;40–43;20	43–43;50
Saturn	6;51	6;30–6;40	5;50

as the argument, κ, lies in the semicircle beginning at 270° from the apogee or in that beginning at 90° from the apogee. Nityānanda, since $R = 1,0$, simply forms $\text{Sin}\,\kappa \times e$ and $\text{Cos}\,\kappa \times e$ without informing his reader that the sexagesimal point must be shifted one place to the left. The square-root of the sum of the squares of these two sides is the distance of the Sun from the earth, SO. Then

$$\text{Sin}\,\delta = \text{Sin}\,\kappa \times \frac{e}{R} \times \frac{R}{SO} = \text{Sin}\,\kappa \times \frac{e}{SO}.$$

Without a hint of irony, Nityānanda states that some extremely clever people advocate the ancient Indian "Method of Sines," in which

$$\text{Sin}\,\delta = \text{Sin}\,\kappa \times \frac{e}{R}.$$

Ptolemy's solution to the problem of finding geometrically the angle which corrects the anomaly in the lunar epicycle is quite straightforward. In figure 9.2 $P_1O = P_2O = 10;19$ and $P_1C = 49;41$. Since angle 2η is given, in the

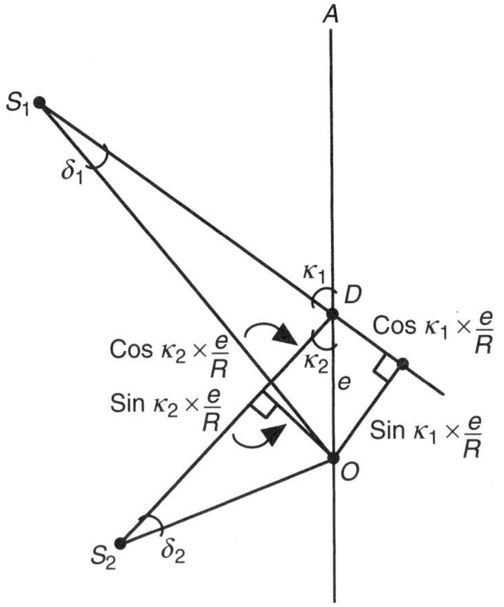

Figure 9.1

right triangle P_2OL_2 the sides P_2L_2 and L_2O can be computed. Similarly, in the right triangle P_1OL_1 the sides P_1L_1 and L_1O are respectively equal to P_2L_2 and L_2O. Then $CL_1^2 = CP_1^2 - P_1L_1^2$; and $CO = CL_1 + L_1O$. Moreover, $CL_2 = CO + OL_2$; and $CP_2^2 = CL_2^2 + L_2P_2^2$. Now, in right triangle CL_2P_2 angle L_2CP_2, which equals angle ACA', can be computed.

Nityānanda's solution, based on the use of the Sine and Cosine functions, is more complex, and its interpretation is made difficult by the terminology that he has invented to name the parts of the figure. $P_1O = OP_2$ he calls the pākṣa epicycle-radius and measures at 10;24,48; P_1C, which is 60−10;24,48 = 49;35,12, is called the radius of the pākṣa eccentric. The elongation of the mean Moon from the mean Sun is the pākṣa argument; it is to be doubled. The Sine of the double elongation is multiplied by the epicycle-radius, P_1O, and divided by the radius of the pākṣa (here one must understand the pākṣa eccentric, P_1C). This produces P_1L_1, the Sine of angle P_1CL_1 measured in the units of a circle whose radius is 60.

$$\text{Sin}_{60}2\eta \times \frac{P_1O}{60} = \text{Sin}_{P_1O}2\eta = P_1L_1. \text{ Also } P_1L_1 = \text{Sin}_{P_1C} \angle P_1CL_1.$$

Therefore,

$$\text{Sin}_{60} \angle P_1CL_1 = \text{Sin}_{60}2\eta \times \frac{P_1O}{60} \times \frac{60}{P_1C} = \text{Sin}2\eta \times \frac{P_1O}{P_1C}.$$

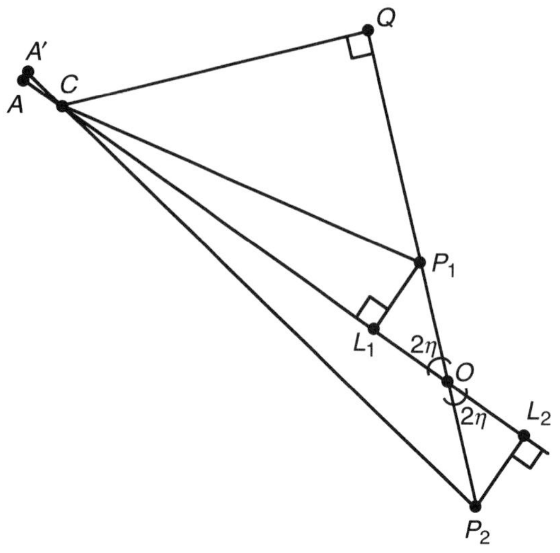

Figure 9.2

For my understanding of the rest of Nityānanda's rules I am indebted to my colleague, Kim Plofker. The first rule is to take the Sine from the argument (that is, 2η) increased or decreased by the arc of the previous result (that is, angle P_1CL_1) as the argument lies between 0° and 180° or 180° and 360° respectively. In figure 9.2 angle $CP_1O = 180° - (2\eta + \angle P_1CL_1)$. Therefore,

$$2\eta + \angle P_1CL_1 = 180° - \angle CP_1O = \angle CP_1Q, \text{ and}$$

$$QC = \mathrm{Sin}(2\eta + \angle P_1CL_1) \times \frac{P_1C}{R}.$$

Right triangles OQC and OL_1P_1 are similar to each other. Therefore,

$$\frac{OC}{QC} = \frac{P_1O}{P_1L_1}, \text{ or } OC = \frac{QC \times P_1O}{P_1L_1}.$$

$$\text{Since } P_1L_1 = \mathrm{Sin}\,2\eta \times \frac{P_1O}{R},$$

$$OC = \mathrm{Sin}(2\eta + \angle P_1CL_1) \times \frac{P_1C}{R} \times P_1O \times \frac{R}{\mathrm{Sin}\,2\eta \times P_1O},$$

which reduces to:

$$OC = \mathrm{Sin}(2\eta + \angle P_1CL_1) \times \frac{P_1C}{\mathrm{Sin}\,2\eta}.$$

Indeed, Nityānanda's instructions are to multiply $\mathrm{Sin}(2\eta + \angle P_1CL_1)$ by the radius of the eccentric, P_1C, and to divide the product by the Sine arising from the original argument—that is, by $\mathrm{Sin}\,2\eta$. The result he correctly states to be the pākṣa hypotenuse, OC.

Now one is instructed to multiply the original $\mathrm{Sin}\,2\eta$ and $\mathrm{Cos}\,2\eta$ by the epicycle-radius, P_1O, and to divide the results by 60; this converts the two Sines from base 60 to base 10;24,48, to the lengths $P_1L_1 = P_2L_2$ and $L_1O = L_2O$ in figure 9.2. Then the sum or difference of OC and LO is CL; and $CL^2 + PL^2 = P_2C^2$, the hypotenuse from the opposite point (abhimukhacihna). Finally,

$$L_2P_2 \times \frac{60}{P_2C} = \mathrm{Sin}\,\angle L_2CP_2;$$

its arc is the correction to the apogee of the epicycle, $\angle ACA'$.

The rest is straightforward, but expressed as far as possible in terms familiar from traditional Sanskrit siddhāntas. The anomaly of the Moon on its epicycle is counted from the corrected epicycle-apogee, and the corrected anomaly, κ', from the uncorrected apogee. The Sine and the Cosine of κ' are multiplied by the radius of the epicycle, r, and divided by the Radius, $R = 60$. The mean pākṣika hypotenuse, CO, is increased or decreased by $\mathrm{Cos}\,\kappa' \times \frac{r}{R}$ and

$$\left(CO + \mathrm{Cos}\,\kappa' \times \frac{r}{R}\right)^2 + \left(\mathrm{Sin}\,\kappa' \times \frac{r}{R}\right)^2 = H^2,$$

where H is the distance of the Moon from the earth. Then the lunar equation is the arc of

$$\mathrm{Sin}\,\kappa' \times \frac{r}{R} \times \frac{R}{H} = \mathrm{Sin}\,\kappa' \times \frac{r}{H}.$$

Following Muslim practice, Nityānanda instructs his readers to reduce the lunar longitude on its orbit to its corresponding longitude on the ecliptic.

The computation of the equation of the center according to Nityānanda is illustrated in figure 9.3, where A is the apogee, E the equant, D the center of the deferent, and O the center of the earth. In order to convert the anomaly at E, κ, to that at D, κ', we have, analogously to the computation of the solar equation, $\mathrm{Sin}\,\delta_1 = \mathrm{Sin}\,\kappa \times \frac{e}{R}$ and $\kappa' = \kappa \pm \delta_1$. Then CH is $\mathrm{Sin}\,\kappa'$ and HD is $\mathrm{Cos}\,\kappa'$. In the right triangle CHO, $HO = HD + e$, and $CO^2 = CH^2 + HO^2$. Then,

$$OG = \mathrm{Sin}\,\delta_2 \times \frac{CO}{R} = \mathrm{Sin}\,\kappa \times \frac{2e}{R} \text{ from which } \mathrm{Sin}\,\delta_2 = \mathrm{Sin}\,\kappa \times \frac{2e}{CO}.$$

The equation of the anomaly is computed in a manner similar to that used in traditional Indian siddhāntas. The difference is that the distance of the center of the epicycle from the earth is now CO rather than R; in this way the influence of the first, manda, equation on the second, śīghra, equation is accounted for. In figure 9.4 $CA = CP = r$ is the radius of the epicycle and $\angle ACP = \gamma$,

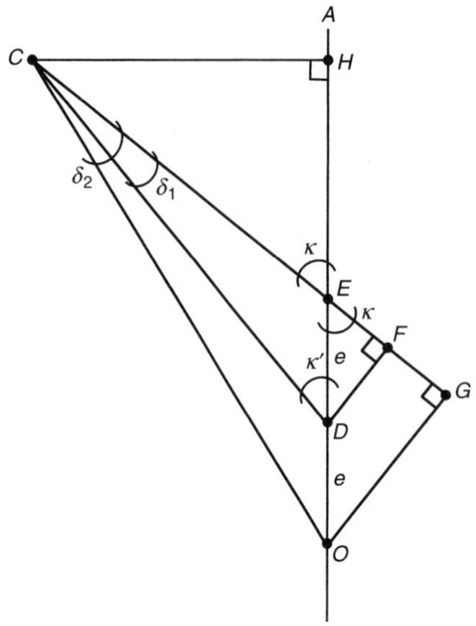

Figure 9.3

the anomaly. Then

$$BP = \text{Sin}\,\gamma \times \frac{r}{R} \text{ and } BC = \text{Cos}\,\gamma \times \frac{r}{R}. \text{ Then}$$

$$BO = CO \pm BC;\ PO^2 = BO^2 + BP^2;\ \text{and Sin}\,\delta = BP \times \frac{R}{PO}.$$

Nityānanda's algorithm for computing the equation of the center of Mercury is, of course, more complex. It can be understood with the help of figure 9.5.

Here $OE = ED = DM = DN = e$, $NC = R$, and $\angle MDN = \angle DEC = \angle OEH = \kappa$. Nityānanda begins by subtracting κ from 180° (or 180° from κ); this yields $\angle NDE$. Since in triangle NDE the perpendicular, DF, from D to NE bisects $\angle NDE$,

$$\angle NDF = \angle FDE = \frac{180° - \kappa}{2}.$$

Then $2\left(\text{Sin}\frac{180° - \kappa}{2} \times \frac{e}{R}\right) = NE.$

Further, since in the right triangle EDF:

$$\angle EDF = \frac{180° - \kappa}{2},\ \angle DEF = 90° - \frac{180° - \kappa}{2} = \frac{\kappa}{2};$$

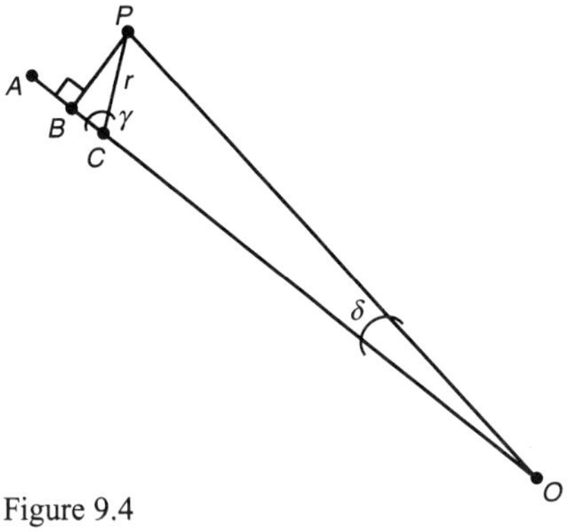

Figure 9.4

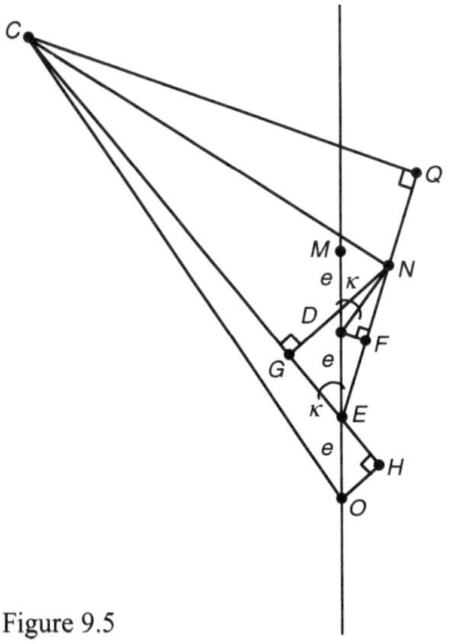

Figure 9.5

therefore $\angle CEF = \dfrac{3\kappa}{2}$.

Then $NG = \text{Sin}\dfrac{3\kappa}{2} \times \dfrac{NE}{R}$.

In triangle ECN $\angle CNE = 180° - \left(\dfrac{3\kappa}{2} + \angle NCE\right)$;

and $\text{Sin}\angle NCE = NG = \text{Sin}\,\dfrac{3\kappa}{2} \times \dfrac{NE}{R}$.

Thus $\angle CNQ = 180° - \angle CNE = 180° - \left(180° - \left(\frac{3\kappa}{2} + \text{arc}\left(\text{Sin}\frac{3\kappa}{2} \times \frac{NE}{R}\right)\right)\right)$;
and $\text{Sin}\angle CNQ = CQ = \text{Sin}\left(\frac{3\kappa}{2} + \text{arc}\left(\text{Sin}\frac{3\kappa}{2} \times \frac{NE}{R}\right)\right)$.

But right triangle *CQE* is similar to right triangle *NGE*; therefore, according to Nityānanda,

$$CE = CQ \times \frac{NE}{NG \times \frac{NE}{R}} = \text{Sin}\left(\frac{3\kappa}{2} + \text{arc}\left(\text{Sin}\frac{3\kappa}{2} \times \frac{NE}{R}\right)\right) \times \frac{NE}{NG} \times \frac{R}{NE}$$
$$= \text{Sin}\left(\frac{3\kappa}{2} + \text{arc}\left(\text{Sin}\frac{3\kappa}{2} \times \frac{NE}{R}\right)\right) \times \frac{R}{NG}.$$

It is clear that $HO = \text{Sin}\,\kappa \times \frac{e}{R}$ and $EH = \text{Cos}\,\kappa \times \frac{e}{R}$;
and $CH = CE + EH$. Then $CO^2 = CH^2 + HO^2$; and $\text{Sin}\,\delta = HO \times \frac{R}{CO}$.

Having provided this much, Nityānanda omits the rules for computing Mercury's equation of the anomaly since they do not differ from those for the other planets.

In all of these algorithms for computing planetary equations Nityānanda has provided only the final bare rules—presumably derived from a Persian commentary on Ulugh Beg's zīj—but no hint of the geometrical rationales that lie behind them. Later on he describes the Muslim system of internesting planetary spheres, and, in the section where he provides the algorithms for these, in most manuscripts there are spaces left blank for diagrams; however, in only one of the half-dozen or so manuscripts that I have been able to inspect, Alwar 2005, which was copied in 1846, are any diagrams actually drawn, and they completely misrepresent the geometry that we can reconstruct from the rules.

Therefore, though the algorithms are mostly correct and though Nityānanda, as we demonstrated at the beginning of this chapter, is doing his best to persuade the traditionalists that the so-called *Romakasiddhānta* is not very different from the *Sūryasiddhānta*, he, or the scribes of the *Sarvasiddhāntarāja* if they simply failed to copy his diagrams, has or have failed to provide sufficient information for an expert in the *Sūryasiddhānta* to understand how the *Romaka* arrives at its results. This problem is made worse by Nityānanda's introduction of a new vocabulary, based on Sanskrit, for referring to the elements of the Ptolemaic models without defining these new terms; it had been the practice of other scholars attempting to describe Muslim models in Sanskrit to transliterate the Arabic/Persian technical terms, and to define them. Thus, Nityānanda refers to the various elements of the lunar crank-mechanism by applying the adjective pākṣa (pertaining to a half-month) to the normal Sanskrit terms: pākṣakarṇa (the distance of the center of the Moon's epicycle from the earth), pākṣakendra (the double elongation of the Moon from the Sun), pākṣapara (the radius of the concentric circle bearing the center of the Moon's deferent), and pākṣapratimaṇ ḍalavyāsārddha (the radius of the Moon's deferent). The term "opposite point" (nuqṭat al-muḥādhā) is translated as "abhimukhacihna." Curiously, the equant (markaz muʿaddal al-masīr) is called by Nityānanda the "abhicārabindu" or "point of magic."

The most noteworthy later use of chapter 3 of Nityānanda's *Sarvasiddhāntarāja* is in a manuscript I found in the Pauṇḍarīka Collection in the Palace Library at Jaipur.[20] This contains the earliest version, datable to 1726 or 1727, of Jagannātha's *Siddhāntakaustubha* in which, on ff. 20v–25v, Jayasiṃha's guru[21] paraphrases or copies out all the verses of this chapter containing the algorithms and the parameters. He only substitutes the equivalent "sammukha" and "unmukha" for "abhimukha;" even "abhicāra" is preserved as the name of the equant. Some of the terminology applied by Nityānanda to the elements of the lunar crank-mechanism combining the adjective "pākṣa" or "pākṣika" with common Sanskrit astronomical terms such as "karṇa" (hypotenuse) and "kendra" (anomaly) appear in about 1732 in Kevalarāma's *Dṛkpakṣasāriṇī*,[22] the poetic version, and in about 1734 in the Sanskrit prose version of de La Hire's *Tabulae astronomicae*[23] and in the *Yavanacandracchedyakopayoginī*;[24] all three were products also of Jayasiṃha's court. The *Sarvasiddhāntarāja*, then, provided Jagannātha with the Sanskrit words to describe Ulugh Beg's planetary models, and Jagannātha's successors with some of the terminology with which they wrote of European astronomy.

NOTES

1. See D. Pingree, "Islamic Astronomy in Sanskrit," *JHAS* 2, 1978, 315–330, and "Indian Reception of Muslim Versions of Ptolemaic Astronomy," *Tradition, Transmission, Transformation*, ed. F. J. Ragep et. al, Leiden 1996, pp. 471–485.

2. D. Pingree, *Census of the Exact Sciences in Sanskrit* (henceforth *CESS*), Series A, vols. 1–5, Philadelphia 1970–1994, A3, 173b–174a; A4, 141a–141b; and A5, 184a; "The Indian Reception," pp. 476–480; and "Amṛtalaharī of Nityānanda," *Sciamus* 1, 2000, pp. 209–217.

3. *The Shah Jahan Nama of 'Inayat Khan*, translated by A. R. Fuller and edited and completed by W. E. Begley and Z. A. Desai, Delhi 1990, p. 13.

4. ʿAbd al-Ḥamīd Lāhūrī, *Pādshāh Nāma*, ed. K. Ahmad and ʿAbd al-Raḥīm, *BI* 56, 2 vols., Calcutta 1867–1868, vol. 1, p. 97.

5. Ibid., vol. 1, pp. 286–287.

6. *The Shah Jahan Nama*, p. 35.

7. Manuscripts 266 and 268 in D. Pingree et al., *A Descriptive Catalogue of the Sanskrit Astronomical Manuscripts Preserved at the Maharaja Man Singh II Museum in Jaipur, India,* forthcoming.

8. Named in manuscripts 266 and 267 in Nāgarī transliterations are: ʿĀẓam Khān of Baṅgāla, ʿAbdallāh Khān of Paṭana, Daī Ṣāḥib of Vānārasī, Iʿtiqad Khān of Dillī, Khān Dauran of Ujjayinī, Mahabat Khān Khān-i-khānān of Burhānapura, Ujīra Khān (?) of Lāhora, Jaʿfar Khān of Kāśmīra, an unnamed person of Multāna, and Nityānanda himself.

9. Manuscript 266 (Khasmohor 4960).

10. Manuscript 267 (Museum 23).

11. The fourth copy, manuscript 269 (Khasmohor 4962), belonged to Jagannātha–perhaps Jagannātha Samrāṭ.

12. Alwar 2627 and Anup 5332 and 5333; and SOI 9410.

13. See the verses cited in *CESS* A3, 174a.

14. Benares 34466 and Wellcome V.36.

15. *CESS* A4, 254b–255b, and A5, 239b.

16. *CESS* A6 (in preparation).

17. *CESS* A4, 358b.

18. See "The Indian Reception," pp. 477–478.

19. H. Hunger and D. Pingree, *The Astral Sciences of Mesopotamia*, Leiden 1999, pp. 229 and 237.

20. Manuscript 45 (Puṇḍarīka (jyotiṣa) 5).

21. *CESS* A3, 56a–58a; A4, 95a; and A5, 113b–114a.

22. Edition in preparation by S. Ikeyama and D. Pingree.

23. Edition in preparation by D. Pingree.

24. Edition in preparation by D. Pingree.

10

On the Lunar Tables in Sanjaq Dār's *Zīj al-Sharīf*

Julio Samsó

1 Introductory Remarks

Nothing is known about the characteristics of the earliest known Maghribī *zīj*, Ibn Abī l-Rijāl al-Qayrawānī's *Ḥall al-ʿaqd wa-bayān al-raṣd* (beginning of the eleventh century), which seems to have been lost. Two centuries later, Abu 'l-ʿAbbās Ibn Isḥāq al-Tamīmī al-Tunisī (fl. Tunis and Marrakesh ca. 1193–1222), left an unfinished set of tables which survive in a unique manuscript of Hyderabad, discovered in 1978 by D. A. King.[1] The predominant influence in Ibn Isḥāq's *zīj* was that of the Andalusian school represented by Ibn al-Zarqālluh (d. 1100), Ibn al-Kammād (fl. Cordova 1116–17) and Ibn al-Hāʾim (fl. 1205). This influence continued along the thirteenth and fourteenth centuries through several "editions" of the *zīj* of Ibn Isḥāq such as the one prepared by the anonymous compiler of the aforementioned collection extant in the Hyderabad manuscript (ca. 665 H./1266–680 H./1281), the *Minhāj* of Ibn al-Bannāʾ of Marrakesh (1256–1321),[2] and the two *zīj*es composed by the Tunisian-Andalusian astronomer Muḥammad ibn al-Raqqām (d. 1315).[3] Andalusian influence is also present in two other fourteenth century *zīj*es, written by two astronomers of Constantine, in the Central Maghrib, who were active in Fez: the one compiled by Abu 'l-Ḥasan ʿAlī ibn Abī ʿAlī al-Qusanṭīnī, the canons of which were written in verse,[4] and the *Zīj al-Muwāfiq* of Abu 'l-Qāsim ibn ʿAzzūz al-Qusanṭīnī (d. 1354),[5] partially based on observations made in Fez with an armillary sphere ca. 1344.

All the aforementioned *zīj*es share a certain number of characteristics among which we should mention that their mean motion tables are sidereal, and that they contain tables based on the theory of trepidation that enable the user to calculate the amount of precession for a given date and, thus, obtain the tropical longitudes of heavenly bodies. Trepidation implies a variation in the obliquity of the ecliptic and these *zīj*es also include tables, based on the model designed by Ibn al-Zarqālluh, which implies the existence of cycles that regulate the diminution and the expected future increase of the obliquity of the ecliptic. The analysis of a limited number of new sources dated between the fifteenth and the early seventeenth centuries (three commentaries on the poem on timekeeping written in 1391 by al-Jadhārī,[6] and the *Kitāb al-adwār fī tasyīr al-anwār*, written by Abū ʿAbd Allāh al-Baqqār in 821/1418)[7] give some evidence that observations were made in the Maghrib in the 13th and 14th centuries. The result of these observations was that astronomers realized that the observed precession of the equinoxes was much larger than the values that could be computed with trepidation tables and that the obliquity of the ecliptic stubbornly continued to diminish below the limits fixed in obliquity tables. The situation being thus, we find a new contact with the East which—at least in the case of al-Andalus—had been interrupted in the eleventh century. This brings to the Maghrib of the late fourteenth century the *Tāj al-azyāj* of Muḥyī al-Dīn al-Maghribī (d. 1283) and the *Zīj al-Jadīd* of Ibn al-Shāṭir (d.1375), and there is evidence of adaptations of these *zīj*es for their use in specific cities of the Maghrib. These *zīj*es compute directly tropical longitudes, are based on constant precession, reject the existence of cycles regulating the obliquity of the ecliptic and they offer results that agree with observations much better than those computed with the tables of the Andalusian school. In spite of this the Andalusian tradition stayed alive until the nineteenth century and coexisted with the "new" Eastern *zīj*es. The reason is clearly explained by the astrologer al-Baqqār: horoscopes were cast using sidereal longitudes and my impression is that astrologers used Ibn al-Bannā's *Minhāj* and other similar sources, while astronomers and conscientious *muwaqqit*s prefered oriental *zīj*es.[8]

The last Eastern *zīj* which was introduced in the Maghrib was Ulugh Begh's *Zīj-i Sulṭānī* (1393–1449),[9] usually called *al-Zīj al-Jadīd* in Maghribī sources. I have no evidence that this *zīj* was known in the Maghrib before the end of the seventeenth century, but it is obvious that it became very popular during the eighteenth and nineteenth centuries. The Arabic translation of the introduction to this *zīj*, made by Ḥasan b. Muḥammad, known as Qāḍī Ḥasan al-Makkī (fl. 17th c.) reached Morocco, for the Ḥasaniyya Library preserves a manuscript of this work dated in 1291/1874.[10] Interestingly, however, there were, at least, two Tunisian recensions of this *zīj* prepared by Muḥammad al-Sharīf, called Sanjaq Dār al-Tūnisī and by ʿAbd Allāh Ḥusayn Quṣʿa b. Muḥammad b. Ḥusayn al-Ḥanafī al-Tūnisī.[11] Both authors are undated but MS 16650 of the Tunis National Library (Khaldūniyya collection) contains a copy of Quṣʿa's recension—under the title *Ghunyat al-ṭālib fī taqwīm al-kawākib*—where it is stated that the work (not the copy) was finished on Thursday 22nd Shawwāl 1091 H./15 November 1680 (which was a Friday), and that the tables have been adapted to the longitude of Tunis (41;45° from the western meridian).[12]

As for Sanjaq Dār's *Zīj al-Sharīf*—also called *al-Zīj al-Mukhtaṣar fī ʿilm al-taʿdīl wa l-taqwīm*—I have been able to study two manuscripts of this work, extant in MSS 1816 of the National Library in Tunis[13] (MS Ta hereafter) and 18104 of the library of the late Tunisian scholar Ḥasan Ḥusnī ʿAbd al-Wahhāb, now also in the Tunis National Library[14] (MS Tb hereafter). MS Tb is dated in 1107/1695–1696 (fol. 62 r).[15] In this *zīj* the mean positions of the Sun, Moon and planets are given, in intervals of thirty years, between 1080/ 1669–1670 and 1290/1873–1874; on the other hand, the positions (?) of the lunar mansions are given, according to the observations of Ulugh Begh, for the end of year 1090/ 1679–1680 (MS Ta, fol. 36r; MS Tb, fol 90 r) and the day, hour and minute of the spring equinox is given between years 1090/1679 and 1205/1790. All this evidence points to a work made in the late seventeenth century,[16] and this agrees with the presumed date for Quṣʿa's *Ghunya*: commentaries on the introduction of this *zīj* were written towards the end of the seventeenth or the beginning of the eighteenth century. One of them was written by Aḥmad b. Muḥammad Bū Daydaḥ al-Qādīrī al-Qayrawānī (fl. before 1150/1737–1738), who praised, in his introduction, the importance of Sanjaq Dār's *zīj* but added that this work was difficult to understand and to use. For that reason commentaries had been written by his masters *al-ḥājj* Abū ʿAbd Allāh Muḥammad al-Qalī and, before him, by Abu 'l-Ḥasan ʿAlī b. Māmī al-Ḥanafī, known as Karbāṣo (or Karbāṣa) (before 1163/1749–1750).[17] Al-Qālī's commentary was lost in the time of al-Qadīrī, while the work of Karbāṣo contained many mistakes and al-Qadīrī wrote, for that reason, his *Ḥāshiya ʿalā sharḥ al-Sharīf al-Tunisī Sanjaq Dār li-zīj al-Sulṭān Ulugh Beg,*[18] which, apparently, is not a very interesting work.

Sanjaq Dār's *zīj*, like Quṣʿa's *Ghunya*, is computed for 41;45° of longitude and it contains tables of oblique ascensions for a latitude of 36;50° (MS Ta, fol. 28 r; MS Tb, fol. 82 r), id. for the division of the houses for the same latitude (MS Ta, fols. 28 v–31 r; MS Tb, fols. 82 v–85 r), id. of half of daylight for the same latitude (MS Ta, fol. 33 v; MS Tb, fol. 87 v), lunar longitude at sunset for a latitude 36;40° (MS Ta, fol. 35 r; MS Tb, fol. 89 r), lunar parallax in longitude and latitude also for 36;40° (MS Ta, fol. 36 v; MS Tb, fol. 90 v) and ascendants of anniversaries for the same latitude (MS TA, fol. 45 v; MS Tb, fol. 98 r)): 36;50° is the modern value for the latitude of Tunis and it does not appear attested in any other historical source, while 36;40° is one of the three values for the latitude of Tunis given by al-Khwārizmī[19] and it is systematically used by the compiler of the Hyderabad recension of the *zīj* of Ibn Isḥāq. Apart from this, it is interesting to note that the prologue of the *Zīj al-Sharīf* states that the author is going to follow the *ḥabṭaq* method, which means that planetary equations are computed in a simplified way using double argument tables. This type of tables is documented in the East from the time of Ibn Yūnus (d. 1009),[20] but I do not know of any instance in which they appear in the Maghrib before the *Zīj al-Sharīf* for the computation of lunar or planetary equations. In spite of this, Ibn al-Kammād (fl. Cordova 1116) used double argument tables for the computation of time from mean to true syzygy[21] and for the calculation of solar eclipses,[22] and the Hyderabad recension of Ibn Isḥāq's *zīj* contains a set of double argument planetary latitude tables, based on Ptolemaic parameters, attributed to an otherwise unknown Ibn al-Bayṭār.[23] In the equation tables of the *Zīj al-Sharīf* one enters the table in the vertical sense with the mean anomaly of the Moon or the planet (tabulated from 0° to 360° with 6° intervals) and in the horizontal sense with the *zodiacal sign* of the *markaz* (the double elongation in the case of the Moon, and the mean longitude of the planet computed from the apogee): the total equation (center + anomaly) will be read directly in the intersection of the two and will be added to the mean longitude of the planet.

The present chapter intends to begin the study of this set of double argument equation tables with the case of the Moon, which seems particularly accessible. The *Zīj al-Sharīf* contains two different sets of tables for the computation of the solar and lunar longitudes: on the one hand, mean motions of the Sun, Moon, solar apogee, lunar nodes, and double elongation calculated to the precision of minutes (MS Ta, fols. 10 r and v; MS Tb, fols. 63 v–64 r); a table of the solar equation (MS Ta, fol. 10 r; MS Tb, fol. 63 v), also calculated to the precision of minutes and with an interval of 6° of the argument (tabular maximum 3;52° for arguments 264°–270°, the table being obviously displaced vertically 1;56°); the aforementioned double argument table for the lunar equation to the precision of minutes (MS Ta, fols. 11 r and v; MS Tb, fols. 65 v–66 r).

The second set of solar and lunar tables is far more precise. The solar mean motion tables (MS Ta, fol. 23 r; MS Tb, fol. 78 r) give the mean motion of the Sun and of the solar apogee (to the precision of thirds), for hours, days, lunar months, lunar years *mabsūṭa*, and the positions of the *markaz* (solar longitude from the apogee) for the end of years 1080–1290 H. in 30-year intervals. Here, as in the rest of the *zīj*, the system of intercalation is the same as that used by Ulugh Beg, in which years 2, 5, 7, 10, 13, 15, 18, 21, 24, 26, and 29 of the 30-year cycle are leap years (*kabīsa*). The corresponding solar equation table (MS Ta, fol. 24 r; MS Tb, fol. 78 r) is calculated for every degree of the argument from 0° to 359°, to the precision of thirds, it has a vertical displacement of 1;55,53,12°,[24] and it reaches a tabular maximum of 3;51,46,24° for an argument of 268°. As for the Moon, the mean motion tables (MS Ta, fols. 24 v–25 r; MS Tb, fols. 78 v–79 r) give the corresponding mean motions in longitude (*wasaṭ*), anomaly (*khāṣṣa*), double elongation (*markaz*) and nodes (*jawzahar*) for the same periods, computed to the precision of thirds (longitude) or seconds (the rest). Three other tables (MS Ta, fols. 25 v–26 v; MS Tb, fols. 79 v–80 v) tabulate more or less standard lunar equations which will be described in detail below. It is easy to see that—with the exception of the double argument equation tables of the first set and the lunar equations of the second—the first set is just the result of a rounding of the second one to the precision of minutes.

2 Mean Solar and Lunar Motions[25]

All the mean motion values derive from Ulugh Beg's *Zīj-i Sulṭānī*.[26]

- *Precession/motion of the apogees*: the analysis of the table of single years and that of positions for 1080–1290H. gives two different results

    0;0,0,8,27,14,19 (years)

and a value between

    0;0,0,8,27,14,*24,31,41,45,32*

and

    0;0,0,8,27,14,*28,52,55,30,50*

for the table of periods. This corresponds,
precisely, to a precession of 1° in 70 Persian years.

- *Sun*: two slightly different parameters appear to have been used.

    0;59,8,11,10,28,20° (days, months, positions 1080–1290H.)

    0;59,8,11,10,28,37° (single years)

- *Lunar longitude*: the different sets of tables are mutually compatible and the mean motion parameter used lies between

    13;10,35,1,47,53,*43,49,40,23,58*°

and

    13;10,35,1,47,53,*45,38,5,22,28*°

- *Lunar anomaly*: the underlying parameter is

    13;3,53,55,54,24°

- *Double elongation*: the underlying parameter is

    24;22,53,*23,46,8,59,46,47,36*°

It is easy to check that this daily parameter does not correspond to the double value of the difference between the mean motion in longitude of the Moon and that of the Sun. For the computation we begin subtracting the daily motion of the apogee from the rounded value of the lunar mean motion in longitude:

13;10,35,1,47,53,45° (±0,1) – 0;0,0,8,27,14,27° (±0,2) =
13;10,34,53,20,39,18° (±0,3).[27]

and then,

13;10,34,53,20,39,18° (± 0,3) – 0;59,8,11,10,28,20° =
12;11,26,42,10,10,58° (± 0,3)

The corresponding mean motion in double elongation will be

24;22,53,*24,20,21,56*° (± 0,6)

This inconsistency is already present in Ulugh Beg's tables and it may be the reason (?) of the incoherence we find in the tables of positions for years 1080–1290. Table 10.1 collects all the information on positions of the solar center (col. 1), solar apogee (col. 2), lunar longitude (col. 4), and double elongation (col. 7): the values transcribed are those of the two manuscripts of the *Zīj al-Sharīf*. A control of errors has been made in all the cases: amounts between parentheses in col. 1 correspond to the difference (expressed in thirds) between the tabular and the recomputed value. Columns 3, 5, 6, 8 have been calculated from the corresponding tabular values and column 9 contains line-by-line differences in column 8.

Table 10.2 corresponds to positions also of the solar center (*markaz*, col. 1), solar apogee (col. 2), lunar longitude (col. 4), and double elongation (col. 7) for years 841 H.–871 H., at one year intervals, as they appear in Ulugh Beg's *Zīj-i Sulṭānī*. An asterisk (*) marks the leap-years (*kabīsa*). Here, once more, we find a disagreement between the values of the double elongation computed from the solar and lunar positions and those appearing in the table. It is easy to check, however, that the positions in the *Zīj al-Sharīf* derive from those in the *Zīj-i Sulṭānī*. To establish this, I have compared two sets of positions: those corresponding to 841 H. *nāqiṣa* (which means midday of the last day of 840) and those corresponding to the last day of 1080: the interval between these two dates should be 85048 days but the actual interval used for the recomputation is one day less (?). On the other hand a correction has been introduced to account for the difference in geographical longitude between Samarqand (longitude 99;16° in Ulugh Beg's *Zīj*) and Tunis (41;45° in the *Zīj al-Sharīf*): the difference in longitude (57;31°) corresponds to $3^h\ 50^m\ 4^s$ or $0;9,35,10^d$. The positions used in the two *zīj*es are:

	Solar *markaz*	Solar apogee	Lunar longitude	Double elongation
841	18;26,0,13°	90;30,4,48°	115;51,56,28°	25;12,49°
1080	321;32,2,24°	93;49,47,58°	51;1,29,2°	2;26,56°

For n = number of revolutions in each case, we can see that:

• *Solar* markaz:
$18;26,0,13° + 0;59,8,11,10,28,20 * 85047;9,35,10^d = 360° * n + 321;32,2,27,9°$

• *Solar apogee*:
$90;30,4,48° + 0;0,0,8,27,14,26° * 85047;9,35,10^d = 93;49,47,57,28°$

• *Lunar longitude*:
$115;51,56,28° + 13;10,35,1,47,53,45° * 85047;9,35,10^d = 360° * n + 51;1,29,2,25°$

• *Double elongation*:
$25;12,49° + 24;22,53,23,46,9° * 85047;9,35,10^d = 360° * n + 2;26,55,2°$

Table 10.1
Positions in the *Zīj al-Sharīf*

	(1)	(2)	(3) = (1) + (2)	(4)	(5) = (4) – (3)	(6) = (5) * 2	(7)	(8) = (7) – (6)	(9)
	Sun (center)	Apogee	[Solar Long.	Lunar Long.	[Elongation	[Double el.	Double el. (tab.)		Diff.
1080	321;32,2,24	93;49,47,58	55;21,50,22	51;1,29,2	355;39,38,40	351;19,17,20	2;26,56	11;7,39	
1110	359;31,30,21	94;14,45,53	93;46,16,14	89;18,12,39	355;31,56,25	351;3,52,50	2;9,51	11;5,58	0;1,41
1140	37;30,58,18(+1)	94;39,43,48	132;10,42,6	127;34,56,17	355;24,14,11	350;48,28,22	1;52,44[1]	11;4,16	0;1,42
1170	75;30,26,13	95; 4,41,42	170;35,7,56	165;51,39,54	355;16,31,58	350;33,3,56	1;35,40	11;2,36	0;1,40
1200	113;29,54,9(–1)	95;29,39,36	208;59,33,45	204; 8,23,32	355;8,49,47	350;17,39,34	1;18,34	11;0,54	0;1,42
1230	151;29,22,5(–1)	95;54,37,31	247;23,59,36	242;25, 7, 9	355;1,7,33	350;2,15,6	1;1,29	10;59,14	0;1,40
1260	189;28,50,2(–1)	96;19,35,26	285;48,25,28	280;41,50[2],46	354;53,25,18	349;46,50,36	0:44,22[3]	10;57,31	0;1,43
1290	227;28,17,58(–1)	96;44,33,20	324;12,51,18]	318;58,34,24	354;45,43,6]	349;31,26,12]	0:27,18	10;55,52	0;1,39

1. MS Tb 45"
2. Ms. Ta 56"
3. MS Tb 24"

Table 10.2
Positions in Ulugh Beg's Zīj-i Sulṭanī

	(1)	(2)	(3) = (1) + (2)	(4)	(5) = (4) – (3)	(6) = (5) * 2	(7)	(8) = (7) – (6)	(9
	Sun (Center)	Apogee	[Solar long.	Lunar long.	[Elongation	[Double el.	Double el.(tab.)		Diff.
841	18;26,0,13	90;30,4,48	108;56,5,1	115;51,56,28[1]	6;55,51,27	13;51,42,54	25;12,49	11;21,6	
842	7;20,18,19	90;30,54,41	97;51,13,0	100;18,37,5[2]	2;27,24,5	4;54,48,10	16;15,51	11;21,3	0;0,3
843*	357;13,44,16	90;31,44,42	87;45,28,58	97;55,52,43[3]	10;10,23,45	20;20,47,30	31;41,47	11;21,0	0;0,3
844	346;8,2,12	90;32,34,35	76;40,36,47	82;22,33,20	5;41,56,33	11;23,53,6	22;44,49	11;20,53	0;0,7
845	335;2,20,7	90;33,24,27	65;35,44,34	66;49,13,57	1;13,29,23	2;26,58,46	13;47,51	11;20,52	0;0,1
846*	324;55,46,14	90;34,14,29	55;30;0,43	64;26,29,35	8;56,28,52	17;52,57,44	29;13,47	11;20,49	0;0,3
847	313;50,4,10	90;35,4,21	44;25,8,31	48;53,10,11	4;28,1,40	8;56,3,20	20;16,49	11;20,46	0;0,3
848*	303;43,30,17	90;35,54,22	34;19,24,39	46;30,25,50	12;11,1,11	24;22,2,22	35;42,45	11;20,43	0;0,3
849	292;37,48,13	90;36,44,15	23;14,32,28	30;57,6,26	7;42,33,58	15;25,7,56	26;45,47	11;20,39	0;0,4
850	281;32,6,9	90;37,34,9	12;9,40,18	15;23,47,3	3;14,6,45	6;28,13,30	17;48,49	11;20,36	0;0,3
851*	271;25,34,16	90;38,24,9	2;3,58,25	13;1,2,41	10;57,4,16	21;54,8,32	33;[1]4,45[4]	11;20,36	0;0,0
852	260;19,50,[5]11	90;39,14,2	350;59,4,13	357;27,43,18	6;28,39,5	12;57,18,10	24;17,46	11;20,28	0;0,8
853	249;[14][6],8,7	90;40,3,54	339;54,12,1	341;54,23,54	2;0,11,53	4;0,23,46	15;20,49	11;20,25	0;0,3
854*	239;7,36,14	90;40,53,56	329;48,30,10	339;31,39,33	9;43,9,23	19;26,18,46	30;46,45	11;20,26	–0;0,1
855	228;1,5[4][7],10	90;41,43,48	318;43,37,58	323;58,20,9	5;14,42,11	10;29,24,22	21;49,47	11;20,23	0;0,3
856*	217;55,[22][8],7	90;42,33,50	308;37,55,57	321;35,35,[4]8[9]	12;57,39,51	25;55,19,42	37;15,43	11;20,23	0;0,0
857	206;49,36,[10]13	90;43,23,42	297;32,59,55	306;2,16,25	8;29,16,30	16;58,33,0	28;18,45	11;20,12	0;0,9
858	195;43,54,8	90;44,13,34	286;28,7,42	290;28,57,2	4;0,49,20	8;1,38,40	19;21,47	11;20,8	0;0,4
859*	185;37,20,15	90;45,3,36	276;22,23,51	288;6,12,39	11;43,48,48	23;27,37,36	34;47,43	11;20,5	0;0,3
860	174;31,38,11	90;45,53,29	265;17,31,40	272;32,53,16	7;15,21,36	14;30,43,12	25;50,45	11;20,2	0;0,3
861	163;25,56,7	90;46;43,22	254;12,39,29	256;59,33,52	2;46,54,23	5;33,48,46	16;53,48	11;19,59	0;0,3

862*	153;19,22,14	90;47,33,23	244;6,55,37	254;36,49,31	10;29,53,54	20;59,47,48	32;19,43	11;19,55	0;0,4
863	142;13,40,10	90;48,23,15	233;2,3,25	239;3,30,7	6;1,26,42	12;2,53,24	23;22,45	11;19,52	0;0,3
864	131;7,58,6	90;49,13,8	221;57,11,14	223;30,10,44	1;32,59,30	3;5,59,0	14;25,48	11;19,49	0;0,3
865*	121;1,24,12	90;50,3,9	211;51,27,21	221;7,26,22	9;15,59,1	18;31,58,2	29;51,43	11;19,45	0;0,4
866	109;55,42,8	90;50,53,2	200;46,35,10	205;34,6,59	4;47,31,49	9;35,3,38	20;54,46	11;19,42	0;0,3
867*	99;49,8,15	90;51,43,3	190;40,51,18	203;11,22,37	12;30,31,19	25;1,2,38	36;20,41	11;19,38	0;0,4
868	88;43,26,11	90;52,32,56	179;35,59,7	187;38,3,14	8;2,4,7	16;4,8,14	27:23,43	11;19,35	0;0,3
869	77;37,44,7	90;53,22,49	168;31,6,56	172;4,43,50	3;33,36,54	7;7,13,46	18;26,46	11;19,32	0;0,3
870*	67;31,10,14	90;54,12,50	158;25,23,4	169;41,59,29	11;16,36,25	22;33,12,50	32;52,41	11;19,28	0;0,4
871	56;25,28,10	90;55,2,43	147;20,30,53]	154;8,40,5	6;48,9,12	13;36,18,24]	24;55,44	11;19,26	0;0,2

1. 4^s 25;51,56,11 in the Aleppo ms.
2. 4^s in the Aleppo ms.
3. 4^s in the Aleppo ms.
4. 4' in the Aleppo ms.
5. 50 in the Alepo ms. The correct value should be 52".
6. [14] is the correct value: the Aleppo ms. has 40.
7. 52" in the Aleppo ms.
8. 48" in the Aleppo ms.
9. 18 thirds in the Aleppo ms.
10. 36" in the Aleppo ms. The correct value should be 40".

3 Solar and Lunar Equations

As I have previously stated, the *Zīj al-Sharīf* contains two displaced tables of the solar equation, the first being the result of a rounding, to the approximation of minutes, of the second one, which was copied from Ulugh Beg's *Zīj-i Sulṭānī*. Table 10.3 contains excerpts of the aforementioned second solar equation table of the *Zīj al-Sharīf*: the amounts between parentheses correspond to the differences, in thirds, between tabular and recomputed values, an eccentricity of 2;1,20^p having been used for the recomputation.[28]

The set of lunar equation tables also derive from Ulugh Beg's *Zīj-i Sulṭānī*. The lunar equation of the center, called *taʿdīl awwal* ("first equation") in the *Zīj al-Sharīf* (excerpts in table 10.4), although displaced vertically 13;15,34°, is a standard table calculated for an eccentricity of 12;33,22^p, the radius of the deferent being 60^p. The four tables (see excerpts below in tables 10.5, 10.6,

Table 10.3
Solar Equation

0	1;55,53,12°	180	1;55,53,12°
10	1;36,24,49	190	2;16,41,50 (+1)
20	1;17,28,37 (+1)	200	2;36,48,32
30	0;59,36,8	210	2;55,33,19 (+2)
40	0;43,17,44	220	3;12,19,28
50	0;29,1,47	230	3;26,35,19
60	0;17,14,5	240	3;37,55,7 (+1)
70	0;8,17,7	250	3;45,59,46 (+1)
80	0;2,29,14	260	3;50,37,13
		268	3;51,46,24
		269	3;51,[45],29[1]
90	0:0,3,57	270	3;51,[4]2,27[2]
		271	3;51,37,19
92	0;0,0,0		
100	0;1,9,11	280	3;49,17,10
110	0;5,46,38 (–1)	290	3;43,29,17
120	0;13,51,17 (–1)	300	3;34,32,19
130	0;25,11,5	310	3;22,44,37
140	0;39,26,57	320	3;8,28,40
150	0;56,13,17	330	2;52,10,16
160	1;14,57,52	340	2;34,17,47 (–1)
170	1;35,4,34 (–1)	350	2;15,21,35

1. 45' is legible in MS Tb fol. 78 r.
2. 42' is also legible in MS Tb, fol. 78 r.

Table 10.4
Lunar Equation of the Center

0	13;15,34°	180	13;15,34°
10	14;44,3	190	9;43,4 (–1)
20	16;[1]2;2[8][1]	200	6;30,10 (–1)
30	17;40,8 (+1)	210	3;51,33 (–1)
40	19;6,57 (+1)[2]	220	1;54,[4]3[3]
50	20;31,57	230	0;40,26
60	21;53,58	240	0;5,1 (–1)
		246	0;0,0
70	23;11,11	250	0;2,34
80	24;21,6	260	0;26,32
90	25;20,20	270	1;10,[4]8[4]
100	26;4,36	280	2;10,2
110	26;28,34	290	3;19,57
114	26;31,8		
120	26;26,7 (+1)	300	4;37,10
130	25;50,42	310	5;59,11
140	24;36,25	320	7;24,13 (+1)
150	22;39,35 (+1)	330	8;51,0 (–1)
160	20;0,58 (+1)	340	10;18,48
170	16;48,4 (+1)	350	11;47,5

1. 16;12,20° in MS Tb, fol. 79 v.
2. 19;6,55° in MS Tb, fol. 79 v.
3. 43" are legible in MS Tb fol. 79 v.
4. 48" are clearly legible in MS Tb, fol. 79 v.

10.7, 10.8) designed for the computation of the equation of anomaly are not so standard:[29] the first, called *taʿdīl thānī* (second equation), is a table displaced vertically 7;37,28°, but, if we subtract this constant from the tabular values, we will discover that the two halves of the table are not symmetrical. The solution to the difficulties posed by this table can be found in the instructions, given in the canons, for the computation of the lunar equation of anomaly (MS Ta, fol. 3v; MS Tb, fols. 56 v–57 r):[30]

- With the mean motion tables obtain the mean longitude (λ_m) and the mean anomaly (α_m) of the Moon as well as the *markaz* (double elongation, $2f$).
- Enter the table of the lunar equation of the center (*taʿdīl awwal*) with $2f$ and obtain the equation of the center (η).
- Add $\eta + \alpha_m = \alpha_v$ (true anomaly, *al-khāṣṣa al-muʿaddala*).
- Enter with α_v the table of the "second equation" (*taʿdīl thānī*) and obtain γ.

Table 10.5
Taʿdīl Thānī (“Second Equation”)
First Table for the Computation of the Lunar Equation of Anomaly

0	7;37,28°	180	7;37,28°
10	6;27,24 (–1)	190	8;34,7 (+1)
20	5;18,48 (–1)	200	9;28,32 (+1)
30	4;13,7	210	10;18,38 (+1)
40	3;11,49	220	11;2,38
50	2;16,23 (–1)	230	11;39,5[1]
60	1;28,22 (–1)	240	12;6,59
70	0;49,[1]7[2] (–2)	250	12;25,43 (+2)
80	0;2[4],40[3]	260	12;34,59 (–1)
		265	12;36,10 (+1)
90	0;3,54	270	12;3[5],3[4] (+1)
98	0;0,0 (–6)		
100	0;0,1[8][5]	280	12;26,13 (+1)
110	0;10,53 (–2)	290	12;[9],10 (+1)[6]
120	0;36,23 (–2)	300	11;44,39
130	1;16,57 (–1)	310	11;13,39 (+1)
140	2;12,0 (–1)	320	10;37,6
150	3;20,13	330	9;56,8 (+1)
160	4;39,21 (–1)	340	9;11,47
170	6;6,21	350	8;25,11

1. 11;39,35° in MS Tb, fol. 80 r.
2. 0;49,47° in Ta (fol. 26 r); 0;49,17° in Tb (fol. 80 r).
3. 0;20,40° in MS Tb, fol. 80 r.
4. 12;34,3 in MS Tb, fol. 26 r; 12;35,3° in MS Ta, fol. 80 r.
5. 0;10,13° in MS Tb, fol. 80 r.
6. 12;10,9° in MS Ta, fol. 26 r; 12;9,10° in MS Tb, fol. 80 r.

• Enter with α_v the table of the “variation” (*ikhtilāf*) and obtain $\Delta\gamma$.

• Enter with the *markaz* (*2f*) the first table of the “minutes of the anomalies” (*daqāʾiq al-ḥiṣaṣ*) if $0° < \alpha_v < 180°$ and obtain m_1. If $180° < \alpha_v < 360°$, then do the same with the second table of the “minutes of the anomalies” and obtain m_2.

• Multiply $m{\cdot}\Delta\gamma$.

• Add $\gamma + m{\cdot}\Delta\gamma$ and you will obtain the “corrected equation” [of anomaly] (*taʿdīl muḥkam*).

• The true longitude of the Moon in its “inclined sphere” (*falak māʾil*) will be: $\lambda_v = \lambda_m + \gamma + m{\cdot}\Delta\gamma$.

Instructions are given, after this, to calculate the equation of time and to obtain the true longitude of the Moon on the ecliptic.

Table 10.6
Ikhtilāf al-Qamar ("Variation of the Moon")
Second Table for the Computation of the Lunar Equation of Anomaly

0	0;0°	90	2;35,59 (–1)
10	0;22,21 (+1)	100	2;39,39 (–1)
		102	2;39,49
20	0;44,21 (+1)	110	2;38,20
30	1;5,41 (–1)	120	2;31,34 (+2)
40	1;26,1	130	2;18,54 (+1)
50	1;44,54	140	2;0,18 (+1)
60	2;1,55 (+1)	150	1;36,4 (–2)
70	2;16,29 (+1)	160	1;7,3
80	2;28,3 (–1)	170	0;34,28 (–1)

Table 10.7
Daqā'iq al-Ḥiṣṣa ("Minutes of the Anomaly")
First Function of Interpolation for the Computation of the Lunar Anomaly

0/360	60;0'	90/270	33;12'
30/330	56;46 (–1)	120/240	17;24 (+1)
60/300	47;24 (+1)	150/210	4;49 (+2)
		179/181	0;1 (–1)

Table 10.8
Daqā'iq al-Ḥiṣṣa ("Minutes of the Anomaly")
Second Function of Interpolation for the Computation of the Lunar Anomaly

0/360	0;0'	90/270	26;48'
30/330	3;14 (+1)	120/240	42;36 (–1)
60/300	12;36 (–1)	150/210	55;11 (–2)
		179/181	59;59 (–3)

Following the logic of the instructions I have just summarized, we can see that the table of the "second equation" (see table 10.6) computes two different functions:

- For $0° < \alpha_v < 180°$, the funcion involved is the lunar equation of anomaly for the minimum distance of the center of the lunar epicycle from the center of the Earth ($\gamma_{R\text{-}e}$). The table is displaced vertically 7;37,28° and, as the equation of anomaly is negative for $0° < \alpha < 180°$, it gives:

  $7;37,28° - \gamma_{R\text{-}e}$.

The table can be recomputed using a radius of the epicycle of 6;17,46°,[31] an eccentricity of 12;33,22^p and a radius of the deferent of 60^p. The choice of the constant of displacement is inadequate and it corresponds to an error in the computation of the maximum value of the table, which should be 7;37,34° instead of 7;37,28°. The mistake was already present in Ulugh Beg's *zīj* and 7;37,28° is explicitely mentioned by Mīram Chalabī, in his commentary on Ulugh Beg's canons,[32] as the amount for the displacement and for the maximum equation.[33]

The table of the "variation" (*ikhtilāf*) is a standard table of differences between the equation of anomaly for minimum ($\gamma_{\text{R-e}}$) and for maximum ($\gamma_{\text{R+e}}$) distances:

$$\gamma_{\text{R-e}} - \gamma_{\text{R+e}}.$$

The interpolation function involved for the aforementioned values of the true anomaly (m_1) decreases monotonically from 60′ (for $2f = 0°$) to 0′ (for $2f = 180°$). It can be recomputed using the expression:

$$m_1 = (max\ \gamma_{2f} - max\ \gamma_{R\text{-}e}) / (max\ \gamma_{R\text{-}e} - max\ \gamma_{R+e})$$

in which $max\ \gamma_{R\text{-}e} = 7;37,28°$, and

$$max\ \gamma_{R+e} = 4;58,42°.$$

It is easy to see the logic of the system for the computation of the lunar equation of anomaly at syzygies in which $2f = 0°$, $m_1 = 60'$ and $\gamma(\alpha_v) = \gamma_{R+e}$:

$$7;37,28° - \gamma(\alpha) = 7;37,28 - \gamma_{R\text{-}e} + 60'(\gamma_{R\text{-}e} - \gamma_{R+e}) = 7;37,28° - \gamma_{R+e}.$$

At quadratures, $2f = 180°$, $m_1 = 0'$ and $\gamma(\alpha_v) = \gamma_{R\text{-}e}$:

$$7;37,28° - \gamma(\alpha) = 7;37,28 - \gamma_{R\text{-}e} + 0'(\gamma_{R\text{-}e} - \gamma_{R+e}) = 7;37,28° - \gamma_{R\text{-}e}.$$

- For $180° < \alpha_v < 360°$, the funcion involved is the lunar equation of anomaly for the maximum distance of the center of the lunar epicycle from the center of the Earth (γ_{R+e}). The table is also displaced vertically 7;37,28° and, as the equation of anomaly is positive for $180° < \alpha < 360°$, it gives:

$$7;37,28° + \gamma_{R+e}.$$

The interpolation function involved for the aforementioned values of the true anomaly (m_2) increases monotonically from 0′ (for $2f = 0°$) to 60′ (for $2f = 180°$). It can be recomputed using the expression:

$$m_2 = (max\ \gamma_{2f} - max\ \gamma_{R+e}) / (max\ \gamma_{R\text{-}e} - max\ \gamma_{R+e}).$$

Again, at syzygies, $2f = 0°$, $m_2 = 0'$ and $\gamma(\alpha_v) = \gamma_{R+e}$:

$$7;37,28° + \gamma(\alpha) = 7;37,28 + \gamma_{R+e} + 0'(\gamma_{R\text{-}e} - \gamma_{R+e}) = 7;37,28° + \gamma_{R+e}.$$

At quadratures, $2f = 180°$, $m_2 = 60'$ and $\gamma(\alpha_v) = \gamma_{R\text{-}e}$:

$$7;37,28° + \gamma(\alpha) = 7;37,28 + \gamma_{R+e} + 60'\ (\gamma_{R\text{-}e} - \gamma_{R+e}) = 7;37,28° + \gamma_{R\text{-}e}.$$

4 Ḥabṭaq Tables of the Combined Lunar Equations

The *Zīj al-Sharīf* contains, as we have seen, a set of double argument tables to calculate the lunar equation of anomaly (MS Ta, fols. 11 r and v; MS Tb, fols. 65 v and 66 r): see samples below, in tables 10.9 and 10.10. The arguments are the *markaz* (double elongation), at intervals of 30°, and the mean anomaly (at intervals of 6°): 720 values of the equation of anomaly are computed and I have recalculated all of them. On the whole, the calculator did a good job and errors have usually an amount of 1′, with the only exception of the equations calculated for a double elongation of 240°, in which errors are systematic and reach a maximum of ± 8'.[34] It has been established that these *ḥabṭaq* tables have been calculated using the lunar equation tables that I have just described and that derive from Ulugh Beg's *Zīj-i Sulṭānī*. For the recomputation of the *ḥabṭaq* tables I have used Benno van Dalen's *Table Analysis* and, especifically, the sub-programme called *Table calculator*. The procedure used for the recomputation has been the following one:

Table 10.9
Ḥabṭaq Table for the Lunar Equation of the Anomaly for $0^{\circ} < \alpha < 180^{\circ}$

Markaz (double elongation)							
Mean anomaly	0	60	120	180	240	300	330
0	6;35 (+1)	5;44	4;53	6;5	7;37	7;13	6;54
12	5;40 (+1)	4;47	3;45	4;44	6;23 (+2)	6;11	5;57
24	4;49	3;55	2;45	3;28	5;10 (+3)	5;12	5;3
36	4;4	3;11	1;54	2;20	4;2 (+4)	4;17	4;15 (+1)
48	3;27	2;37	1:16	1;23	3;1 (+5)	3;30	3;33
60	2;59	2;15	0;53	0;39	2;10 (+7)	[2];51[1]	3;1
72	2;43	2;6	0;46	0;10	1;30 (+7)	2;23	2;39
84	2;39	2;11	0;57 (−1)	0;0 (+5)	1;4 (+8)	2;8	2;30
96	2;4[8][2]	2;31	1;28	0;10	0;54 (+8)	2;6 (−1)	2;35
108	3;11	3;6	2;17 (−1)	0;41	1;2 (+8)	2;20	[2];53[3]
120	3;47	3;56	3;25	1;34 (+1)	1;28 (+7)	2;49	3;24 (−1)
132	4;34 (−1)	4;57	4;47	2;47 (+1)	2;13 (+6)	3;32 (−1)	4;9
144	5;32	6;8	6;20	4;17	3;16 (+5)	4;29	5;5
156	6;36 (−1)	7;24	7;56 (−1)	6;0	4;34 (+4)	5;36	6;9
168	7;44 (−1)	8;41	9;33 (−1)	7;49	6;2 (+1)	6;50	7;49

1. MS Ta 3;51°
2. MS Ta 2;43°
3. MS Ta 3;53°

Table 10.10
Ḥabṭaq Table for the Lunar Equation of the Anomaly for 180° < α < 360°

Markaz (double elongation)							
Mean an.	0	60	120	180	240	300	330
180	8;52	9;54	11;3	9;38	7;37 (–1)	8;7	8;29
192	9;55 (–1)	10;59	12;19	11;[1]9[1] (+1)	9;12 (–3)	9;22	9;[3]6[2] (–1)
204	10;51	11;53	13;20	12;46 (+1)	10;41 (–5)	10;31	10;37
216	11;36 (–1)	12;33	14;2	13;55 (+1)	11;59 (–6)	11;31	11;28
228	12;10	12;59	14;24	14;43	13;2 (–7)	12;[1]7[3]	12;7
240	12;30	13;9	14;28	15;9	13;47 (–8)	12;50 (+1)	12;32 (–1)
252	12;36	13;5	14;14	15;14	14;13 (–8)	13;7 (+1)	12;44
264	12;30 (+1)	12;47	13;45	15;0	14;21 (–8)	13;9 (+1)	12;42
276	12;11	12;17	13;2	14;28	14;15[4] (–3)	12;57 (+1)	12;27
288	11;41	11:36	12;7	13;41	13;45 (–7)	12;32 (+1)	12;0
300	11;2	10;47	11;3	12;41	13;5 (–6)	11;55	11;23
312	10;16	9:51	9;53	11:31 (–1)	12:13 (–5)	11;10 (+1)	10:39 (+1)
324	9;24	8;51	8;39 (+1)	10;14	11;12 (–4)	10;17 (+1)	9;47
336	8;29	7;49	7;22	8;52 (–1)	10;7	9;18	8;51
348	7;32 (+1)	6;46	6;6	7;28 (–1)	8;53	8;16	7;53

1. MS Ta 11;59°.
2. MS Ta 9;56.
3. 12;7 in MS Ta, in which 17' appears in a marginal correction.
4. 14;15° in MS Ta; 14;11° in MS Tb.

1. For 0° < α_v < 180°.

1.1 Enter a table $T1(c_m)$ of the lunar equation of center η as a function of the mean centrum, for e = 12;33,22^p, displaced vertically 13;15,34°, for arguments comprised between 0° and 330°, with intervals of 30°. The actual values entered were those that appear above in table 10.4.

From here on all the steps are given for one particular value of c_m.

1.2 Enter a table $T2(c_m)$ of the function of interpolation m_1 for arguments 0, 30, 60...360°. The values entered are those of table 10.7 above.

1.3 Add

$$\eta + \alpha_m = \alpha_v$$

to obtain the true lunar anomaly by adding, for each particular value of c_m,

$$\alpha_m + T1(c_m), \text{ for } \alpha_m = 0, 6, 12...174°.$$

The result obtained will be table $T3(\alpha_m)$.

1.4 Calculate a table $T4(\alpha_m)$ of the equation of anomaly for minimum distance ($\gamma_{R\text{-}e}$), the arguments being the true anomalies obtained from $T3(\alpha_m)$, for each particular value of c_m. The table has to be displaced 7;37,28° and we must, therefore, calculate

$$7;37,28° - \gamma_{R\text{-}e}.$$

The expression used is:

$$7;37,28-(atan((6;17,46*\sin(T3))/(47;26,37 + 6;17,46*\cos(T3)))),$$

in which $T3$ means $T3(\alpha_m)$.

1.5 Calculate a table $T5(\alpha_m)$ of the equation of anomaly for maximum distance (γ_{R+e}), also displaced 7;37,28°, again using as arguments the values of $T3(\alpha_m)$, for each particular value of c_m:

$$7;37,28 - (atan((6;17,46*\sin(T3))/(72;33,22+6;17,46*\cos(T3)))),$$

$T3$ being $T3(\alpha_m)$.

1.6 The final table for each value of the mean anomaly, and still for our particular value of c_m, will be:

$$T4 + ((T5\text{-}T4)*T2),$$

in which

$T2$ is $T2(\alpha_m)$,
$T4$ is $T4(\alpha_m)$,
$T5$ is $T5(\alpha_m)$.

2. For $180° < \alpha_v < 360°$

2.1 As in 1.1 above: $T1(c_m)$.

2.2 Enter a table $T2(c_m)$ of the function of interpolation m_2 for arguments 0, 30, 60...360°. The values entered are those of table 10.8 above.

2.3 As in 1.3 above for α_m = 180, 186, 192...354°:

$T3(\alpha_m)$, for each particular value of c_m.

2.4 Calculate 7;37,28 – γ_{R+e} for the values of $T3(\alpha_m)$, using the same expression as above in 1.5:

$T4(\alpha_m)$.

2.5 Calculate 7;37,28 – $\gamma_{R\text{-}e}$ for the values of $T3(\alpha_m)$, using the same expression as above in 1.4:

$T5(\alpha_m)$.

2.6 As in 1.6 above, calculate

$$T4 + ((T5–T4)*T2).$$

Acknowledgments

The present chapter has been prepared within a research program on "The circulation of astronomical ideas in the Mediterranean between the twelfth and fourteenth centuries," sponsored by the Dirección General de Investigación Científica y Técnica of the Spanish Ministry of Education and Culture. Benno van Dalen's programs CALH (for the conversion of dates) and TA ("Table Analysis"), as well as Honorino Mielgo's ATMM (for the analysis of mean motion tables) have been extensively used. Information about manuscripts in the National Library of Tunis and the Ṣabīḥiyya Library of Salé (Morocco) was given to me by Dr. Mercè Comes and by my research graduate student Hamid Berrani. Benno van Dalen read two successive drafts of this chapter and corrected some errors I had made. He also tried to improve my hopeless mathematical inconsistencies and sent me photocopies of Sédillot's publications on Ulugh Beg's *zīj*. My gratitude to all of them.

Notes

1. See f. ex. David A. King, "An Overview of the Sources for the History of Astronomy in the Medieval Maghrib." *Deuxième Colloque Maghrébin sur l' Histoire des Mathématiques Arabes*. Tunis, 1988, pp. 125–157. See also a recent survey of this manuscript in A. Mestres, "Maghribī Astronomy in the 13th Century: A Description of Manuscript Hyderabad Andra Pradesh State Library 298," in J. Casulleras and J. Samsó (eds.), *From Baghdad to Barcelona. Studies in the Islamic Exact Sciences in Honour of Prof. Juan Vernet*. Barcelona, 1996, I, 383–443. A complete edition of the canons of the Hyderabad compilation, together with a commentary and a partial edition of the numerical tables was presented by Angel Mestres as a Ph.D. dissertation (University of Barcelona, January 2000).

2. The canons were edited by J. Vernet, *Contribución al estudio de la labor astronómica de Ibn al-Bannā*ʾ. Tetuán, 1952. The solar tables were studied by J. Samsó and E. Millás, "Ibn al-Bannāʾ, Ibn Isḥāq and Ibn al-Zarqālluh's Solar Theory" in J. Samsó, *Islamic Astronomy and Medieval Spain*. Variorum. Aldershot, 1994, no X (35 pp.). For the planetary tables see J. Samsó and E. Millás, "The Computation of Planetary Longitudes in the *Zīj* of Ibn al-Bannāʾ," *Arabic Sciences and Philosophy* 8 (1998), 259–286.

3. See E. S. Kennedy, "The Astronomical Tables of Ibn al-Raqqām, a Scientist of Granada," *Zeitschrift für Geschichte der Arabisch-Islamischen Wissenschaften* 11 (1997), 35–72. On these *zīj*es a doctoral dissertation by Muḥammad ʿAbd al-Raḥmān (Institute for the History of Arabic Science, Aleppo), *Ḥisāb aṭwāl al-kawākib fī l-Zīj al-Shāmil fī Tahdhīb al-Kāmil li-Ibn al-Raqqām*, was presented in the University of Barcelona (September 1996).

4. E. S. Kennedy and David A. King, "Indian Astronomy in Fourteenth Century Fez: the Versified Zīj of al-Qusunṭīnī," *Journal for the History of Arabic Science* 6 (1982), 3–45.

Reprint in D. A. King, *Islamic Mathematical Astronomy*, Variorum Reprints, London, 1986, no. VIII (second revised edition by Variorum, Aldershot, 1993).

5. J. Samsó, "Andalusian Astronomy in 14th Century Fez: *al-Zīj al-Muwāfiq* of Ibn ʿAzzūz al-Qusanṭīnī," *Zeitschrift für Geschichte der Arabisch-Islamischen Wissenschaften*, 11 (1997), 73–110. See also J. Samsó, "Horoscopes and History: Ibn ʿAzzūz and his retrospective horoscopes related to the battle of El Salado (1340)," in Lodi Nauta and Arjo Vanderjagt (eds.), *Between Demonstration and Imagination. Essays in the History of Science and Philosophy Presented to John D. North*, Brill, Leiden-Boston-Köln, 1999, 101–124.

6. See Kennedy and King, "Indian Astronomy," p. 9; King, "Overview," pp. 131–132 and 144.

7. See Juan Vernet, "Tradición e innovación en la ciencia medieval," *Oriente e Occidente nel Medioevo: Filosofia e Scienze*, Accademia dei Lincei, Roma, 1971, pp. 741–757; reprint in Vernet, *Estudios sobre Historia de la Ciencia Medieval*, Barcelona-Bellaterra, 1979, pp. 173–189 (see pp. 188–189). See now Montse Díaz-Fajardo, *La teoría de la trepidación en un astrónomo marroquí del siglo XV. Estudio y edición crítica del* Kitāb al-adwār fī tasyīr al-anwār (*parte primera*) *de Abū ʿAbd Allāh al-Baqqār.* Barcelona, 2001.

8. See J. Samsó, "An Outline of the History of Maghribī Zījes from the End of the Thirteenth Century." *Journal for the History of Astronomy* 29 (1998), 93–102; see also Samsó, "Astronomical Observations in the Maghrib in the Fourteenth and Fifteenth Centuries," *Science in Context* 14 (2001), 165–178.

9. See L. P. E. A. Sédillot, *Prolegomènes des Tables Astronomiques d'Oloug-Beg*. 2 vols., Paris, 1847 and 1853.

10. Muḥammad al-ʿArabī al-Khaṭṭābī, *Fahāris al-Khizāna al-Ḥasaniyya* vol. III (Rabat, 1983), pp. 193–195 (no. 227); see also D. A. King, *A Survey of the Scientific Manuscripts in the Egyptian National Library*. Winona Lake, Indiana, 1986, p. 164 (G99). MSS 11420 and 12616 of the National Library contain versions of Ulugh Beg's *zīj* for the coordinates of Cairo (long. 55° from the western meridian).

11. Ms. Cairo DM814 is a copy of this recension dated in 1325/1907–1908: cf. King, *Survey* p. 142 (F46). A third (?), anonymous, Tunisian recension is preserved in two mss. of the Ḥasaniyya Library in Rabat: 2650 (undated) and 2148 (dated in 1218/1803). See Khaṭṭābī, *Fahāris* 335–338 (no. 406–408). Ekmeleddin Ihsanoğlu *et al., Osmanli Astronomi Literatürü Tarihi* (*History of Astronomy Literature during the Ottoman Period*). Istanbul, 1997, p. 314 (no. 175) and pp. 347–348 (no. 213).

12. Another copy of the same *zīj* is extant in MS 5990 of the same library. The copy was finished in 1229 H./1814 by Muḥammad b. ʿAbd Allāh al-Khayyārī.

13. Copy dated in 1242/1826. Mercè Comes obtained a microfilm of that manuscript and called my attention to it.

14. See ʿAbd al-Ḥafīẓ Manṣūr, *al-Fihris al-ʿāmm li 'l-makhṭūṭāt. I. Raṣīd maktabat Ḥasan Ḥusnī ʿAbd al-Wahhāb*, Tunis, 1975, p. 398.

15. See also King, *Survey,* p. 143 (F53) who lists other manuscripts of the same work: ms. Cairo Ṭalʿat Riyāḍa 319,1 (pp. 1–16, ca. 1300 H.) contains the mean motion tables of this *zīj*. A complete copy is in ms. Paris B.N. ar. 2536 (48 fols., ca. 1150 H.).

16. The incomplete copy extant in MS Tb (fol. 104 r) gives an example of the computation of the solar longitude and the date used is the 16th Jumādā I 1099 H/ 19.3.1688.

17. Driss Lamrabet, *Introduction à l'histoire des mathématiques maghrebines* (Rabat, 1994), p. 143 (no. 506), mentions Ibn Māmī and places him tentatively after the sixteenth century. One copy (dated 1163 H/ 1749–1750) of his work is extant in Cairo N. L. (King, *Survey* p. 143, F54).

18. Tunis Nat. Library, col. Ḥ. Ḥ. ʿAbd al-Wahhāb 18104, dated in 1301/1883–1884: see ʿA. Ḥ. Manṣūr, *Fihris* p. 397, and in Tunis Nat. Library MS 2770. Another copy of the same work is extant in MS 1042 of the Ṣabīḥiyya Library in Salé (Morocco). Two copies of this work, dated in 1150/1737–1738 and in 1169/1755–1756, are also extant in Cairo N. L.: see King, *Survey* p. 144, F55. I have another reference to a summary of Sanjaq Dār's *Mukhtaṣar*, mixed with materials derived from Ibn al-Shāṭir's *zīj*, in MS Tunis N. L. 5608. Finally, the Ṣabīḥiyya Library also contains two summaries (*Mukhtaṣar*s) of Ulugh Beg's *zīj* in MSS 1167 and 1168, the author being Muḥammad b. Abī l-Fatḥ al-Ṣūfī.

19. E. S. Kennedy and M. H. Kennedy, *Geographical Coordinates of Localities from Islamic Sources* (Frankfurt, 1987), pp. 362–363.

20. D. A. King, "A Double Argument Table for the Lunar Equation Attributed to Ibn Yūnus," *Centaurus* 18 (1974), 129–146. Reprinted in King, *Islamic Mathematical Astronomy* no. V. For other double argument tables for the computation of lunar and planetary equations see D. A. King, "On the Astronomical Tables of the Islamic Middle Ages," *Colloquia Copernicana* 13 (1975), 37–56 (cf. pp. 44–45, 55), reprinted in *Islamic Mathematical Astronomy* no. II; Claus Jensen, "The Lunar Theories of al-Baghdādī," *Archive for History of Exact Sciences* 8 (1971–1972), 321–328; Mark J. Tichenor, "Late Medieval Two-Argument Tables for Planetary Longitudes," *Journal of Near Eastern Studies* 26 (1967), 126–128, reprinted in E. S. Kennedy, Colleagues and Former Students, *Studies in the Islamic Exact Sciences* (Beirut, 1983), 126–128; G. Saliba, "The Double Argument Lunar Tables of Cyriacus," *Journal for the History of Astronomy* 7 (1976), 41–46 and "The Planetary Tables of Cyriacus," *Journal for the History of Arabic Science* 2 (1978), 53–65.

21. See a similar table in the *Zīj al-Sharīf* MS Ta fol. 31 v; MS Tb, fol. 85 v.

22. J. Chabás and B. R. Goldstein, "Andalusian Astronomy: *al-Zîj al-Muqtabis* of Ibn al-Kammâd," *Archive for History of Exact Sciences* 48 (1994), 1–41 (see pp. 14 and 23).

23. I owe this information to Angel Mestres.

24. This displacement is explicitely mentioned by Maḥmūd b. Muḥammad b. Qāḍīzādah al-Rūmī, known as Mīram Chalabī (d. 1524) in his commentary to the *zīj* of Ulugh Beg: cf. Sédillot II (1853), p. 140.

25. Mean motions have been calculated using Honorino Mielgo's ATMM. See

H. Mielgo, "A Method of Analysis for Mean Motion Astronomical Tables," in J. Casulleras and J. Samsó (eds.), *From Baghdad to Barcelona. Studies in the Islamic Exact Sciences in Honour of Prof. Juan Vernet* (Barcelona, 1966), I, 159–179.

26. I have checked ms. Aleppo, Waqfiyya 1307. A microfilm of this manuscript was obtained from the Institute for the History of Arabic Science (Aleppo), microfilm no. 507.

27. I am following here Benno van Dalen's notation: the two digits after the ± refer to the last two digits of the parameter.

28. This eccentricity is mentioned by Mīram Chalabī in his commentary of Ulugh Beg's canons: see Sédillot II (1853), p. 142.

29. Mīram Chalabī mentions an equivalent eccentricity of $10;23^p$ for a radius of the deferent of $49;37^p$. See Sédillot II (1853), p. 147.

30. The same set of instructions appears in Ulugh Beg's canons: see Sédillot II (1853), pp. 138–153.

31. Mīram Chalabī mentions a radius of the epiycle of $5;12^p$ for a radius of the "inclined sphere" of 60^p. He seems to be making a mistake here: $5;12^p$ is probably a rounded value and it corresponds to a radius of the deferent of $49;37^p$. $6;17,18^p$ would be the corresponding epicycle radius for a deferent radius of 60^p. See Sédillot II (1853), p. 158.

32. See Sédillot II (1853), p. 151.

33. We have seen that Mīram Chalabī also ascribes to Ulugh Beg a radius of the epicycle of $5;12^p$ (for a deferent radius of $49;37^p$), equivalent to $6;17,18^p$ (for $R = 60^p$), instead of our $6;17,46^p$. With this radius the maximum value of the equation of anomaly for minimum distance would be:

$$\gamma_{max} = \text{atan}\ (6;17,18/(60–12;33,22)) = 7;33,0°.$$

34. Benno van Dalen, after reading a draft of this chapter, suggested that the errors practically disappear if one assumes that Sanjaq Dār made a mistake and used interpolation coefficients 20;24 (instead of 17;24) and 39;36 (instead of 42;36). He is absolutely right. After recalculating the column for mean anomaly 240°, using these erroneous coefficients, the only differences (in the sample values) are:

12	6;23(+1)	216	11;59(–1)
168	6;2(–1)	276	14;15(+4)
180	7;37(–1)	336	10;7(+3)
192	9;12(–1)	348	8;53(+1)
204	10;41(–1)		

VI

Science and Medicine in the Maghrib and al-Andalus

11

A Panorama of Research on the History of Mathematics in al-Andalus and the Maghrib between the Ninth and Sixteenth Centuries

Ahmed Djebbar

A Survey of the Research before the 1980s

For the period 1834 to 1980 I have surveyed more than one hundred works exclusively or partially devoted to the mathematics and astronomy in the Muslim West. This number is modest considering the length of the period, the important historical role of these regions, and the number of works on the same subjects written in the same period in other parts of the Islamic world.[1] One reason for this may be the political conditions of the societies of these regions at the end of the nineteenth and beginning of the twentieth century, and the consequent effects on their cultures.

Nevertheless, in spite of their modest number, and the different aims and levels of these works, they can be said to have contributed as a whole to writing the first drafts of the history of science in al-Andalus and the Maghrib in the Middle Ages.

Two kinds of works can be immediately identified, each having a specific goal. The first is founded on the traditional criteria for scholarly research, as developed in the nineteenth century, whose first concern is to make known the facts of the subject in a disinterested way. The second, a product mainly of the Maghrib and Egypt during the first half of the twentieth century, has a clearly stated nationalistic goal: to nurture a sense of pride in the heritage of the country and the importance of its contribution to science.[2]

Even this second kind, with its rhetorical form and political aim, has also contributed to a better knowledge of the scientific heritage of the Muslim West. For, in a field that is almost virgin territory, any data can be helpful, and, as long as the researcher can separate the wheat from the chaff, he or she can find in this nationalistic literature, if not secure factual information, at least some hints (sometimes quite important) that may reveal new documents or suggest new ways of approach.

For this reason it has seemed useful to make a rapid inventory of the significant publications in this field before 1980, in order to put recent research (including my own) in perspective. In addition, although focusing on the mathematical aspects of the scientific tradition of the Muslim West, my presentation will also include the essential elements relating to the astronomical tradition of this region, in both its theoretical and practical aspects. For astronomy uses, implicitly or explicitly, mathematical concepts, objects, and instruments belonging to various scientific traditions (Greek, Indian, Eastern Muslim), which fed the science of al-Andalus and the Maghrib. Moreover, some mathematicians in these regions, such as Maslama al-Majrīṭī and Ibn al-Bannā, published astronomical writings, and it would seem unfair to neglect this aspect of their scientific activities.

The Nineteenth Century

Although the Muslim West was mentioned in the earliest works in the history of mathematics or of astronomy, by J. F. Montucla,[3] C. Bossut,[4] and J. B. Delambre[5] in France, M. Pelayo in Spain,[6] and Hammer-Purgstall in Prussia,[7] the first substantial investigations concerning its scientific tradition were the work of J.-J.-E. and L.-A. Sédillot[8] in astronomy and of F. Woepcke in mathematics.[9] These works formed part of their authors' interest in the Arabic scientific heritage as a whole.[10]

Astronomy

In astronomy, the translation of the first volume of al-Marrākushī's work (ca. 661/1262),[11] *Jāmiʿ al-mabādiʾ wa l-ghāyāt fī ʿilm al-mīqāt* [the Collection of Principles and Goals in the Science of the Determination of Time][12] appeared in 1834 and was followed, ten years later, by an analysis of the whole treatise.[13] These publications stimulated interest in the history of sciences in the Maghrib, even though the contents of the work in question reflect the scientific tradition of the East before the thirteenth century, rather than that of Marrakech, the city where al-Marrākushī was born. However, one had to wait until the second half of the nineteenth century to see a new interest in the astronomy of the Muslim West, which concerns two great figures of medieval Spain: az-Zarqālluh (= az-Zarqiyāl = az-Zarqālī) (d. 493/1100) of Toledo and the Castilian king Alphonso X (1252–1284), who perpetuated the Arab scientific tradition of Spain by his knowledge and his patronage. Steinschneider's[14] works on az-Zarqālluh and those of M. Rico y Sinobas[15] on the *Libros del Saber* stimulated new research in the first half of the twentieth century, which was aided by the advance in the history of astronomy as a whole in the nineteenth century, together with the study of the history of mathematics.

Mathematics

The publication of a translation of the *Muqaddima* [Introduction] of Ibn Khaldūn was the first stimulus for research into the history of Western Islamic mathematics by F. Woepcke, A. Marré, M. Steinschneider, and H. Suter.[16] All the investigations by these authors of the Maghribi tradition in arithmetic and algebra, were more or less related to passages on the classification of the sciences in the work of Ibn Khaldūn. Woepcke, for example, noticed the importance of the chapter in the *Muqaddima* on mathematics. He published extracts from it in 1854[17] and a complete translation two years later.[18]

Since these passages also influenced an important part of the research that will be mentioned below, I will also quote them and comment on their contents in the light of the research of these four historians. Because different interpretations of these passages have arisen out of the previous translations by Quatremère,[19] de Slane,[20] Woepcke, Rosenthal,[21] and Monteil,[22] I will give here my own translation, referring to a future study for the justification of my readings.[23]

After devoting section 19 of chapter VI of the *Muqaddima* to a general presentation of the rational sciences, Ibn Khaldūn discusses, in section 20, entitled *Sciences of Number*, the sciences of arithmetic, algebra, commercial transactions, and inheritances. Here are the main passages of the book, which is dated 779/1377:[24]

> The Sciences of number: the first of them is arithmetic, and it is the knowledge of the properties of numbers from the viewpoint of composition, either according to a successive "progression" or by duplication; as, for example, if the numbers follow one another while exceeding each other by the same number, then the sum of their two extremes is equal to the sum of any pair of numbers whose distance from the two extremes is the same, and it is equal to double of the middle [term] if these numbers are odd in number. For example: the sequence of the [integer] numbers, the sequence of the even numbers, the sequence of the odd numbers. . . .
>
> This science is the first part of mathematics and the best established, and it is used in the proofs of calculation. The ancient and modern scientists have publications on it, and the majority of them integrate it into mathematics without devoting [specific] writings to it. This is what Ibn Sīnā did in his books *ash-Shifā'*[25] and *an-Najāt*,[26] as well as others among the ancient scientists. As far as the moderns are concerned, they gave it up because it is not practiced and because its usefulness is in the proofs and not in the calculations; and it is for this reason that they abandoned it after they had extracted the essence in arithmetical proofs, as Ibn al-Bannā did in the book *Rafʿ al-ḥijāb* [The Raising of the Veil], as well as others. . . .

> And among the branches of the science of number, [there is] the art of calculation. It is a practical art [relating] to the calculation of numbers by joining and separation. Joining takes place in numbers in a single act [of adding], and this is [then] addition, and by repetition [of the same added portion], i.e., one increases a number [by its own value] as many [times as there are] units in another number; and this is multiplication. Separation also takes place for numbers, either in a single act, like cutting off of a number from [another] number and knowing of the remainder, and this is then subtraction; or cutting of a number into a determined number of equal parts; and this is division. . . .
>
> And among the best writings concerning the [subject] at this time in the Maghrib, there is al-Ḥaṣṣār's small book. Ibn al-Bannā from Marrakech has on the [subject] a useful summary which establishes the rules of its operations. Then he commented on it in a book, which he entitled *Rafʿ al-ḥijāb*, which is obscure for a beginner because of the firmly-structured proofs it contains. It is a book of great value and we saw high-level professors giving much consideration to it; it is a book which indeed deserves that.
>
> The author, God bless him, followed Ibn Munʿim's book, *Fiqh al-ḥisāb* [The Science of Calculation] and al-Aḥdab's [book] *al-Kāmil* [The Complete <Book>]. He summarised their demonstrations and changed them—because of the use of [numerical] symbols—into clear and abstract justifications which are the secret and the essence of representation by [numerical] symbols.[27] All these [matters] are obscure,[28] but the obscurity arises [only] from the method of demonstration [which] is specific to the mathematical sciences, since their problems and their operations are themselves quite clear. However if one wanted to explain them, it would be necessary to give the justifications for these operations. And there are, in understanding them, difficulties which one does not find in the resolution of problems.[29]

In these extracts of the *Muqaddima*, European specialists in Arabic sciences found a whole research program, namely the study of the mathematical works quoted by Ibn Khaldūn. With one or two exceptions, their publications were confined to the topics that were suggested to them by their own readings of the quoted passages.

An ambiguous passage in the section on calculation[30] directed Woepcke toward his research of symbolism in the mathematical writings of the Muslim West. Thus he discovered some symbols in the work of al-Qalaṣādī (d. 892/1486), an Andalusi mathematician who studied in the Maghrib and taught in Tlemcen and Tunis.[31] In 1854, Woepcke presented a note on his discovery to the Academy of Sciences.[32]

In this note, Woepcke announced the discovery, in al-Qalaṣādī's handbook "*of a well developed algebraic notation among the Arabs of the West*," as well as a "*multiplication table of the algebraic powers*" in a Persian manuscript.

A few years later, Woepcke published al-Qalaṣādī's book which contained this symbolism and which is entitled *Kashf al-asrār ʿan ʿilm ḥurūf al-ghubār* [Disclosure of the Secrets of the Science of the Dust-Numerals].[33] He also translated an anonymous Maghribi handbook on calculation[34] as well as extracts from several commentaries on *Talkhīṣ aʿmāl al-ḥisāb* [Abridgement of the Operations of Calculation] of Ibn al-Bannā (d. 721/1321).[35] In connection with al-Qalaṣādī, we must also mention a note by M. A. Cherbonneau on his writings and an article by G. Eneström on a method of approximation in his *Kashf al-asrār*.[36]

Although Ibn al-Bannā's *Talkhīṣ* was mentioned in the *Muqaddima* and in almost all other mathematical Maghribi texts studied between 1850 and 1860, neither Woepcke nor any of the other researchers mentioned above devoted an article to the contents of this handbook itself. In 1864, Marré published it with a somewhat literal and sometimes strange translation.[37] This translation does not seem to have aroused much interest in Ibn al-Bannā and his work on the part of the researchers of this time, with the exception of Steinschneider who devoted a brief note to him in 1877.[38]

An erroneous reading of another passage of the *Muqaddima* was the origin of research on al-Ḥaṣṣār (5th/12th c.), one of the three other mathematicians mentioned by Ibn Khaldūn. In this passage, one reads "*Kitāb al-Ḥaṣṣār aṣ-ṣaghīr*" which one can translate as "*the book of the small al-Ḥaṣṣār*" or "*the book of the small saddle*" or "*al-Ḥaṣṣār's small book*." On the basis of this sentence and its various interpretations, patient research was carried out by two of the most important historians of sciences in Europe at the end of nineteenth century: H. Suter and M. Steinschneider. In 1874, the latter provided a satisfactory interpretation of the passage on the basis of Hebrew texts. At the same time, he revealed the existence of a Hebrew translation of the 'small book' of al-Ḥaṣṣār made by Moses Ibn Tibbon in 1271.[39]

In 1893, the discovery of an anonymous manuscript and its comparison with this Hebrew translation enabled him finally to give the true title of the book as *Kitāb al-bayān wa t-tadhkār* [Book of the Demonstration and the Recollection] and to provide new information on the complete name of its author.[40]

Suter investigated the biography of al-Ḥaṣṣār and the remainder of his mathematical production. He conjectured the existence of a second treatise by Ḥaṣṣār, of which the *Bayān wa t-tadhkār* was only an abridged version. After long and unsuccessful investigations to confirm this conjecture, he published a German translation of the *Bayān*.[41]

Suter presented the results of his research on the other mathematicians mentioned by Ibn Khaldūn in his biobibliographic work which he completed at the turn of the century.[42] This work can be regarded as the final result of half a century of research, edition and translation, in Europe and in Egypt, of the sources of Arab Science in a wide sense.

For the primary sources of the Muslim West, Suter initially used the first eight volumes of the famous *Bibliotheca Arabica Hispana*[43] which were published between 1883 and 1892, and al-Maqqarī's *Nafḥ aṭ-ṭīb* [Diffusion of the Perfume].[44] He also used the Eastern biobibliographic works from before the twelfth century which had been edited by European scholars there.[45]

It is natural to wonder whether, at the same time, one can detect in the nineteenth-century intellectual circles of the Maghrib, a similar interest for the history of the scientific heritage of the Muslim West in general. The most recent biobibliographic publications by Maghribi researchers, especially al-Manūnī, suggest a negative answer.[46] However, there are two phenomena that deserve further study.

The first is the continuation of the traditional scholarly activities of publishing and teaching. In mathematics and astronomy, this included the continuation of curricula in the institutions of higher education of Qarawiyyūn in Fez and of Zaytūna in Tunis, which had not changed since the sixteenth century, except in the reduction of their volume and the progressive lightening of their contents.[47] The authors belonging to this traditionalist trend continued to write handbooks prompted by the earlier writings of the Maghribi tradition. In the field of historical research, there were numerous publications on the political, religious, and cultural history of the area of which none, to my knowledge, is devoted to aspects of the scientific history of the medieval period, or even of the Ottoman period (10th–13th/16th–19th c.). The only initiative concerning the scientific heritage was associated with the arrival, in 1865, of lithography (which was set up initially in Meknès, then definitively in Fez).[48] *The Recension of Euclid's Elements* by Naṣīr ad-Dīn aṭ-Ṭūsī was lithographed,[49] not with the aim of putting it at the disposal of researchers in the history of science, but rather to be used as a reference work for the teaching of elementary geometry at the Qarawiyyūn in Fez.[50] The scientific heritage continued to provide material for teaching until the end of the nineteenth century.

The second phenomenon is the publication of translations, adaptations or popularizations of European writings about sciences (geometry, logarithms) or about military technology. This is related to initiatives taken in the Ottoman empire in order to acquire scientific and technological knowledge from Europe.[51] Among the translated works were Legendre's *Elements of Geometry* and Lalande's *Treatise of Astronomy*.[52] For reasons that are still unclear, the initiators of this trend do not seem to have been interested in researches on the history of the scientific heritage of the Muslim West.

THE TWENTIETH CENTURY BEFORE 1980

In the twentieth century, research on the history of mathematics and astronomy in al-Andalus and in the Maghrib increased significantly but it continued along the same lines as the work of the pioneers of the second half of the nineteenth century which we mentioned above. In the biobibliography of Andalusi mathematicians, the works of Suter, H. Hankel,[53] and C. Brockelmann[54] were supplemented by the publication of *Kitāb ikhbār al-ʿulamāʾ bi-akhbār al-ḥukamāʾ* [The Book which Informs the Scholars on the Life of the Wise] of Ibn al-Qifṭī (d. 646/1248)[55] in an abridged version, and by the important discovery of the *Kitāb ṭabaqāt al-umam* [Book of the Categories of Nations] of Ṣāʿid al-Andalusī (d. 461/1068). This small work (on profane sciences and philosophy) was written by a specialist in astronomy and it contains invaluable information on the contemporaries of the author and their works.[56] On the basis of these new sources, J. A. Sánchez-Pérez compiled, in, 1921, a biobibliography on the history of mathematics and astronomy in al-Andalus.[57]

This work contains data on 191 mathematicians and astronomers, but in contrast to astronomical and astrological writings, Sánchez-Pérez could name very few extant copies of mathematical works produced in al-Andalus between the ninth and the fifteenth centuries, these being the *Commentary* of Ibn Zakariyyāʾ al-Gharnāṭī (d. 809/1406) on Ibn al-Bannā's *Talkhīṣ*, Ibn Badr's *Abridged Book in Algebra* (7th/13th) and two writings of Ibn Muʿādh al-Jayyānī (d. 460/1067): his *Commentary* on Book V of Euclid's *Elements* and his treatise of spherical Trigonometry, *Kitāb majhūlāt qisiyy al-kura* [The Book of the Unknown Arcs of the Sphere].[58] At the same time, Sánchez-Pérez gave a long list of the lost mathematical works by the scientists who worked in al-Andalus between the ninth and fifteenth centuries. This difference is even more striking because, except for Ibn Muʿādh's book on trigonometry, the mathematical works identified by the time of Sánchez-Pérez are minor writings, compared with the works of the tenth to twelfth centuries. Thanks to information provided by Ṣāʿid al-Andalusī and by mathematicians of the thirteenth to fourteenth centuries, such as Ibn Munʿim, Ibn al-Bannā, and Ibn Zakariyyāʾ al-Gharnāṭī, we know that dozens of mathematicians of the tenth to twelfth centuries, including Ibn as-Samḥ, az-Zahrāwī, and al-Muʾtaman, had published works of a high level on geometry, calculation and the theory of numbers. Ibn Khaldūn also said, in his *Muqaddima*, with reference to the *Book of Algebra* by the Egyptian Abū Kamil (d. 930), that it "*was commented on by many Andalusis*" and that they "*excelled*" in their commentaries, whereas only the rather average work by Ibn Badr has reached us.[60]

No work similar to that of Sánchez-Pérez was written in this time, on the scientific tradition of the Maghrib. Two contributions were made by H. P.-J. Renaud (1881–1945) in the years 1932–1933: an appendix to Suter's book[61] and an article on the edition of scientific works in the Maghrib prior to 1880.[62]

The Contents of Research

Astronomy and Astrology

For these two important fields of science in al-Andalus,[63] the most important research undertaken in the first half of the 20th century was that by Millas Vallicrosa, in particular on az-Zarqālluh's works.[64] This was followed by contributions by J. Vernet, G. J. Toomer, W. Hartner, B. Goldstein, J. Samsó, D. A. King, R. Lorch, and others[65] on certain aspects of the astronomical production of az-Zarqālluh, and on other astronomers, such al-Majrīṭī and al-Biṭrūjī (d. 581/1185).

The research in the first half of the twentieth century showed that Andalusi scientists designed new astronomical instruments, both instruments of observation[66] and universal astrolabes.[67] It also appeared that attempts were made by Ibn Bājja (d. 533/1138), Ibn Ṭufayl (d. 581/1185), Ibn Rushd (d. 595/1198) and al-Biṭrūjī, to establish a new cosmology. On the basis of the criticism of the Ptolemaic model by Ibn al-Haytham,[68] these scientists tried to replace the Ptolemaic model by a model that was, in their opinion, more compatible with the physical world.[69] The originality of the Andalusi mathematicians was also revealed through their development of astronomical tables, such as the tables of Ibn Muʿādh and the Toledan tables.[70]

The research between 1900 and 1980 also produced four essential insights. First, it showed that a number of scientific works by Greek, Indian or ninth-century Arabic authors, such as al-Khwārizmī, Ḥabash al-Ḥāsib, and al-Battānī, were available in al-Andalus at an early stage.[71] Second, it revealed that a Latin astrological tradition influenced the birth and the development of astrology in al-Andalus, which had been previously thought to depend only on the traditions of the Eastern Islamic world.[72] There are also connections with the local tradition of sundials prior to the advent of Islam in Spain.[73] Third, it showed the important role of the group of scientists associated with Alphonso X in the transmission of science from Arabic into Latin.[74] This transmission was studied with specific examples, such as the universal astrolabe.[75] Fourth, it demonstrated the important role of mathematical methods, which are often implicitly present in works on the theoretical aspects of astronomy, astronomical instruments and sundials.

Mathematical methods must also have been used for the computation of the extant astronomical tables. Mathematical methods are discussed explicitly in Ibn Muʿādh al-Jayyānī's trigonometrical treatise *Kitāb majhūlāt qisiyy al-kura*. This book is a description of trigonometrical procedures for the solution of astronomical problems.[76] His approach is independent of that of his immediate predecessor in Asia, al-Bīrūnī, in his *Maqālīd ʿilm al-hayʾa* [The Keys of Astronomy].[77] Ibn Muʿādh develops a procedure for determining all the elements (sides and angles) of an unspecified spherical triangle from three given elements. He uses the polar triangle for the resolution of the case where three angles are given, in a way independent of Abū Naṣr ibn ʿlrāq (4th/l0th c.) in the Eastern Islamic world. He is, finally, the first known Andalusi mathematician to have calculated a table of tangents.

In the historical research of the first half of the twentieth century, the place of the astronomical tradition of the Maghrib is quite modest by comparison with that of the astronomy in al-Andalus. This astronomical tradition could not compete with that of al-Andalus, but nevertheless deserves some attention.

The first work in this tradition to be studied was Ibn al-Bannā's *Minhāj aṭ-ṭālib fī taʿdīl al-kawākib* [The guide of the student for the correction of the star movements], of which a partial Spanish translation was published by J. Vernet,[78] who also listed, four years later, the astronomical manuscripts of Ibn al-Bannā.[79] In the 1930s, H. P.-J. Renaud started to work on particular astronomical subjects. He published biographical notes on two astronomers, Ibn al-Bannā[80] and Abū Miqraʿ,[81] then he conducted research on the astronomical and astrological traditions in the Western Maghrib as a whole,[82] on the obliquity of the ecliptic[83] and the appearance of the lunar crescent.[84] He published the only known calendric treatise of the Maghribi tradition, whose contents also show the close connection between the tradition in the Maghrib and that of al-Andalus.[85]

An interest for the study of the scientific heritage of the Muslim West started relatively early in certain intellectual spheres in the Maghrib. From the 1950s on as we will see in detail in the following chapter devoted to mathematics, the results of this interest were modest in quantity and uneven in quality. The history of astronomy does not seem to have benefited from the passion for the Arabic sciences of the Middle Ages which was aroused by the phenomenon of *Nahḍa* [rebirth] preached by the Muslim reformers of the beginning of the century, such as Shaykh ʿAbduh and Jamāl ad-Dīn al-Afghānī.[86] I have not found any trace of original historical work devoted to a particular astronomer of al-Andalus or the Maghrib, other than notes in manuscript catalogues and in general works on the cultural and scientific history of the former Islamic empire. However, some of these general works contain important, and sometimes new, information on particular astronomers of the Muslim West or on their contribution. Examples are Gannūn's book, which is limited to the cultural history of the Western Maghrib; Ṭūqān's book, which is devoted to the entire history of Arabic astronomy and mathematics; and the book of al-ʿAzzāwī dealing with the history of astronomy in the Islamic world, from the arrival of the Mongols until the second decade of the twentieth century.[89] These books contributed in a double way to a better knowledge of the history of Islamic science in general. The works by Gannūn and Ṭūqān were widely distributed and went through several editions.[90] They changed the attitude among the intellectuals of the Arabic countries, and in the Islamic world as a whole, toward the cultural and scientific heritage of the civilization of their ancestors.[91] In addition, the new information in these works was of interest to the biobibliographers and the Western specialists in the history of science, who integrated it into their own publications. For example, Sarton said that the book of Ṭūqān was indispensable for researchers in the history of Arabic astronomy and mathematics. He also published a positive review of this work.[92] In 1942, C. Brockelmann wrote to A. Gannūn in connection with his book:

> I received your invaluable book entitled 'the Moroccan genius in the Arabic literature'; you have honoured me by sending it. I began to read it and I benefited much from it in connection with what has escaped my research up to now on the history of the Moroccan literature. I hope to use its invaluable contents for my profit and the profit of my Orientalist colleagues, in the Supplement to my first book on the history of Arabic literature which is now published in the city of Leiden.[93]

Finally, thanks to the research of Ṣāliḥ Zakī, Western specialists have been informed about the existence of the mathematician Ibn Ḥamza (16th century), and his book *Tuḥfat al-aʿdād* [The Ornament of Numbers], in which Zakī claims to have found some form of the concept of the logarithm.[94]

Mathematics

In the early twentieth century, research on the history of mathematics in al-Andalus and the Maghrib was limited to six authors and some specific topics, with an emphasis on biobibliographic research and editions of texts.

Three of these authors are Maghribi, namely al-Būnī (d. 622/1225), Ibn al-Bannā (d. 721/1321) and Ibn al-Yāsamīn (d. 601/1204). The fourth author, al-Qalaṣādī, is Andalusi but spent most of his life in the Maghrib. The origin of the two remaining mathematicians, al-Ḥaṣṣār and Ibn Badr (7th/13th c.) is uncertain. Judging from the contents of their mathematical writings and from their names (which are carried by a number of Andalusi intellectuals mentioned in the biobibliographic works), one may guess that they were of Andalusi origin.

This historical research concentrated mainly on the science of calculation, algebra, and, to a lesser extent, elementary geometry with plane figures. I will also mention astrology and the science of inheritances as applications of mathematics, even though the research in these two subjects was more modest in quantity.

Ibn al-Bannā and his work were the subject of the greatest number of studies. Renaud published two articles on him, first a very concise article, giving precise details on his date of birth,[95] and secondly, a much longer one, in which he reveals that Ibn al-Bannā was the author of many works in different fields.[96] In his articles Renaud corrects certain errors according to which Ibn al-Bannā's origin was Saragossa or Granada, and which were based on confusing[97] or unbased assertions, for which Casiri, the author of the catalog of Arabic manuscripts of the Escurial, seems to have been responsible.[98] Later, A. Gannūn[99] and A. al-Fāsī[100] published biobibliographies of Ibn al-Bannā, but these did not bring in any new elements beyond what Renaud's researches had already revealed.

Only one of the mathematical writings of Ibn al-Bannā was the subject of a detailed study, namely the *Talkhīṣ*, a handbook on the science of calculation. This work was first translated (but not edited) by Marré, then edited, translated a second time into French, and analyzed by Mohamed Souissi.[101] In this work, Souissi used other Maghribi mathematical writings, such as Ibn al-Bannā's *al-Maqālāt al-arbaʿ* [The Four Epistles][102] and the commentaries on the *Talkhīṣ* by Ibn al-Bannā's student al-Huwārī (7th/13th centuries), and al-Qalaṣādī.[103] This comparative study of the sources themselves confirmed the existence of a certain continuity in the subjects and methods used in the mathematical teaching in the Maghrib in the fourteenth and fifteenth centuries.[104]

We know today that these same topics and methods already occurred in al-Ḥaṣṣār's handbook *al-Bayān wa t-tadhkār*, of which Suter had published a detailed analysis in 1901. A comparison between this handbook and the *Talkhīṣ* in Marré's translation could have revealed the relationships between the mathematical traditions of twelfth and the fourteenth centuries. As far as I know, no new research was done on this subject during the period which interests us here.

A new opportunity for the realization of this comparative study arose in 1964 when A. Gannūn revealed the existence of a twelfth-century treatise, written by Ibn al-Yāsamīn, with a rich mathematical content which is important for a better knowledge of the mathematical tradition of the Muslim West as a whole.[105] But the article, which was written in Arabic, apparently went unnoticed. Twenty years later, Ibn al-Yāsamīn continued to be known only for his algebraic poem, *al-urjūza al-yāsamīniyya* [the Poem of al-Yāsamīn].[106] This *urjūza* did not get a better fate, even though a certain number of manuscript copies had been mentioned by Suter and Brockelmann. But in 1916, an important algebraic work, Ibn Badr's *Mukhtaṣar fī l-jabr*, was edited and translated into Spanish by Sánchez-Pérez. As we will see in the next section, this text throws light on the connections between the algebraic tradition of the Muslim West and the algebra of al-Khwārizmī and Abū Kāmil (d. 318/930).[107]

In geometry, the only contribution was the publication of a facsimile with English translation of Ibn Muʿādh al-Jayyānī's treatise on the theory of proportions of Euclid's *Elements* Book V.[108] This text is important both for the history of this theory in the Islamic world and for the history of transmission between the Eastern and the Western Islamic world. But its publication did not inaugurate new research in the mathematical tradition of the Maghrib and al-Andalus.

Publications on applications of mathematics are also rare and limited to the presentation of the contents of a manuscript. Two texts in the science of inheritances drew the attention of Sánchez-Pérez, probably more for their sociological interest than for the mathematical content.[109] The construction of magic squares was studied through al-Būnī's writings,[110] but the work of al-Būnī, who lived in Egypt, is representative of the Eastern Islamic tradition.

The Contents of the Published Mathematical Texts

The research between 1900 and 1980 shows a predominant interest among historians in the mathematical works themselves, through editions, translations and analyses of texts. This has made it possible to make a preliminary overview of the mathematics that was taught or used in the Maghrib from the twelfth century onwards, although nothing precise had yet been revealed on the teaching of geometry. Thanks to the handbooks analyzed by Suter, Woepcke and Souissi, we know that the following subjects in arithmetic were taught in the Maghrib, between the twelfth and fourteenth centuries:

- the positional decimal system and its representation by the *ghubār* symbols.[111]
- the seven traditional arithmetical operations, i.e., duplication, halving, addition, subtraction, multiplication, division, and extraction of the exact or approximate

square root of a number, with various algorithms for integers and with extensions of some of these operations to fractions and to quadratic or biquadratic irrationals.

- Arithmetic formulas for the summation of finite series of natural, odd, or even numbers, squares, cubes, as well as the finite geometrical series.
- Operations specific to fractions, such as reduction and conversion.

The second part of these handbooks of arithmetics is devoted to solutions of equations by arithmetical methods, that is, the rule of three and the method of false position (in the Muslim West called *ʿamal al-kaffāt* [method of scales]), then by the six canonical equations of al-Khwārizmī.[112]

In addition, irrational numbers are found in a poem by Ibn al-Yāsamīn, entitled *Urjūza fī l-judhūr* [Poem on the Roots],[113] and in al-Qalaṣādī's letter entitled *Risāla fī dhawāt al-asmāʾ wa l-munfaṣilāt* [Epistle on the Binomials and the Apotomes].[114] One finds here the tradition of Euclid's *Elements* Book X, rephrased arithmetically, as well as some extensions of this tradition discovered in the Eastern Islamic word between the ninth and the eleventh century.

The historical studies that were published during this period on the numeral systems used in the Muslim West all deal with the question of origin, usually in respect to the positional decimal system. Probably for cultural reasons, this subject inspired a series of articles on the origin of the symbols which were used for this system in the Maghrib.[115] These articles contributed nothing new to the history of mathematics, because they were written from cultural and ideological motivations by scholars who had no expertise in the scientific aspects of the subject.

At the beginning of this century, one also observes a revival of a non-positional numeral system which had been used for centuries, in some administrations of the Western Maghrib. These were the so-called *numbers of Fez* using 27 different signs. The system was also known in Maghrib under the name of *ḥurūf az-zimām* [figures of the account book]. The calculation which was associated with it was called *ḥisāb rūmī* [Byzantine calculation]. A number of old texts, dealing with this system and its applications, by al-ʿUqaylī, al-Fāsī, and Sakrīj,[116] appeared in a lithographed edition. These publications might explain the interest around the 1920s for this numeral system on the part of researchers of different backgrounds, such as G. S. Colin, C. Pellat, and J. A. Sánchez-Pérez.[117] This explanation needs to be supported by further research in the cultural history of the Maghrib in the first half of twentieth century. It would also be interesting to know if this revival represents a return to the scientific heritage similar to, but on a lower level than, the development in China in the end of eighteenth and throughout the nineteenth centuries.[118]

The chapters on algebra in the arithmetical handbooks and the small poem of Ibn al-Yāsamīn which I have mentioned, are limited to the definition of algebra, its objects, and its methods: the unknowns, the numbers, the four arithmetical operations applied to these objects when they stand alone or are combined in polynomial expressions, the six canonical equations, the operations of balancing and, finally, the method of solution.

In the *Muqaddima*, Ibn Khaldūn says on algebra:

> the first one who wrote on this science is Abū ʿAbdallāh al-Khwārizmī and, after him, Abū Kāmil Shujāʿ ibn Aslam; then people followed his method. His book on the six equations is among the best-written on the [subject]. Many [are those], among the people of al-Andalus, [who] commented on it well. And among the best commentaries is al-Qurashī's book. And we heard that one of the greatest masters in mathematics among the people of the East, was led to [a greater number] of equations than these six species, bringing them to more than twenty, and that he determined valid procedures for their resolution, followed by geometrical demonstrations.[119]

This passage has been translated and analyzed many times by historians of sciences since Woepcke. The passage showed that, in the Muslim West, algebra did not remain in the condition in which al-Khwārizmī had left it, and as we find it in the Maghribi handbooks of the fourteenth to fifteenth centuries analyzed by Woepcke, Suter, and Marré. But Ibn Khaldūn's assertions were partially confirmed by the publication of Ibn Badr's *Mukhtaṣar fī l-jabr* [Concise Work on Algebra].[120] This work contains all the material of al-Khwārizmī's book but in a different presentation and with extensions and new applications. Ibn Badr treats the same topics as those studied in the Islamic East before al-Karajī. He goes beyond al-Khwārizmī because he generalizes the rule of the powers, discusses the necessary terminology, and introduces irrational numbers. He also solves the same types of problems which one encounters in Abū Kāmil's work: problem of "tens," "capitals," "meetings," "cereals," the "pursuit" problem, etc. But the number of examples solved in each category is less and Ibn Badr generally selected the problems which do not involve too many technical complications.[121] Fifty years later, an English translation of Abū Kāmil's treatise on algebra was published on the basis of the Hebrew version of Mordechaï Finzi.[122] Abū Kāmil's work is important for the transmission of algebra from the East to the West, but its publication did not lead to a re-examination of the algebraic texts from the Muslim West.

In number theory, Ibn Khaldūn was not of great help for historians of science since he said merely that the mathematicians

> had given it up after they had drawn the main points for arithmetic demonstrations, as Ibn al-Bannā did in the book *Rafʿ al-ḥijāb*, as well as others.[123]

The last sentence revealed the title of a work whose contents were promising, but this became available only in the 1980s. The last word of this sentence kept its secrets even after the publication of Ṣāʿid al-Andalusī's *Kitāb ṭabaqāt al-umam*. These "other" mathematicians who wrote on number theory seem to have been Andalusis. Ṣāʿid al-Andalusī, who designates them by "*scientists in the theory of numbers*"[124] seems to differentiate them from the "*scientists in calculation.*" Knowledge of the titles of their works could have made it possible to solve the question of their origin but Ṣāʿid seldom mentions the works of the scientists he introduces. Among the titles he quotes, only one could be connected to an Andalusian tradition in arithmetic, namely the *Kitāb ṭabīʿat al-ʿadad* [Book on the Nature of the Number],[125] but up to now, we do not know its contents.

Ibn Khaldūn seems to refer to an arithmetical tradition in the Maghrib in his discussion of certain types of integers and of the figurate numbers of the *Arithmetica* of Nicomachus of Gerasa.[126] The existence of such a tradition was confirmed in 1976 by the publication of an anonymous text, attributed to Ibn al-Bannā and dealing with the determination of amicable numbers.[127]

Orientations and Results of Research after 1980

Before 1980, research on the mathematical tradition of al-Andalus and the Maghrib focused on arithmetic and algebra, and was based on the known writings of three authors: al-Ḥaṣṣār, Ibn al-Bannā, and al-Qalaṣādī (12th, 13th–14th, 15th century respectively).

In spite of the major contributions of the first researchers in this field, not all questions raised by the publication of Ibn Khaldūn's *Muqaddima* had been answered, and some interpretations concerning the mathematicians or the mathematical tradition of the Muslim West were unsatisfactory. In addition, the modest place of mathematics in published sources suggested that profitable research was still possible. Several questions were still open.

The first question was the beginning of mathematical activity in the Muslim West. The second concerned the circulation of scientists, ideas, and techniques between al-Andalus and the Maghrib, in relation to the cultural and political history of the area.

A third question related to the contents of the various handbooks for teaching which had already been published and analyzed. Obviously these contents did not reflect the level of mathematics in the Eastern Islamic world, and they were well below the level of some handbooks published in Baghdad a few centuries earlier. The historians tried to find an explanation for this established fact, or to qualify it by discovering new sources.

In addition to these general questions, there were also more specific ones. The first concerned the mathematical symbolism which Woepcke found in the work of al-Qalaṣādī. What were the origins of this symbolism, its development, and, especially, its role in teaching? The second was whether other intellectual activities in Western Islamic cities could have influenced the thought of the mathematicians in relation to some mathematical concepts or definitions.

In addition, the biobibliographic or historical works suggested two equally important topics: the history of the infrastructure of teaching, and the place of the community of mathematicians and their activities in the context of the intellectual life and the cultural and religious practices of the cities of al-Andalus and the Maghrib.

The Mathematical Topics That Were Studied

Combinatorics

Research on this subject began with the discovery in 1978 of Ibn al-Bannā's treatise entitled *Tanbīh al-albāb ʿalā masāʾil al-ḥisāb* [Warning to Intelligent People on the Problems of Calculation].[128] The analysis of this treatise showed the interests of the author in combinatorics as well as his contributions to this field. This contribution seemed to be based on earlier work by another mathematician, Ibn Munʿim, whose book *Fiqh al-ḥisāb* [the Science of Calculation] had been mentioned by Ibn Khaldūn in his *Muqaddima*.

Further progress was made by the analysis of another previously unpublished work of Ibn al-Bannā, *Rafʿ al-ḥijāb*. The chapter on figurate numbers and the summations of finite sequences of integers, as well as some information drawn from late commentaries on Ibn al-Bannā's *Talkhīṣ*, made it possible to supplement the combinatorial part of *Tanbīh al-albāb* and to place these parts in a context of interests and practices which seem specific to the Maghrib.[129]

The subsequent discovery of Ibn Munʿim's *Fiqh al-ḥisāb* and the study of its important chapter 11 showed that the Maghribi contributions in combinatorics, prior to those of Ibn al-Bannā, were important both in quantity and in quality. Not less than nineteen pages of this work are exclusively devoted to combinatorial definitions, theorems, and techniques.[130]

The *Fiqh al-ḥisāb* showed the utility for a historian of science of a sufficient knowledge of the political, cultural, and ideological environment where the mathematical activities took place. Thus it was possible to refute several assumptions about Ibn Munʿim's life and activities which had prevailed since the 1940s and which had been a source of debate between European historians of sciences such as Steinschneider, Suter, Woepcke, and Renaud.

The technical contents of the combinatorial practice in the two above-mentioned sources can be summarized as follows. The combinatorial activities of the medieval Arabic tradition are not a direct continuation of the Greek and Indian traditions, but their origin is closely related to an Arabic cultural reality.[131] For they originated not from the traditional mathematical themes, but in linguistics and thus with the first studies on the Arabic language and the lexical morphological terminological discussion which they prompted.

At a later stage, one observes inside the mathematical tradition of the Islamic East the first signs of what one could call a combinatorial approach, in connection with simple enumerations in the solution of geometrical or algebraic problems, which did not require the establishment of formulas or general processes. But, paradoxically, the mathematicians of the Maghrib who dealt with combinatorics, were not inspired by this later tradition.

Ibn Munʿim tells us that his project is to mathematize a lexicographical problem inherited from the period of al-Khalīl Ibn Aḥmad (d. 797). He initially established a rule enabling him to determine all the possible combinations of n colors in groups of p. For that, he has to construct the arithmetical triangle according to both an inductive and a combinatorial approach. Then he demonstrates, according to the same double approach, the formulas for the permutations of a set of letters, with or without repetitions, and the formulas giving, by recurrence, the number of possible readings of the same word of n written letters, taking into account all the unwritten signs used in a given language (e.g., vowels and sukūns for Arabic). He also gives a formula for arrangements without repetitions of n letters in groups of p, taking account of the unwritten signs (vowels, etc.) that can accompany these letters, and he solves a certain number of problems on the maximum number of combinations with repetitions. These results enable the author, by means of a series of tables, to solve the initial problem of the enumeration of all the words from one to ten letters which can be formed with the letters of an alphabet, taking into account all the possible repetitions of the letters in a word and the (unwritten) signs which can be placed on these letters.

The results in Ibn Munʿim's work were new, and he also derived them by a new method, which involved a purely combinatorial reasoning. In the history of combinatorics, the *Fiqh al-ḥisāb* is, at the same time, the end of a stage of calculation by means of tables, and the beginning of a new stage, the substitution of arithmetic formulas for these tables. These formulas are subsequently used to solve mathematical problems in various fields.

The contributions of Ibn al-Bannā develop the scope of Ibn Munʿim. In the *Tanbīh al-albāb*, Ibn al-Bannā establishes the famous arithmetical formula permitting a direct computation of the number of combinations of n objects in groups *p* without having to construct the arithmetical triangle, as his predecessor had done.

In his *Raf*ᶜ *al-ḥijāb*, one finds combinatorial theorems similar to those in the *Tanbīh al-albāb*, but presented as complements to two chapters on the theory of numbers inherited from the arithmetical tradition of Nicomachus: on the summations of integers and on the figurate numbers respectively.

In later Maghribi mathematical writings, combinatorial approaches appear in very different fields, such as magic squares, inheritances, grammar, and even religion.[132]

The fact that all later authors use the same combinatorial terminology, and that none of them claim authorship for results or applications, reinforce the continuity of the combinatorial tradition since Ibn Munᶜim.[133]

Algebra

The first research in the history of Maghribi algebra after 1980 was related to the diffusion in the Muslim West of the six canonical equations of degree one or two. The available texts were Ibn al-Yāsamīn's *Urjūza*, Ibn Badr's *Ikhtiṣār fī l-jabr*, both already published, and the still unpublished *Kitāb al-uṣūl* of Ibn al-Bannā. To these texts, which are exclusively devoted to algebra, one can add the algebraic chapters which usually conclude the handbooks of calculation.

This research showed that the canonical equations were generalized and that changes were made in the classification of these equations, in close connection with some developments of the algebra of polynomials.[134] We have already mentioned specific developments of notational symbolism of the Muslim West discovered by Woepcke and Renaud.[135] Traces of this symbolism were found in Ibn al-Yāsamīn's *Talqīḥ al-afkār* [The Fecondation of Spirits],[136] and through specific testimonies, it has been possible to follow the evolution of these symbols and abbreviations into an instrument for solving arithmetical and algebraic problems.[137]

Based on these specific studies on equations and notational symbolism, I wrote a general assessment of the research on the available algebraic sources from the Muslim West.[138] A few years later, Saidan published a critical edition and analysis of the only work on algebra from the Maghrib which has reached us, namely Ibn al-Bannā's *Kitāb al-uṣūl wa l-muqaddamāt fī l-jabr wa l-muqābala* [the Book of the Bases and of the Preliminaries in Restoration and Balancing].[139] This is a work in the algebraic tradition of Abū Kāmil, and it gives an idea of the level of teaching of algebra in some cities in the Maghrib, such as Bougie, Marrakesh, and Fez.

Since the publication of Ibn Khaldūn's *Muqaddima*, we know that there was an algebraic tradition in this area.[140] This information was partially confirmed by Ibn al-Yāsamīn's *Urjūza*, Ibn Badr's *Ikhtiṣār*, and finally, by some surviving citations of the lost work of the Andalusi mathematician, Abu 'l-Qāsim al-Qurashī, who taught algebra and the science of inheritance in Bougie, in the twelfth century. Two distinct aspects of this tradition appear in two other categories of writings translated from Arabic or strongly related to the Arabic algebraic tradition of al-Andalus. The first category consists of the chapters of the *Liber Mahameleth* on Algebra[141] and some parts of Abū Bakr's *Liber Mensurationum*, and it contains solutions of practical problems using the method of the algebra of al-Khwārizmī.[142]

The second category consists of texts on the solution of problems of (land) measurement, including Abrahām Bār Ḥiyya's *Liber Embadorum*,[143] parts of Abū Bakr's *Liber Mensurationum* and the *Risāla fī at-taksīr* [Book on Measurement] of Ibn ʿAbdūn (d. after 976).[144] These texts show the existence in al-Andalus of a problem-solving tradition with algebraic aspects anterior to al-Khwārizmī, and related to the Babylonian tradition.

The analysis of the available sources shows that the mathematical tradition of the Maghrib was not a faithful preserver of various aspects of the algebraic activity in al-Andalus. In the Maghrib, one finds neither direct or indirect references to three important traditions: algebra associated with the geometry of measurement such as it appears in Ibn ʿAbdūn's handbook, study of the polynomials on the basis of the work done by Sinān Ibn al-Fatḥ in the East[145] and, finally, the geometrical solution of cubic equations of al-Khayyām (d. 526/1131).[146]

The first tradition existed in al-Andalus and its absence in the Maghribi writings can be explained either by the loss of the sources or by the lack of interest on the part of Andalusi mathematicians from the twelfth century onwards. Future research may reveal which of these alternatives is correct. The second tradition is partially present in the Maghribi texts, but the algebraic writings of Eastern authors, like Ibn al-Fatḥ, al-Karajī (d. 414/1023) and al-Samawʾal (d. 571/1175) seem to have been unavailable in the West. The silence of the mathematical sources and Ibn Khaldūn's testimony suggest that the third tradition did not take root in al-Andalus and the Maghrib.[147]

The contents of Western algebraic writings prior to the thirteenth century can be summarized as follows. The oldest witness in al-Andalus of the "algebraic" tradition in measurement problems seems to be Ibn ʿAbdūn's handbook. Many problems are set out and solved according to the procedures in Babylonian texts, without any reference to the six canonical equations of al-Khwārizmī's *Algebra* and without the algebraic terminology of the ninth century. Textual similarities to problems in the *Liber Mensurationum* and the *Liber Embadorum* suggest that this tradition of measuring survived in Andalus until the beginning of the twelfth century.

In the tradition of al-Khwārizmī and Abū Kāmil, Abū Bakr's *Liber Mensurationum* contains problems of measurement that are similar and sometimes identical to those treated by Ibn ʿAbdūn and Bār Ḥiyya.[148] The author of the *Liber Mahamalet* refers explicitly to Abū Kāmil's Algebra, the *Kitāb al-kāmil*, but no other work is quoted.

Three texts were influenced by the tradition after Abū Kāmil. The mathematical poem of Ibn al-Yāsamīn summarizes the algorithms for the solution of the six canonical equations and accompanies them with some operations on quadratic irrationals. Ibn Badr's *Ikhtiṣār al-jabr wa l-muqābala* [the Summary of Restoration and Balancing] is a summary of algebra in the tradition of al-Khwārizmī and Abū Kāmil, with some further additions. However, an analysis of the methods and the problems treated by Ibn Badr reveals some characteristics that are difficult to relate to what is known about the algebraic traditions of the East prior to al-Khayyām. Ibn al-Bannā's *Kitāb al-uṣūl* [Book of Elements] is the last important work on Algebra of the Muslim West, and later algebraic works, in the Maghrib and in Egypt, derive their inspiration entirely from the problems and the methods of this book, or reproduce them with explicit references.[149] The *Book of Elements* is in two parts. The first part deals with numbers; it is a summary of Books VII–X of Euclid's *Elements* with additions such as the division by irrational expressions of the form $n + \sqrt{m} + \sqrt{p}$. The second part is about the solution of various types of problems using algebraic methods. First Ibn al-Bannā treats problems whose solutions are integers or fractions, and he concludes with problems whose solutions are irrational numbers. This part of the book is inspired by Abū Kāmil's problems, and sometimes use exactly the same coefficients, but follow a different form of presentation.[150] The additions by Ibn al-Bannā are problems of algebra which do not occur in Abū Kāmil's book but which fit into the same tradition, and problems on the theory of numbers, such as the representation of an integer as to the sum of two squares of integers or rational numbers.

Ibn al-Bannā also dealt with algebra in two other works, the *Talkhīṣ* and the *Rafʿ al-ḥijāb*. Although the *Talkhīṣ* contains only the rules and the basic algorithms of algebra, without demonstration and application, it was the Arabic mathematical work that received the most commentaries in the Maghrib, between the fourteenth and the sixteenth century. The *Rafʿ al-ḥijāb* contains some new elements, in particular purely algebraic methods for setting out and calculating the solutions of quadratic equations, without reference to the geometrical demonstrations found in the *Algebra* of al-Khwārizmī and Abū Kāmil.

After Ibn al-Bannā's *Kitāb al-uṣūl*, three kinds of algebraic texts were written in the Maghrib: commentaries on earlier works, independent observations, and chapters in arithmetical works. Their contents are not always a simple repetition of the methods in the earlier works.

All the algebraic commentaries which have reached us concern Ibn al-Yāsamīn's algebraic poem. Some were written in the Maghrib.[151] The analysis of these texts does not disclose anything new, other than that algebraic symbolism was intensively used in the mathematical teaching in the Maghrib during the fourteenth to fifteenth centuries.

The algebraic chapters of two other texts are important for showing the use of this symbolism. The first is Ibn Qunfudh al-Qasanṭīnī's *Ḥaṭṭ an-niqāb ʿan wujūh aʿmāl al-ḥisāb* [the Lowering of the Veil on the Various Operations of Calculation] which contains, in particular, a symbolic expression of an equation whose second term is zero.[152] The second work is *Rashf ar-ruḍāb min thughūr aʿmāl al-ḥisāb* [Sucking the Nectar from the Mouths of the Operations of Calculation] by al-Qaṭrawānī, a mathematician who lived in Tunis probably at the end the fourteenth century or the beginning of the fifteenth. This is the only extant mathematical text of the Maghrib in which the calculation of the square and cubic roots of an abstract polynomial is explained. The interest of this fact does not lie in the results themselves, since the extraction of the square root of a polynomial of arbitrary degree had been explained by as-Samawʾal (d. 1175) in Baghdad, and since the extraction of the cube root could be carried out by any mathematician who understood the technique.[153] The inventors of the method in the Eastern Islamic world used the symbolism of tables to support of the algorithm, but al-Qaṭrawānī uses the Maghribi letter symbols to write out the data, the operations and the results. Al-Qaṭrawānī's book is a good illustration of the circulation of ideas and techniques among various scientific centers in the Islamic world.

Arithmetic

Since 1990, the history of arithmetic in al-Andalus and the Maghrib has been the subject of intensive research in the Maghrib itself. This has resulted in the editions and studies of the following works written between the twelfth and fourteenth century: Ibn al-Bannā's *Rafʿ al-ḥijāb*,[154] Ibn Qunfudh's *Ḥaṭṭ an-niqāb*,[155] Ibn al-Yāsamīn's *Talqīḥ al-afkār*,[156] al-ʿUqbānī's *Sharḥ at-Talkhīṣ*,[157] and al-Ḥaṣṣār's *Kitāb al-bayān wa t-tadhkār*.[158] In addition, studies have been made of the methods of false positions[159] and arithmetical algorithms[160] in the tradition of al-Andalus and the Maghrib. The relationship between arithmetic and the science of Islamic inheritances has been studied on the basis of the *Kitāb al-istiqṣāʾ wa t-tajnīs fī ʿilm al-ḥisāb* [the Book of the Investigation and the Classification in Calculation] of al-Ḥubūbī (4th/10th c.),[161] *al-Mukhtaṣar* [the Abridged <book>] of al-Ḥūfī (d. 1192)[162] and *Sharḥ Mukhtaṣar al-Ḥūfī* [Comment on al-Ḥūfī's abridged <book>] of al-ʿUqbānī (d. 1408).[163] Another research project has dealt with aspects of arithmetic practiced in al-Andalus between the ninth and the eleventh century and their presence in Maghribi writings, or in Andalusi works written after the eleventh century that were distributed and taught in the Maghrib. These investigations started with the study of the arithmetical parts of three important works which had never been analyzed before: in chronological order, the first three chapters of Ibn al-Yāsamīn's *Talqīḥ al-afkār*, chapters I–VII of the first part and the entire second part on fractions of Ibn Munʿim's *Fiqh al-ḥisāb* and, lastly, Book I of Ibn al-Bannā's *Rafʿ al-ḥijāb*.[164] This increased knowledge of the practice of arithmetic in the Maghrib has led to the identification of the first volume of the full version of al-Ḥaṣṣār's work, *al-Kitāb al-kāmil fī ʿilm al-ghubār*. Suter conjectured the existence of such a work and he distinguished it from the *Kitāb al-bayān* of the same author.[165]

Since research is still in its initial stage, it does not allow us to describe the arithmetic tradition of the Muslim West in detail, but it is already possible to sketch briefly the major outlines and the characteristic aspects of this tradition.

As in the East, the Indian positional decimal system, with modified number symbols of Indian origin, dominated Andalusi and Maghribi teaching. But two other systems of numeration and calculation were used in al-Andalus and in the Maghrib. The *Ḥisāb al-yad* (Finger Reckoning) does not seem to have been marginal since mathematicians of some importance devoted writings to it, as they also did in the Eastern Islamic world.[166] The second system is the Rūmī calculation (see above, p. 321). Evidence in some mathematical texts shows that this system was used at least since the twelfth century. Authors of Andalusi origin or Andalusi formation, like al-Ḥaṣṣār, Ibn Munʿim and Ibn al-Yāsamīn, discuss the system at length. In mathematical texts the system is presented in the form of a definition with comment, or in the form of an independent chapter. Ibn al-Bannā even devoted a whole treatise to it.[167]

The arithmetical algorithms in the extant works are mostly the same as those already used in the East. Presentation and usage change in the course of time. In the works prior to the thirteenth century, there are two separate sections on the operations of doubling and halving respectively, multiplication is treated before the addition, and an important place is given to the chapter dealing with fractions, which sometimes occupies up to half of the work.[168] From the thirteenth century onward, these characteristics disappear from the handbooks. From now on fractions are represented with the bar separating numerator from denominator (as we do nowadays) but no author claims this innovation.

The Theory of Numbers

Ibn al-Bannā's *Rafʿ al-ḥijāb* was the first work to show historians that number theory was studied in the Maghrib. The work deals with problems of the Euclidean tradition, such as the study of prime numbers and perfect numbers, and problems from the arithmetical tradition of Nicomachus, on figurate numbers, summation of a finite series of integer numbers, and amicable numbers.[169] An original contribution of Ibn al-Bannā is the integration of combinatorial theorems into the theory of numbers. He expressed these theorems by means of a finite series of figurate numbers in the tradition of Nicomachus, or by means of the finite series of powers of integers.

It was the discovery of Ibn Munʿim's *Fiqh al-ḥisāb* that raised the veil on some aspects of the Andalusi tradition. In chapter VIII, Ibn Munʿim establishes the formula for the sum of a series of integers, such as the sum of the first *n* even, odd, even-even or even-even-odd numbers, and the sum of the first *n* squares and cubes. In this chapter, Ibn Munʿim also uses analysis and synthesis to find results usually demonstrated by induction.[170] Chapter IX follows the tradition of Andalusi writings of the eleventh to twelfth centuries on various summations of rows or columns of the table of the figurate numbers.[171] Chapter X is devoted to perfect and amicable numbers. The author gives here the calculation of Fermat's couple of amicable numbers (17296 and 18416), which al-Ḥaṣṣār also gave in the lost second volume of his *Kitāb al-kāmil*.[172]

The second source of information on the Andalusi tradition in the theory of numbers is in the chapter on arithmetic of the *Kitāb al-istikmāl*, which was identified at the beginning of the 1980s. It is an abridged version of what was, in the Arabic mathematical tradition after the ninth century, the core of the theory of numbers, namely Books VII, VIII, and IX of Euclid's *Elements*,[173] the Arithmetical Introduction of Nicomachus[174] and the *Risāla fī l-aʿdād al-mutaḥābba* [Letter on the Amicable Numbers] of Thābit Ibn Qurra (d. 901).[175]

Geometry

What was known before 1980 about the geometrical tradition of the Muslim West was limited to the problems of measurement and preliminaries for astronomy. The importance of this discipline was confirmed by Ṣāʿid al-Andalusī, the eleventh century historian of sciences who discussed the activities and works of Andalusi authors who were specialists in geometry[176] and who practiced this discipline before or in the eleventh century. Ibn Khaldūn also mentions important geometers who lived in the twelfth to thirteenth centuries.[177] But until the end of the 1970s, no important geometrical work of the Muslim West had been identified and studied, except al-Jayyānī's *Kitāb majhūlāt qisiyy al-kura*.

Some Latin and Hebrew sources are important evidence of one aspect of the of Arabic geometrical tradition of al-Andalus, because they are translations of Arabic texts or compilations based on Arabic material. The field covered by these sources is limited, since they are mainly[178] concerned with topics related to astronomy or land measurement.[179]

The publications after 1980 fall into three categories: identification of lost works, the transmission of Greek and Arabic geometrical works from East to West, and the teaching of geometry.

In 1984, I identified and analyzed an anonymous text containing information on the original contributions of the Andalusi mathematician Ibn Sayyid. These are an extension of the Greek material on conics, and concern what would be called much later the study of curves of degree higher than 2. Ibn Sayyid outlined a classification of one category of these plane curves and their use to solve geometrical problems of the Greek tradition which would be expressed in modern terms by equations of degree equal to or higher than 5.[180]

The discovery, in 1984, of most of the geometrical chapters of the *Kitāb al-istikmāl* of al-Muʾtaman, by J. P. Hogendijk, answered the questions of Sarton in his *Introduction to the History of Science*.[181] The work throws new light on the Andalusi geometrical tradition[182] and on the transmission of the geometrical writings of the Greek corpus, such as Euclid's *Elements*, Apollonius's *Conics*,[183] or Menelaus's *Spherics*[184] in Arabic from the Muslim East to the West. Al-Muʾtaman also used geometrical works of the tenth century such as Ibrāhīm Ibn Sinān's treatise on the area of the parabola, and even of the eleventh century, such as the works of Ibn al-Haytham on optics and on analysis and synthesis.[185] Al-Muʾtaman, or his Andalusi predecessors, also made original contributions to geometry, such as the theorem of Ceva[186] and the construction of two mean proportionals between two given lines.[187]

However, the manuscripts of the *Kitāb al-istikmāl* are incomplete, and they are a preliminary draft by al-Muʾtaman of a work in two parts, of which the second part was apparently never published. Information on the missing sections and on the chapters of the proposed second part of al-Muʾtaman's project became available ten years later, thanks to the discovery of two copies of a manuscript of the thirteenth century. These manuscripts were written in Asia by Ibn Sartāq, a mathematician of the East who had carefully studied the published version of al-Muʾtaman's book and who had made a new version that was not very different from the original one.[188]

Another important problem in the transmission of mathematical texts relates to the Arabic manuscript copies of Euclid's *Elements* preserved in the Escurial and in Rabat respectively. A comparative analysis showed that the geometers of al-Andalus and the Maghrib might have had at their disposal the second al-Ḥajjāj version of the Arabic translation of the *Elements*, which was dedicated to the Caliph al-Maʾmūn.[189]

In the field of the elementary teaching of geometry on the measurement of figures several treatises (*rasāʾil*) were published in the 1980s.[190] But much work remains to be done before we can have an adequate idea of this teaching. To begin with, the published texts will have to be compared with those which are not yet published, such as the geometrical chapter of Ibn al-Yāsamīn's *Talqīḥ al-afkār*[191] and Ibn ʿAbdūn's *Risāla fī t-taksīr*.[192]

Conclusion

The following points that arise from this brief survey, may indicate what course to follow in investigating both material that has yet to be disclosed, and texts which have been discovered, but not yet analyzed.

First, there is the problem of breaks of continuities in the transmission from the East to the West. Then, there are problems relating to the subject matter of calculation, geometry, and algebra.

In the science of calculation, the questions are: Why did authors prior to the thirteenth century give so much space to fractions in their teaching manuals? What made Ibn al-Bannā reduce this space, and why did he not hit upon the idea of reducing all fractions to one form? And why did the science of calculation become the main element of mathematics (in all its aspects) in the Maghrib in the post-Almohad period? The last question involves the position of mathematics in society as a whole, and the possible negative influences of the environment on scientific activity. But, in the light of what we now know on the cultural history of Maghrib, we should also question the role of this environment in the protection of contents and scientific level, which one could designate as minimal.

In geometry the survey has revealed various Arabic versions of Euclid's *Elements*, which need to be carefully compared with each other, and Andalusi and Maghribi geometrical texts which need to be fitted into the Euclidean tradition.[193]

In algebra, the most puzzling factor is the silence of Andalusi biobibliographers concerning the presence of the subject in the curricula of the scientists they mention. Moreover, they do not mention which works on algebra were transmitted from the East to the West. It is only through the Latin and Hebrew translations made from the twelfth century onward that we know that al-Khwārizmī's *Mukhtaṣar* and Abū Kāmil's *Algebra* must have been available in al-Andalus. More research is needed to find these works, and their traces, in Arabic sources in the region.

Finally, it has been shown that, in addition to research on the mathematical texts themselves, investigations using the biobibliographical writings from the Muslim West have proved useful. For they have resulted in several publications relevant to the history of the Andalusi and Maghribi mathematical traditions, ranging from writings on the biobibliography of mathematicians in the Middle Ages,[194] through the mathematical activities in each of the three main geographical regions of the Maghrib,[195] to the study of the situation in a specific period in a specific region.[196] But these studies, while providing a useful starting point, also show how much more research remains to be done.

ACKNOWLEDGMENTS

I should like to thank Jan P. Hogendijk for his helpful advice and suggestions during the preparation of this paper, and C. Burnett for correcting the English translation.

NOTES AND REFERENCES

1. There is not yet, to my knowledge, any database of the publications on the history of mathematics and astronomy in the Islamic world, but the existing partial bibliographies confirm this disproportion. See D. A. King: The Exact Sciences in Medieval Islam: some remarks on the Present State of Research, *M E. S. A.*, 4 (1980), 17 pp.; J. L. Berggren: History of Mathematics in the Islamic World: The Present State of the Art, *Middle East Studies Association Bulletin*, 19, 1 (1985), 9–33; D. A. King: *An overview of the sources for the history of Astronomy in the medieval Maghrib*, 2e Colloque Maghrébin sur l'Histoire des Mathématiques Arabes, Tunis, 1–3 Décembre 1988, Actes du Colloque: Tunis, Université de Tunis I-I.S.E.F.C.-A.T.S.M., 1990, 125–157; J. L. Berggren: Mathematics and her Sisters in Medieval Islam: A Selective Review of Work Done from 1985–1995, *Historia Mathematica* 24 (1997), 407–440.

2. A. Djebbar: Les scientifiques arabes face à leur patrimoine, *Maghreb-Machrek*, Paris, Documentation française, 105 (1984), 48–64.

3. J. F. Montucla: *Histoire des Mathématiques*, Paris, 1799–1802.

4. C. Bossut: *Histoire générale des mathématiques*, Paris, 1810.

5. J. B. Delambre: *Histoire de l'Astronomie du Moyen-âge*, Paris, 1819.

6. M. Pelayo: *La ciencia española*, Madrid, 1887–1889.

7. J. von Hammer-Purgstall: *Literaturgeschichte der Araber*, Vienna, 1853–1855.

8. J.-J.-E. Sédillot (1777–1832) graduated at the Ecole Polytechnique and then at the Ecole des Langues Orientales (founded in 1795). He was appointed as an associate astronomer by the Bureau des Longitudes (created especially for the study of the history of astronomy in Eastern civilizations). He completed the translation of Ibn Yūnus's *Zīj* [Tables] started by Caussin de Perçeval, before starting to work on other important translations, namely Ulugh Beg's *Zīj* and the *Book of the Principles and the Goals* by al-Ḥasan al-Marrākushī. The last work was completed by his son L.-A. Sédillot.

9. After studying in mathematics in Berlin and Arabic in Bonn, Franz Woepcke (1826–1864) was sent to Paris when he was 24 years old, by Alexander von Humbolt who had directed him toward research on Arabic mathematics. He was the first to publish the contributions of al-Karajī and al-Khayyām in algebra and the use of symbolic notation in Maghribi mathematical writings.

10. It would be of interest to study the motivations of the first publications in this field (which date back to the very first year of the nineteenth century)—especially the translation in 1804 by Caussin de Perceval of the first chapters of Ibn Yūnus's *Ḥākemite tables*—and to consider the possible connections with the expedition to Egypt. See J. Dhombres: L'image du monde arabe dans le bilan des activités scientifiques dressé par l'Institut de France sous l'Empire, *Sciences et Techniques en Perspective*, 20 (1992), 155–163; J. Dhombres: *Scientific Motivation for and Mood From the Experience of the Egyptian Expedition*, XXth International Congress of History of Sciences, Symposium "Science, Technology, and Industry in the Ottoman World," Liège (Belgium), 20–26 July 1997, Turnhout, Brepols Publishers, 2000, 91–99; J. Dhombres, J. B. Bebert and J. Fourier: *La chaleur mathématisée*, Paris, Berlin, 1998, chap. 5, *L'Égypte*.

11. In this paper a notation such as 661/1262 means 661 in the (Islamic) Hijra era and 1262 in the Christian era.

12. J. J. Sédillot (trans.): *Traité des instruments astronomiques des Arabes composé au treizième siècle par Aboul Hassan Ali de Maroc*, Paris, Imprimerie royale, 1834. Reprinted: Frankfurt, Institut für Geschichte der arabisch-islamischen Wissenschaften, 1984.

13. L. A. Sédillot: *Mémoire sur les instruments astronomiques des Arabes*, Paris, 1844. Reprinted in: *Al-Marrākushī Abū ʿAlī al-Ḥasan ibn ʿAli ibn ʿUmar* (7th/13th.), Frankfurt, Institut für Geschichte der arabisch-islamischen Wissenschaften, 1998, 45–312.

14. Moritz Steinschneider was born March 30, 1816. He graduated in Prague and in Berlin. After teaching in these two cities, he was given a post in the Berlin library. He died in Berlin on January 24, 1907. For az-Zarqālluh, see his: Études sur Zarkali, astronome arabe du XIe siècle, et ses ouvrages, *Bullettino di Bibliografia e di Storia delle scienze matematiche e fisiche*, 14 (1881), 171–182; 16 (1883), 493–513; 17 (1884), 765–794; 18 (1885), 343–360; 20 (1887), 1–36, 574–604.

15. M. Rico y Sinobas: *Libros del Saber de Astronomía del Rey D. Alfonso X de Castilla*, Madrid, 1863–1867.

16. Heinrich Suter was born in 1848 near Zürich in Switzerland. He studied mathematics and received a doctorate in the history of mathematics in 1871. At the age of 40 years he began to learn Arabic and for the next thirty years he did research on Arabic

mathematics and astronomy. For a detailed biography of H. Suter by J. Ruska, see *Isis* 5 (1923), 409–417.

17. F. Woepcke: Notice sur des notations algébriques employées par les Arabes, *Journal Asiatique*, 5e série, 4 (1854), 369–372.

18. F. Woepcke: Traduction d'un chapitre des Prolégomènes d'Ibn Khaldūn relatif aux sciences mathématiques, *Atti dell'Accademia Pontifica dei Nuovi Lincei*, 10 (1856–57), 236–248. Reprint in F. Woepcke: *Études sur les mathématiques arabo-islamiques*, Frankfurt, Institut für Geschichte der arabisch-islamischen Wissenschaften, 1986, vol. 1, 711–723.

19. A. M. Quatremère: *La Muqaddima*, Paris, 1858.

20. M. de Slane: *Prolégomènes*, Paris, 1844–1862.

21. F. Rosenthal: *The Muqaddimah, an Introduction to History*, 3 vols, New York, 1958.

22. V. Monteil (trans.): *Discours sur l'Histoire universelle*, Paris, Sindbad, 1978, 3 vols. All references in this paper are to the Arabic edition: Ibn Khaldūn: *al-Muqaddima*; in *Kitāb al-ʿibar* [The Book of Examples], Beirut, Dār al-Kitāb al-Lubnānī, Maktabat al-Madrasa, 1983, vols. 1–2.

23. A. Djebbar: *Ibn Khaldūn et les Mathématiques, à travers la classification des sciences de la Muqaddima*, to appear.

24. This is the year in which Ibn Khaldūn began writing the *Muqaddima*. He finished the first version by 1382, when he offered a copy to the Prince of Tunis. During the next twenty years Ibn Khaldūn introduced into his work additions and corrections, as he says in the autograph manuscript now in Istanbul (Ms. Istanbul, Atıf Efendi, 1936).

25. Ibn Sīnā included in his compendious work on philosophy *ash-Shifāʾ* four sections dealing with the four disciplines of the quadrivium: number theory, geometry, astronomy, music. For geometry, he simply wrote a summary of Euclid's *Elements* (including the three arithmetical books). His treatise on number theory is a work of synthesis. See Ibn Sīnā: *Kitāb ash-Shifāʾ* [Book of Cure], geometry and arithmetic, ed. A. I. Sabra and A. L. Maẓhar, Cairo, 1975–1976.

26. Ibn Sīnā also included in his work *Kitāb an-Najāt* (which is a summary of *Kitāb ash-Shifāʾ*) chapters on the four mathematical disciplines of the quadrivium. These chapters were later translated into Persian and enclosed in Ibn Sīnā's *Danesh-nameh* [the Book of Science]. See Ibn Sīnā: *Kitāb an-Najāt* [Book of Salvation], Le Caire, 1903; M. Achena and H. Massé (trans.): *Avicenne, Le livre de science* [Danesh-nameh], Paris, Les Belles Lettres, 1958, 91–270.

27. "*Lakhkhaṣa barāhīnahā wa ghayyarahā ʿan ikhtilāfi l-ḥurūfi fīhā ilā ʿilalin maʿnawiyyatin ẓāhiratin hiya sirru l-ishārati bi l-ḥurūfi wa zubdatuhā.*" Ibn Khaldūn: *al-Muqaddima*, op. cit., pp. 897–898.

28. Ibn Khaldūn uses the term *mughlaq*.

29. Op. cit., note 22, vol. 2, 896–898.

30. Woepcke translates the passage of note 27 as follows: "*Il résuma les démonstrations de ces deux ouvrages et autre chose encore en fait de ce qui concerne l'emploi technique des signes dans ces démonstrations, servant à la fois pour le raisonnement abstrait et pour la représentation visible (figurée), ce qui est le secret et l'essence de l'explication (des procédés du calcul) au moyen des signes.*" See op. cit., note 18, 239.

31. On al-Qalaṣādī, see M. Souissi: ʿĀlim riyāḍī andalusī tūnusī al-Qalaṣādī [al-Qalaṣādī, an Andalusi-Tunisian mathematical scientist], *Bulletin de l'Université de Tunis*, 9 (1972), 33–49; A. S. Saidan: *al Qalaṣādī*. In *Dictionary of Scientific Biography*, Ch. C. Gillispie (ed.), New York, Scribner's Sons, 1970–1980, vol. 11, 229–230. From now on, the *Dictionary of Scientific Biography* will be quoted as: *D. S. B.*; A. Djebbar: al-Qalaṣādī, un savant andalo-maghrébin du XVe siècle, *Revue Arabe des Technologies*, Paris, 9 (1990), 12–23.

32. F. Woepcke: Note sur des notations algébriques employées par les Arabes, *Comptes Rendus de l'Académie des Sciences*, 39 (1854), 162–165. Reprinted in F. Woepcke: *Études sur les mathématiques arabo-islamiques*, op. cit., vol. 1, 641–644.

33. F. Woepcke: Traduction du traité d'Arithmétique d'Aboul Haçan Ali Ben Mohammed Alkalçadi, *Extraits des Atti dell'Accademia Pontificia de Nuovi Lincei*, 12 (1858–1859), 230–275, 399–438, in: *Recherches sur plusieurs ouvrages de Léonard de Pise, découverts et publiés par M. le Prince Balthasar Boncompagni*, Rome, Imprimerie des sciences mathématiques et physiques, 1859, 1–66.

34. F. Woepcke: Introduction au calcul gobârî et hawa'î, *Atti dell'Accademia Pontificia de Nuovi Lincei*, 19 (1866), 365–383. Reprint in F. Woepcke: *Études sur les mathématiques arabo-islamiques*, op. cit., vol. 2, 541–559.

35. F. Woepcke: Passages relatifs à des sommations de séries de cubes, extraits de trois manuscrits arabes inédits de la bibliothèque impériale, *Annali di scienze matematiche e fisiche, compilati de Barnaba Tortolini*, 5 (1863), 147–181. Reprint in F. Woepcke: *Études sur les mathématiques arabo-islamiques*, op. cit., vol. 2, 476–510.

36. M. A. Cherbonneau: Notice bibliographique sur Kalaçadi mathématicien arabe du XVe siècle, *Journal Asiatique*, 5e série, XIV (1859), 437–448. G. Eneström: Sur une formule d'approximation des racines carrées donnée par Al Kalaçadi, *Bibliotheca Mathematica*, 236/9 (1886), 222–229.

37. A. Marre: Le Talkhys d'Ibn al-Bannâ, *Atti dell'Accademia Pontificia dei Nuovi Lincei*, 17 (1864), 289–319. For example, the passage: "*al-gharaḍu fī hādhā l-kitāb talkhīṣū aʿmāl al-ḥisāb wa taqrību abwābihī wa maʿānīhī wa ḍabṭi qawāʿidihī wa mabānīhī*" [the goal, in this book, is to summarize the operations of calculation, to make understandable its chapters and its concepts and to fix its rules and its structures], was translated by Marré as follows: "*Le but dans la composition de ce traité est d'analyser succinctement les opérations du calcul, d'en rendre plus facilement accessibles les portes et les vestibules, et d'en établir solidement les fondements et la bâtisse.*" M. Souissi explains the literal character of this translation by the concerns of Marré to show that «*Ibn al-Bannā, fils de maçon, a employé la langage d'un artisan*». See M. Souissi: *Ibn al-Bannā al-Marrākushī, Talkhīṣ aʿmāl al-ḥisāb* [Ibn al-Bannā from

Marrakech, Handbook on the Operations of Calculation], French edition, translation and commentaries, Tunis, Publications de l'Université de Tunis, 1969, 10.

38. M. Steinschneider: Rectification de quelques erreurs relatives au mathématicien arabe Ibn al-Bannā, *Bulletino di Bibliografia e di Storia delle Scienze Matematiche e Fisiche* (Boncompagni) 10 (1877), 313–314.

39. M. Steinschneider: *Hebraïsche Bibliographie*, 80 (1874). Reprint: New York, Olms, 1972, vol. 3, 41.

40. M. Steinschneider: *Die hebräischen Übersetzungen des Mittelalters und die Juden als Dolmetscher*, Berlin, Bibliographisches Bureau, 1893, vol. 2, 557–558.

41. H. Suter: Das Rechenbuch des Abû Zakarîyâ el-Ḥaṣṣâr, *Bibliotheca Mathematica*, serie 3, 2 (1901),12–40.

42. H. Suter: *Die Mathematiker und Astronomen der Araber und ihre Werke*, Leipzig, Teubner, 1900.

43. The *Bibliotheca Arabica Hispanica* consists of the editions made in Madrid and Saragossa by F. Codera and his collaborators including: Ibn Bashkuwāl's *Kitāb al-ṣila* [the Book of the continuation] (1883), al-Ḍabbī's *Bughyat al-multamis fī tārīkh rijāl ahl al-Andalus* [the Desire of the Researcher on the History of the Men of al-Andalus] (1885), *al-Muʿjam* [the Lexicon], Ibn al-Abbār's *at-Takmila li Kitāb aṣ-ṣila* [the Completion to the Book of the Continuation] (1885, 1886–1889) and Ibn al-Faraḍī's *Tārīkh ʿulamāʾ al-Andalus* [History of the Scientists in al-Andalus] (1891–1892).

44. Maqqarī (al-): *Nafḥ aṭ-ṭīb min ghuṣn al-Andalus ar-raṭīb* [Scent of the Perfume of a Tender Branch of al-Andalus], ed. R. Dozy, G. Dugat, L. Krehl and W. Wright, Leiden, 1855–1861.

45. Chapter 13 of Ibn Abī Uṣaybiʿa's book is devoted to the medical doctors of the Muslim West. See Ibn Abī Uṣaybiʿa: *ʿUyūn al-anbāʾ fī ṭabaqāt al-aṭibbāʾ* [Sources of Information on the Categories of the Physicians], A. Müller (ed.), Cairo-Königsberg, 1882–1884. Other equally important works by Ibn Khallikān and Ḥājjī Khalīfa were also available. See Ibn Khallikān: *Wafayāt al-aʿyān* [Necrology of the Famous], trans. M. De Slane, Paris-London, 1843–1871; Ḥājjī Khalīfa: *Kashf aẓ-ẓunūn ʿan asāmi al-kutub wa l-funūn* [Revealing the Doubts about the Names of Books and Arts], trans. G. Flügel, Leipzig, 1835–1858.

46. M. Al-Manūnī: *al-maṣādīr al-ʿarabiyya li tārīkh al-Maghrib* [The Arabic Sources of Moroccan history], Publications de la Faculté des Lettres et des Sciences Humaines-Rabat, Série Études bibliographiques, 1–2; vol. 1, Casablanca, *Muʾassasat Banshara li ṭ-ṭibāʿa wa n-nashr*, 1983; vol. 2, Mohammadia, Maṭbaʿat Fuḍalā, 1989.

47. H. P.-J. Renaud: L'enseignement des sciences exactes et l'édition d'ouvrages scientifiques au Maroc avant l'occupation européenne, Hesperis, XVI (1933), 78–89.

48. A. C. Binebine: *Histoire des bibliothèques au Maroc*, Publications de la Faculté des Lettres et des Sciences Humaines-Rabat, Série Thèses et Mémoires, 17, Casablanca, Imprimerie Najāḥ al-jadīda, 1992. Bencheneb and Levi Provençal: *Essai de répertoire chronologique des éditions de Fès*, Alger, 1922.

49. J. Murdoch: *Euclid.* In D. S. B., op. cit., vol. 4, 414–459.

50. Op. cit., note 47.

51. E. Ihsanoğlu: Ottomans and European Science. In *Science and Empires*, P. Petitjean, C. Jami and A. M. Moulin (edit.), Boston Studies in the Philosophy of Science, vol. 136, Dordrecht, Boston, London, 1992, 37–48. F. Günergun: Introduction of the Metric System to the Ottoman State. In E. Ihsanoğlu (ed.): *Transfer of Modern Science and Technology to the Muslim World*, Istanbul, IRCICA, 1992, 297–316.

52. L. Benjelloun-Laroui: *Les bibliothèques au Maroc*, Paris, Maisonneuve et Larose, 1990, 55–56.

53. H. Hankel: *Zur Geschichte der Mathematik im Altertum und Mittelalter,* Leipzig, 1874, pp. 247–250.

54. C. Brockelmann: *Geschichte der Arabischen Litteratur*, Weimar, 1898–1902.

55. Ibn al-Qifṭī: *Kitāb Ikhbār al-ʿulamāʾ bi akhbār al-ḥukamāʾ* [Book which informs Scholars on the Life of the Wise Men], Lippert (ed.), Leipzig, 1903.

56. Ṣāʿid al-Andalusī: *Kitāb Ṭabaqāt al-umam* [Book of the Categories of Nations], L. Cheikho (ed.), Beirut, Imprimerie Catholique, 1912. Translation: R. Blachère: *Livre des Catégories des nations*, Paris, 1935.

57. J. A. Sánchez-Pérez: *Biografías de matemáticos árabes que florecieron en España*, Madrid, Impr. E. Maestre, 1921.

58. We will return to these works below.

59. One of these lost works in particular intrigued George Sarton, namely the *Kitāb al-istikmāl* of al-Muʾtaman Ibn Hūd. Sarton said about this book: "*It is strange that a work believed to be so important and written by a king should be lost.*" G. Sarton: *Introduction to the History of Science*, Baltimore, Williams and Wilkins, 1927, vol. 1, 759. About the discovery of this work and its importance, see below.

60. Op. cit., vol. 2, 899.

61. H. P.-J. Renaud: Additions et Corrections à Suter "Die Mathematiker und Astronomen der Araber," Isis 18 (1932–1933), 166–183.

62. Op. cit., note 47.

63. In al-Andalus as well as in the Islamic East, astronomy had a more important position than mathematics. From the very beginning of the arabo-islamic civilization, astronomy was related to problems of religious practice, such as the determination of the azimuth of Mecca, the calculation of the times of the daily prayers, and fixing the first day of Ramadan. Astronomy was also used in astrology. The relative importance of astronomy and astrology throughout the Arabic middle ages is reflected by the number of pages devoted to them by the medieval Islamic and modern bibliographers. For example, for the period prior to 1038, F. Sezgin devotes a volume of 521 pages to astronomy and another of 486 pages to astrology, whereas mathematics occupies a single volume of 514 pages: F. Sezgin: *Geschichte des arabischen Schrifttums*, Band

V, *Mathematik* bis ca. 430 H. Leiden, Brill. 1974; Band VI, *Astronomie* bis ca 430 H., 1976; Band VII, *Astrologie* bis ca 430 H., 1978.

64. J. M. Vallicrosa: *Estudios sobre Azarquiel*, Madrid-Granada, 1943–1950.

65. For a detailed bibliography of these contributions, see: J. Samsó: *The Exact Sciences in al-Andalus*; in S. K. Jayyūsī (ed.): *The Legacy of Muslim Spain*, Leiden, Brill, 1994, vol. 2,952–973.

66. J. Samsó: *Tres notas sobre astronomía hispánica en le siglo* XIII. In J. Vernet: *Estudio sobre histoia de la ciencia arabe*, Barcelona, 1980, 175–177. R. Lorch: The Astronomy of Jābir ibn Aflaḥ, Centaurus, 19 (1976), 85–107.

67. J. Samsó and M. A. Catalá: Un instrumento astronómico de raigambre zarqālī: el-cuadrante shakkāzī de Ibn Ṭībughā, *Memorias de la Real Academia de Buenas Lettras de Barcelona*, 13 (1971–1975), 5–31. J. Samsó: *A propos de quelques manuscrits astronomiques des bibliothèques de Tunis: contribution à une histoire de l'astrolabe dans l'Espagne musulmane*, Actas del 11 Colloquio Hispano-Tunecino de Estudios Historicos, Madrid, 1973, 171–190; D. A. King: On the Early History of the Universal Astrolabe in Islamic Astronomy and the Origin of the Term "Shakkāziya" in Medieval Scientific Arabic, *Journal for the History of Arabic Science*, 3 (1979), 244–257.

68. S. Pines: *Ibn al-Haytham's Critique of Ptolemy*, Proceedings of the Xth International Congress of History of Sciences, Ithaca, 1962. Paris, 1964, vol. 1, 547–550.

69. F. J. Carmody: The Planetary Theory of Ibn Rushd, *Osiris*, 10 (1952), 556–586; B. R. Goldstein: *al-Biṭrūjī: Principles of Astronomy*, New Haven-London, 1971.

70. H. Hermelink: Tabulae Jahen, *Archive for History of Exact Sciences*, 2 (1964), 108–112; G. J. Toomer: A Survey of the Toledan Tables, Osiris, 15 (1968), 5–174.

71. J. Vernet and M. A. Catalá: Las obras matemáticas de Maslama de Madrid. In J. Vernet (édit.): *Estudios sobre historia de la ciencia medieval*, Barcelona-Bellatera, 1979, 241–271.

72. J. Samsó: The early Development of Astrology in al-Andalus, *Journal for the History of Arabic Science*, 3, 2 (1979), 228–243. The author specifies there, in connection with the loans from the Latin astrological tradition: "*the aformentioned Arabic text is based on the translation of a Latin astrological work which was known in al-Andalus towards the end of the eighth or beginning of the ninth century, therefore being one more item in the long series of contacts between Isidorian-Latin and Arabic culture in Muslim Spain*" (p. 233).

73. D. A. King: Three Sundials from Islamic Andalusia, *Journal for the History of Arabic Science*, 2 (1978), 358–392; J. Samsó: *The Exact Sciences in al-Andalus*, op. cit., n. 65, vol. 2, p. 956.

74. J. Samsó: Maslama al-Majrīṭī and the Alphonsine Book on the Construction of the Astrolabe, *Journal for the History of Arabic Science*, 4 (1980), 3–8.

75. E. Poulle: Un instrument astronomique dans l'Occident latin: la 'Saphea,' *Studi Medievali*, 10 (1969), 491–510.

76. M. V. Villuendas: *La trigonometria europea en el siglo XI. Estudio de la obra de Ibn Muʿād, El Kitāb mayhūlāt*, Barcelona, Instituto de Historia de la Ciencia de la Real Academia de Buenas Letras, 1979.

77. Al-Bīrūnī: *Kitāb maqālīd ʿilm al-hayʾa, La Trigonométrie sphérique chez les Arabes de l'Est à la fin du Xe siècle*, M. T. Debarnot (ed. and trans.), Damascus, Institut Français de Damas, 1985.

78. Ibn al-Bannā: *Kitāb Minhāj aṭ-ṭālib li taʿdīl al-kawākib* [The guide to the Student for the Correction of the Movements of the Stars], in J. Vernet: *Contribución al estudio de la labor astronómica de Ibn al-Bannā*, Tetouan, 1952.

79. J. Vernet, 1956: *Les manuscrits astronomiques d'Ibn al-Bannā*. In: Proceedings of the VIIIth International Congress of History of Sciences (Milano, 1956), Paris, Hermann, 1958, 297–298.

80. H. P.-J. Renaud: Sur les dates de la vie du mathématicien arabe marocain Ibn al-Bannā' (XIIe–XIIIe s. J.C.), *Isis* 37, 2 (1937), 216–218; H. P.-J. Renaud: Ibn al-Bannā' de Marrakech, sufi et mathématicien (XIIe–XIIIe S. J. C.), *Hespéris* XXV (1938), 13–42.

81. H. P.-J. Renaud and J. S. Colin: Note sur le muwaqqit marocain Abū Muqriʿ-ou mieux Abū Miqraʿ al-Baṭṭīwī, *Hespéris* XXV (1938), 94–96.

82. H. P.-J. Renaud: Astronomie et Astrologie marocaine, *Hespéris* XXIX (1942), 41–63.

83. H. P.-J. Renaud: Déterminations marocaines de l'obliquité de l'écliptique. *Bulletin de l'enseignement public*, 170 (1941), 321–336.

84. H. P.-J. Renaud: Sur les lunes du Ramadan, *Hespéris* XXXII (1945), 51–68.

85. H. P.-J. Renaud: *Le calendrier d'Ibn al-Bannā de Marrakech*, Paris, 1948.

86. In this connection it would be interesting to study the writings of these two Muslim intellectuals, especially those which they published during their stay in Paris.

87. ʿA. Gannūn: *an-Nubūgh al-maghribī fī l-adab al-ʿarabī* [The Moroccan Genius in Arabic Literature], First edition, Tétouan, 1938.

88. Q. H. Ṭūqān: *Turāth al-ʿArab al-ʿilmī fī r-riyāḍiyyāt wa l-falak* [The Arabic Scientific Heritage in Mathematics and Astronomy], First edition, Beirut, Dār ash-shurūq, 1941, 280 pp.

89. ʿA. al-ʿAzzāwī: *Tārīkh ʿilm al-falak fī l-ʿIrāq wa ʿalāqatuhū bi l-aqṭār al-islāmiyya wa l-ʿarabiyya fī l-ʿuhūd at-tāliyya li ayyām al-ʿAbbāsiyyīn* (656–1335/1258–1917) [History of Astronomy in Iraq and its Relationship to the Islamic and Arabic Regions during the Periods of Time Which Followed the Abbassid Period], Baghdad, *Maṭbaʿat al-majmaʿ al-ʿilmī al-ʿirāqī*, 1958.

90. In 1954 the book of Ṭūqān appeared in a second edition which was enlarged from 280 to 450 pages. It was sold out within two years, and was followed in 1963 by a third

edition which included corrections and additions. The book of Gannūn also appeared in revised editions in 1961 and in 1975.

91. To these three works, written in Arabic, it is necessary to add Ṣāliḥ Zakī's book, written in Turkish, the diffusion of which was more limited, but which also reached the European researchers. See Ṣ. Zakī: *Āthār bāqiyya* [Remaining Relics], Istanbul, 1911.

92. G. Sarton: Compte-rendu au Livre de Q. H. Ṭūqān, *Isis* 36 (1944), 140.

93. Op. cit., note 87, third edition, 1975, vol. 2, 6.

94. Op. cit., note 91, vol. 2, 286–291. W. Hartner was the first western specialist who mentioned this mathematician while referring to Ṣ. Zakī and to Q. H. Ṭūqān. See W. Hartner: *Quant et comment s'est arrêté l'essor de la culture scientifique dans l'Islam*? In R. Brunschvig and G. E. Von Grunebaum: *Classicisme et déclin culturel dans l'histoire de l'Islam, Actes du Symposium International d'Histoire de la civilisation musulmane* (Bordeaux, 25–29 Juin 1956), Paris, Maisonneuve et Larose, 1977, 334, note 4. A manuscript of Ibn Ḥamza's work exists in Cairo (Ms. Cairo National Library, Ṭalʿat riyāḍa, turki 1), but to my knowledge, it has not yet been the subject of any thorough study. Since Ibn Ḥamza originated from the city of Algiers, it would be interesting to see how this work relates to the traditions in calculation in the Muslim West and that in the Muslim East.

95. Op. cit., note 80.

96. Op. cit., note 80.

97. There existed another Abu'l-ʿAbbās Ibn al-Bannā originating from Saragossa who wrote a mystical poem entitled: *al-Mabāḥith al-aṣliyya ʿan jumlat aṭ-ṭarīqa aṣ-ṣūfiyya* [Original Researches on the Entire Sufi Way]. See Ms. Alger B. N., 2026, ff. 214b–227b. A commentary on this poem was written by Aḥmad Zarrūq al-Barnūsī, entitled *Sharḥ urjūzat Ibn al-Bannā as-Saraqusṭī* [Comment on the Poem of Ibn al-Bannā as-Saraqusṭī]. See Ms. Alger B. N., 2069 and 2295.

98. M. Casiri: *Bibliotheca Arabico-Hispana Escurialensis*, 2 vols., Madrid, 1760–1770. H. P.-J. Renaud himself says about Casiri's errors: «*Les parties proprement littéraires et historiques de l'œuvre de Casiri sont peut-être acceptables. On ne saurait en dire autant des deux grandes sections qui ont trait à la médecine et aux sciences exactes. Lucien Leclerc s'en était déjà plaint tout au long de son 'Histoire de la médecine arabe,' après avoir fait le voyage de l'Escurial et constaté par lui-même les méprises et les omissions de Casiri. Suter, malgré la prudence justifiée qu'il montre vis-à-vis des assertions de cet écrivain, n'a pas pu toujours les contrôler et les reproduit souvent dans son ouvrage*». H. P.-J. Renaud: Additions et corrections à Suter "Die Mathematiker und Astronomen der Araber," *Isis* 18 (1932–1933), 167.

99. ʿA. Gannūn (no date): *Dhikrayāt mashāhīr rijāl al-Maghrib* [Memories of the Famous Men of Morocco], First edition (assembling forty biographies of Moroccan scientists), Tetouan, Manshūrāt maʿhad Mawlāy al-Ḥasan li l-abḥāth; second edition (increased by ten biographies of scientists), Beirut, Maktabat al-madrasa, Dār al-kitāb al-lubnānī.

100. M. al-Fāsī: Ibn al-Bannā al-ʿadadī al-Marrākushī [Ibn al-Bannā the Arithmetician from Marrakech], *Ṣahīfat maʿhad ad-dirāsat al-islāmiyya*, Madrid, 6, (1–2) (1958), 1–9.

101. Op. cit., note 37.

102. Edited by A. S. Saidan: *Kitāb al-maqālāt fī l-ḥisāb li Ibn al-Bannā* [Ibn al-Bannā's Book on the Epistles of Calculation], Amman, 1984.

103. Op. cit., note 37.

104. M. Souissi evaluates Ibn al-Bannā's handbook in these terms: *"On serait tenté d'affirmer que, chez Ibn al-Bannā', l'arithmétique et certaines questions d'algèbre ont pris leur forme définitive, celle sous laquelle nous les rencontrons de nos jours. Nous ne pouvons, pourtant, soutenir que tout, dans ces questions, est apport propre d'Ibn al-Bannā'. Ce dernier, en effet, nous présente, en cette fin du XIIIe siècle et début du XIVe, une somme des connaissances arabes en arithmétique et en algèbre."* See op. cit., note 37, 33–34.

105. ʿA. Gannūn: *Talqīḥ al-afkār fī l-ʿamal bi rushūm al-ghubār li Ibn al-Yasāmīn* [The Fecondation of Spirits for the Use of the Figures of Dust, of Ibn al-Yāsamīn], *Majallat al-baḥth al-ʿilmī*, Bagdad, 1 (1964), 181–190.

106. B. Ḥimṣī: *Urjūzat Ibn al-Yāsamīn fī l-jabr wa l-muqābala* [Ibn al-Yāsamīn's Poem on algebra], Second International Symposium on the History of Arabic Sciences. Abstracts, Aleppo, I.H.A.S., 1979, 76–78.

107. J. A. Sánchez-Pérez: *Compendio de Álgebra de Abenbéder*, edition, Spanish translation and mathematical analysis, Madrid, Apirica, 1916.

108. E. B. Plooij: *Euclid's Conception of Ratio and his Definition of Proportional Magnitudes as Criticized by Arabian Commentators*, Rotterdam, 1950.

109. J. A. Sánchez-Pérez: *Partición de herencias entre los musulmanes del rito Malequi, con transcription anotada de dos manuscritos aljamiados*, Madrid, 1914.

110. W. Ahrens: Die magischen Quadrate al-Būnīs, *Der Islam* 12 (1922), 157–177. B. Carra De Vaux: Une solution arabe du problème des carrés magiques. *Revue d'Histoire des Sciences*, 1 (1948), 206–212. H. Hermelink: Die ältesten magischen Quadrate höherer Ordnung und ihre Bildungsweise, *Sudhoffs Archiv* 42 (1958), 199–217.

111. These are our present "Arabic" number symbols except for the four and the five, which are slightly modified. They were called in al-Andalus and in the Maghrib *ḥurūf al-ghubār* [letters or figures of dust], and sometimes *rushūm al-ghubār* [signs of dust].

112. Copies of al-Khwārizmī's *Algebra* circulated in al-Andalus and were used for the twelfth-century Latin translations of the book made by Gerard of Cremona and Robert of Chester. See M. Steinschneider: *Die Europäischen Übersetzungen aus dem Arabischen bis Mitte des 17. Jahrhunderts*, Vienna, 1904–1905. Reprint, Graz, 1956, 24, 72. The contents of al-Khwārizmī's *Algebra* are partially present in the handbooks of the Maghrib which have reached us. In his classification of sciences, Ibn

Khaldūn says about algebra: "*the first one who wrote about this art is Abū ʿAbdallah al-Khwārizmī.*" See op. cit., note 22, vol. 2, 899. No copy of al-Khwārizmī's *Algebra* has been found in the Maghrib.

113. M. al-Fāsī: Urjūzat Ibn al-Yāsamīn fī l-judhūr [Ibn al-Yāsamīn's Poem on the Radicals], *Majallat Risālat al-Maghrib*, 1 (1942).

114. M. Souissi: *Un mathématicien tuniso-andalou, al-Qalaṣādī*, in: Actas del II Colloquio Hispano-Tunecino de Estudios Historicos, Madrid, 1973, 147–169; M. Souissi: *Risālat dhawāt al-asmāʾ li Abī l-Ḥasan . . . al-Qalaṣādī*, [al-Qalaṣādī's Letter on the Binomial Numbers], in *Dawr al-ʿulūm aṣ-ṣaḥīḥa fī tanmiyyat al-buldān an-nāmiya, Dirāsāt multaqā ʿAli al-Qalaṣādī*, Béja (Tunisia), 28–30 Mai 1976. Tunis, Ministère des affaires culturelles, 1978, 161–190.

115. A. at-Tāzī: al-arqām al-maghribiyya arqām ʿarabiyya aṣīla [The Moroccan Number Symbols are Authentic Arabic Number Symbols], *Majallat al-Lisān al-ʿarabī*, 2 (1965), 37; M. as-Sarrāj: aṭ-Ṭābiʿ al-ʿarabī fī l-arqām ar-riyāḍiyya [The Arabic Influence on the Mathematical Number Symbols], *Majallat al-Lisān al-ʿarabī*, 3 (1965), 70–64; Abū Fāris: *Dalīl jadīd ʿala ʿurūbat al-arqām al-mustaʿmala fī l-Maghrib al-ʿarabī* [A New Proof Concerning the Arabic Nature of the Number Symbols Used in the Arabic Maghrib] *Majallat al-Lisān al-ʿarabī*, 10, 1 (1973), 231–233; M. H. al-Yasīn: *al-Arqām al-ʿarabiyya fī ḥālihā wa tirḥāliḥā* [Arabic Number Symbols, their Condition and Circulation], *Āfāq ʿarabiyya*, 12 (1980), 42–49.

116. M. al-ʿUqaylī: *Silk farāʾiḍ al-yawāqīt fī l-ḥisāb wa l-farāʾiḍ wa l-mawāqīt* [Pearl Thread of Sapphires on Arithmetic, Inheritance, and Timekeeping], lithography, Fez, 1901; A. Sakrīj: *Irshād al-mutaʿallimīn an-nāsī fī ṣifat ashkāl al-qalam al-fāsī* [Guide for Forgetful Students on the Forms of Fez Writing Signs] (Commentary on the *urjūza* of A. al-Fāsī), lithography (without date).

117. Ch. Pellat (trans.): *Irshād al-mutaʿallim an-nāsī fī ṣifat ashkāl al-qalam al-fāsī de Sakrīj*, Alger, 1917; G. S. Colin: De l'origine grecque des "chiffres de Fès" et de nos "chiffres arabes," *Journal Asiatique*, 222 (1933), 193–215; J. A. Sánchez Pérez: Sobre las cifras rūmīes, *Andalus*, 3 (1935), 97–125; M. E. Viala: *Le mécanisme du partage des successions en Droit musulman, suivi de l'exposé des "signes de Fez,"* Alger, 1917; A. Guiraud: *Jurisprudence et procédure musulmane*, Casablanca, 1925, 109–110.

118. J. C. Martzloff: *Histoire des mathématiques chinoises*, Paris, Masson, 1988, 31–32.

119. Op. cit., note 22, 898–899.

120. Op. cit., note 107.

121. Op. cit., note 107, 24–73.

122. M. Levey: *The Algebra of Abū Kāmil, Kitāb fī l-jabr wa l-muqābala, in a Commentary by Mordechaï Finzi*, Hebrew Text, Translation and Commentary with Special Reference to the Arabic Text, Madison, Milwaukee and London, University of Wisconsin Press, 1966.

123. Op. cit., note 22, 896.

124. Op. cit., note 56, 64–65, 67, 68, sqq.

125. Op. cit., note 56, 69.

126. Op. cit., note 22, 896.

127. M. Souissi: *Un texte d'Ibn al-Bannā' sur les nombres parfaits, abondants, déficients et amiables*, International Congress of Mathematical Sciences, Karachi, 14–20 July 1975; Arabic version in *Bulletin de l'Université de Tunis*, 13 (1976), 193–209.

128. M. Aballagh and A. Djebbar: *Ḥayāt wa mu'allafāt Ibn al-Bannā (ma'a nuṣūṣ ghayr manshūra)* [Ibn al-Bannā's Life and Works (with unpublished texts)], Rabat, Publications de la Faculté des Lettres et des Sciences Humaines, 2001.

129. A. Djebbar: *Enseignement et Recherche mathématiques dans le Maghreb des XIIIe–XIVe siécles*. Paris, Publications Mathématiques d'Orsay, 1980, 81–102, 55–112; hereafter, the title of this publication will be quoted as *E. R. M.*

130. A. Djebbar: *L'analyse combinatoire au Maghreb: l'exemple d'Ibn Mun'im (XIIe–XIIIe siècles)*, Paris, Publications Mathématiques d'Orsay, 1985, 85–101.

131. External influences are, however, suggested by a remark of al-Bīrūnī, on the metrics in Sanskrit poetry. See al-Bīrūnī: *Taḥqīq mā li l-Hind*, Beirut, 'Ālam al-kutub, 1983, 104. E. C. Sachau: *Alberuni's India*, New Delhi, S. Chand & Co., 1888, vol. 1, 147.

132. Op. cit., note 129, 99–112.

133. A. Djebbar: Matériaux pour l'étude des pratiques combinatoires dans le Maghreb médiéval. To appear.

134. Op. cit., note 129, 6–40.

135. Op. cit., note 32, 162–165; H. P.-J. Renaud: Sur un passage d'Ibn Khaldūn relatif à l'histoire des mathématiques, *Hespéris* 31 (1944), 35–47.

136. A. Djebbar: *Quelques aspects de l'algèbre dans la tradition mathématique arabe de l'Occident musulman*, First Maghribin Conference of Algiers on the History of Mathematics Arabic, 1–3 December 1986. Published in the Proceedings of the Conference, Algiers, Maison du Livre, 1988, 99–123. For the edition of *Talqīḥ al-afkār* of Ibn al-Yāsamīn, see T. Zemouli: *Mu'allafāt Ibn al-Yāsamīn ar-riyāḍiyya* [Mathematical Writings of Ibn al-Yāsamīn], M.Sc. thesis in History of Mathematics, Algiers, E. N. S., 1993.

137. Op. cit., note 129, 41–54.

138. Op. cit., note 136, 106–107, 111; M. Zerrouki: *Abū l-Qāsim al-Qurashī, Ḥayātuhū wa mu'allafātuhū ar-riyāḍiyya* [Abū l-Qāsim al-Qurashī, His Life and His Mathematical Writings], *Cahier du Séminaire Ibn al-Haytham* 5 (1995), Algiers, E. N. S., 10–19.

139. A. S. Saidan: *Tārīkh 'ilm al-jabr fī l-'ālam al-'arabī*, II, Kuwait, 1985, 398–613; A. Djebbar: *Mathématiques et Mathématiciens du Maghreb médiéval (IXe–XVIe siècles)*, Thèse de Doctorat, Nantes, 1990.

140. Op. cit., note 22, vol. 2, 898.

141. J. Sesiano: *Le Liber Mahamalet, un traité mathématique latin composé au XIIe siècle en Espagne*, Premier Colloque Maghrébin sur l'Histoire des Mathématiques Arabes, Algiers, 1–3 December 1986; in Proceedings of the Conference, Algiers, Maison du Livre, 1988, 69–98.

142. H. L. L. Busard: L'algèbre au moyen-âge: le "Liber mensurationum" d'Abū Bakr, *Journal des savants*, April–June 1968, 65–124.

143. A. Bar Ḥiyya: *Libre de Geometria*, trans. J. Millás Vallicrosa, Barcelona, 1931.

144. A. Djebbar: *Procédés algébriques et géométrie métrique dans l'Espagne médiévale, à travers un manuel du Xe siécle*, Paris, University of Paris-Sud, Prépublications Mathématiques d'Orsay. In press.

145. Y. Atik: *L'épître d'algèbre de Sinān Ibn al-Fatḥ*, Second Maghribin Conference of History of Arabic Mathematics, Tunis, 1–3 December 1988; in: Proceedings of the Conference, Tunis, Tunis University II.S.E.F.C.-A.T.S.M., 1990, 5–19.

146. A. Djebbar and R. Rashed: *L'œuvre algébrique d'al-Khayyām*, Aleppo, Institute for the History of Arabic Sciences, 1981.

147. Op. cit., note 22, vol. 2, 899.

148. There are also Latin texts translated from Arabic which have not yet been analyzed and which could be connected with this tradition. See: op. cit., note 142, 71.

149. Ibn Al-Majdī: *Ḥāwī l-lubāb wa sharḥ talkhīṣ aʿmāl al-ḥisāb* [Collection of the Essence of, and Commentary on, the Abridged Book on the Operations of Calculation], Ms. London, British Museum 7469, ff. 144a–245a.

150. J. Sesiano: *La version latine médiévale de l'Algèbre d'Abū Kāmil*, in: M. Folkerts and J. P. Hogendijk (ed.): *Vestigia Mathematica, Studies in Medieval and Early Modern Mathematics in Honour of H. L. L. Busard*, Editions Rodopi, Amsterdam-Atlanta, 1993, 315–452.

151. See Ibn Qunfudh al-Qasanṭīnī (d. 810/1407), *Mabādiʾ as-sālikīn fī sharḥ urjūzat Ibn al-Yāsamīn* [Principles for Those Who Undertake the Explanation of Ibn al-Yāsamīn's Poem] and al-Qalaṣādī (m. 892/1486), *Tuḥfat an-nāshiʾīn ʿalā urjūzat Ibn al-Yāsamīn* [The Ornament of the Young People on the Poem of Ibn al-Yāsamīn].

152. Op. cit., note 136.

153. S. Ahmad and R. Rashed: *As-Samawʾal, al-Bāhir fī l-Jabr*, Damas, Maṭbaʿat Jāmiʿat Dimashq, 1972.

154. M. Aballagh: *Rafʿ al-ḥijāb ʿan wujūh aʿmāl al-ḥisāb li Ibn al-Bannā al-Marrākushī* [The Raising of the Veil on the Forms of the Processes of Calculation of Ibn al-Bannā al-Marrākushī], Fès, Faculté des Lettres et des Sciences Humaines, 1994.

155. Y. Guergour: *al-Aʿmāl ar-riyāḍiyya li Ibn Qunfudh* [Mathematical Writings of Ibn Qunfudh], M.Sc. thesis in History of Mathematics, Algiers, E. N. S., 1990.

156. T. Zemouli: *Mu'allafāt Ibn al-Yāsamīn ar-riyāḍiyya* [Mathematical Writings of Ibn al-Yāsamīn], M.Sc. thesis in History of Mathematics, Algiers, E. N. S., 1993.

157. A. Harbili: *L'enseignement des mathématiques à Tlemcen au XIVe siècle à travers le commentaire d'al-ʿUqbānī (m. 811/408)*, Magister in History of Mathematics, Algiers, E. N. S., 1997.

158. M. Zoubeidi: *Le Kitāb al-bayān wa t-tadhkār d'al-Ḥaṣṣār*, M.Sc. thesis in History of Mathematics, Algiers, E. N. S. In preparation.

159. K. Kouidri: *Ṭarīqat al-khaṭa'ayn fī t-taqlīd ar-riyādī al-ʿarabī* [Method of False Position in the Arabic Mathematical Tradition], M.Sc. thesis in History of Mathematics, Algiers, E. N. S., 1999.

160. M. Abdelkader-Khaddaoui: *al-Khawārizmiyyāt al-ḥisābiyya fī t-taqlīd ar-riyāḍī al-maghribī* [Arithmetic Algorithms in the Maghribi Mathematical Tradition], M.Sc. thesis in History of Mathematics, Algiers, E. N. S., 2000.

161. E. Laabid: *Arithmétique et Algèbre d'héritage selon l'Islam, deux exemples: Traité d'al-Ḥubūbī (X^e–XIe s.) et pratique actuelle au Maroc*, Mémoire de Maîtrise, Montréal, Quebec University, 1990.

162. E. Laabid: *L'enseignement mathématique dans le Maghreb extrême à travers l'exemple des partages successoraux*, Thèse de Doctorat, Rabat, Université Mohamed V. In preparation.

163. M. Zerrouki: *ʿIlm al-fārā'iḍ fī l-Maghrib al-wasīṭ min khilāl sharḥ Mukhtaṣar al-Ḥūfī li l-ʿUqbānī* [The Science of Inheritances in the Medieval Maghrib Through the Commentary of al-ʿUqbānī on the *Mukhtaṣar* of al-Ḥūfī], M.Sc. thesis in History of Mathematics, Algiers, E. N. S., 2000.

164. A. Djebbar: *Le traitement des fractions dans la tradition mathématique arabe du Maghreb*; in Paul Benoit, Karine Chemla et Jim Ritter: *Histoire de fractions, fractions d'histoire*. Basel-Boston-Berlin, Birkhäuser Verlag, 1992. Chapter XII, 223–245.

165. M. Aballagh and A. Djebbar: Découverte d'un écrit mathématique d'al-Ḥaṣṣār (XIIth c.), le Livre I du Kāmil, *Historia Mathematica*, 14 (1987), 147–158.

166. J.G. Lemoine: Les anciens procédés de calcul sur les doigts en Orient et en Occident (Note additionnelle sur le comput digital, texte inédit d'Ibn Bundūd), *Revue des Etudes Islamiques*, Paris, 6 (1932), 1–58; J. S. Colin: Un texte inédit d'Ibn Bundūd sur la dactylonomie, *Revue des Etudes Islamiques*, Paris, 6 (1932), 59–60.

167. Y. Guergour: *Sur les différentes numérations utilisées au Maghreb à l'époque ottomane*, XXth International Congress of History of Sciences, Symposium "Science, Technology, and Industry in the Ottoman World," Liège (Belgium), 20–26 July 1997, Turnout, Brepols Publishers, 2000, 67–74.

168. Op. cit., note 164.

169. A. Djebbar: *Kitāb al-uṣūl wa l-muqaddimāt fī l-jabr wa l-muqābala* [The Book of the Bases and the Preliminaries of Algebra], in: Op. cit., note 139, vol. 2.

170. A. Djebbar: *La Théorie des nombres dans le Fiqh al-ḥisāb d'Ibn Munʿim*, 5th International Symposium on the History of Arabic Sciences, Granada, 30 March–4 April 1992; to be published.

171. A. Djebbar: *Les nombres figurés dans la tradition mathématique de l'Andalousie et du Maghreb*, in: Op. cit., note 139, vol. 2.

172. See Ibn Munʿim: *Fiqh al-ḥisāb*, Ms. Rabat B. G., 416 Q, 320.

173. B. Vitrac: *Euclide, Les Éléments*, Paris, Presses Universitaires de France, vol. 2, 1994, 247–493.

174. W. Kutsch: *Kitāb al-Mudkhal ilā ʿilm al-ʿadad, tarjamat Thābit Ibn Qurra* [The Introduction to the Science of Number, the Translation of Thābit Ibn Qurra], Beirut, Catholic Press, 1958. For a French translation, see J. Bertier: *Nicomaque de Gerase, Introduction Arithmétique*, Translation, notes and index, Paris, Vrin, 1978.

175. F. Woepcke: Notice sur une théorie ajoutée par Thâbit Ben Korrah à l'arithmétique spéculative des Grecs, *Journal Asiatique*, 4th séries, 20 (1852), 420–429.

176. Op. cit., note 56, 158–181.

177. Op. cit., note 22, 901–905.

178. An exception is: T. Levy: *Fragment d'Ibn al-Samḥ sur le cylindre et sur ses sections planes.* Edition and French translation; in: R. Rashed (ed.): *Les Mathématiques infinitésimales du IXe au XIe siècle,* vol. 1, *Fondateurs et commentateurs*, London, al-Furqān Islamic Heritage Foundation, 1995, 928–973.

179. In astronomy it is the geometry of the instruments, which one finds, for example, in the *Ṣafīḥa* of az-Zarqālluh and in the *Astrolabe* of al-Majrīṭī. For the geometry of measurement, it is Abū Bakr's *Liber Mensurationum* and the Bār Ḥiyya's *Liber Embadorum* mentioned above.

180. A. Djebbar: *Abū Bakr Ibn Bājja et les mathématiques de son temps*; in: Etudes philosophiques et sociologiques dédiées à Jamal ed-Dine Alaoui, Publications de la Faculté des Lettres et des Sciences Humaines, Dār El Mahraz n° 14, Fez, 1998, 5–26.

181. J. P. Hogendijk: Discovery of an eleventh century Geometrical Compilation: The Istikmāl of Yūsuf al-Muʾtaman Ibn Hūd, King of Saragossa. *Historia Mathematica* 13 (1986), 43–52.

182. J. P. Hogendijk: The Geometrical Part of the *Istikmāl* of Yūsuf al-Muʾtaman ibn Hūd (eleventh century). An Analytical Table of Contents, *Archives Internationales d'Histoire des sciences* 127 (1991), vol. 41, 207–281.

183. On Apollonius, see G. J. Toomer: *Apollonius*; in D. S. B., op. cit., vol. 1, 179–193.

184. J. P. Hogendijk: 'Which version of Menelaus' Spherics Was Used by al-Muʾtaman ibn Hūd in his *Istikmāl*'? Proceedings of the international conference "*Mathematische Probleme im Mittelalter, der lateinische und arabische Sprachbereich,*" Wolfenbüttel (Germany), 18–22 June 1990, M. Folkerts (ed.), Wiesbaden, Harrassowitz Verlag, 1996, 17–44.

185. J. P. Hogendijk: al-Muʾtaman's Simplified Lemmas for Solving "Alhazen's Problem;" in J. Casulleras and J. Samsó (éd.): *From Baghdad to Barcelona, Studies in the Islamic Exact Sciences in Honour of Prof. Juan Vernet*, Barcelona, Anuari de Filologia XIX-Instituto "Millás Vallicrosa," 1996, vol. 1, 59–101.

186. J. P. Hogendijk: 'Le roi géomètre al-Muʾtaman Ibn Hūd et son livre de la perfection (Kitāb al-Istikmāl),' First Maghribin Conference of History of Arabic Mathematics, Algiers, 1–3 December 1986; in Proceedings of the Conference, Algiers, Maison du Livre, 1988, 51–66.

187. J. P. Hogendijk: Four constructions of two mean proportionals between two given lines in the Book of Perfection of al-Muʾtaman Ibn Hūd, *Journal for the History of Arabic Science*, 10 (1992–1993–1994), 13–29.

188. A. Djebbar: La rédaction de l'Istikmāl d'al-Muʾtaman(XIe s.) par Ibn Sartāq un mathématicien des XIIIe–XIVe siècles, *Historia Mathematica*, 24 (1997), 185–192.

189. See J. W. Engroff: *The Arabic Tradition of Euclid's Elements: Book V*, Dissertation Cambridge/Mass., Harvard University, 1980; G. De Young: *The Arithmetical Books of Euclid's Elements*, Dissertation Cambridge/Mass., Harvard University, 1981; S. Brentjes: Varianten einer Ḥaǧǧaǧ-version von Buch II der Elemente; in M. Folkerts and J. P. Hogendijk (eds): *Vestigia Mathematica. Studies in Medieval and Early Modern Mathematics in Honour of H. L. L. Busard*, Amsterdam, 1993, 47–67; A. Djebbar: Quelques commentaires sur les versions arabes des Eléments d'Euclide et sur leur transmission à l'Occident musulman. Proceedings du Colloque International "*Mathematische Probleme im Mittelalter, der lateinische und arabische Sprachbereich*," Wolfenbüttel (Germany), 18–22 June 1990, M. Folkerts (ed.), Wiesbaden, Harrassowitz: Verlag, 1996, 104–111.

190. M. L. al-Khaṭṭābī: Sharḥ al-Iksīr [Commentary on al-Iksīr], *Daʿwat al-ḥaqq*, Rabat, 258 (1987), 77–87; M. L. al-Khaṭṭābī: *Risālatān fī ʿilm al-misāḥa li Ibn ar-Raqqām wa Ibn al-Bannā* [Two Epistles of Ibn ar-Raqqām and Ibn al-Bannā on the Science of Measuring], *Daʿwat al-ḥaqq*, Rabat, 256 (1986), 39–47.

191. A. Djebbar: *Matériaux pour l'étude de la tradition géométrique d'al-Andalus et du Maghreb*. In preparation.

192. Op. cit., note 144.

193. Y. Guergour: *Les Eléments et les Données d'Euclide dans le Kitāb al-istikmāl d'al-Muʾtaman Ibn Hūd*. Thèse de Doctorat, in preparation.

194. D. Lamrabet: *Introduction à l'histoire des mathématiques maghrébines*, Rabat, al-Maʿārif al-Jadīda Press, 1994.

195. A. Djebbar: Quelques éléments nouveaux sur l'activité mathématique arabe dans le Maghreb Oriental (IXe–XVIe s.), 2nd Maghribin Conference on the History of Arabic Mathematics, Tunis, 1–3 December 1988; in *Proceedings of the Conference*, Tunis, University of Tunis-I.S.E.F.C.-G.E.H.M.A.-A.T.S.M., 1990, 53–73; A. Djebbar: *Les activités mathématiques dans les villes du Maghreb Central (IXe–XVIe s.)*, 3rd Maghribin Conference on the History of Arabic Mathematics, Tipaza (Algeria), 2–4

December 1990; in Proceedings of the Conference, Algiers, Office des Publications Universitaires, 1998, 73–115; A. Djebbar: *Quelques aspects de l'activité mathématique dans le Maghreb Extrême (IXe–XVIe)*, 4th Maghribin Conference on the History of Arabic Mathematics, Fez, 2–4 December 1992. To be published in the proceedings of the Conference.

196. M. al-Manūnī: Nashāṭ ad-dirāsāt ar-riyāḍiyya fī Maghrib al-ʿaṣr al-waṣīṭ ar-rābiʿ [Activity of the Mathematical Studies in Morocco of the Fourth Period of the Middle Ages], *al-Manāhil*, Rabat, 33 (1985), 77–115.

12

Another Andalusian Revolt? Ibn Rushd's Critique of al-Kindī's *Pharmacological Computus*

Y. Tzvi Langermann

Introduction

In the opening paragraph of his *al-Kulliyyāt fī al-Ṭibb*, Ibn Rushd alerts his readers that the book shall be controversial: "We shall aspire to those doctrines which conform to the truth, even if this conflicts with the views of the people of the art."[1] In this chapter we shall be concerned with one controversy which Ibn Rushd initiated in his medical textbook, namely a biting critique of al-Kindī's pharmacological computus. In the first part of this chapter we shall very briefly sketch out al-Kindī's theory and, in much greater detail, look at the doctrines advocated by Ibn Rushd and the criticisms which he directs at al-Kindī. In the second section we shall proceed to the central question of this inquiry: to which (if any) historical context (or contexts), does this departure belong?

The Controversy

I Al-Kindī

Léon Gauthier published a book-length study of the theory of al-Kindī, including the Arabic text (from the one surviving manuscript) of the latter's *Fī maʿrifat al-adwiya al-murakkaba*.[2] As the title of his book readily indicates, Gauthier approached his subject with the aim of demonstrating that the work of al-Kindī, and, to a much lesser extent, the critique of al-Kindī given by Ibn Rushd, anticipate advances in European science that are associated with two nineteenth-century figures, E. H. Weber and T. G. Fechner. Needless to say, this sort of orientation in research is unacceptable in our own day. Al-Kindī's treatise raises several issues that are of great interest to the historian of medieval science, especially the early stages of scientific culture in Islamic civilization, and surely warrants an in-depth study of its own. That, however, lies far beyond the purview of this chapter.

The pharmacological computus with which we are concerned can be summarized in a few sentences.[3] Classical pharmacology had classified drugs into four degrees—more precisely, it had characterized some as temperate, and classified those that were not so into four degrees. These degrees signify the drug's potency in terms of the four elemental qualities: heat, cold, wetness, dryness. Al-Kindī's contribution lies in the claim that the intensity of drugs increases geometrically with the increase in degree, according to the "double ratio;" (*nisbat al-ḍiʿf*); thus a drug in the first degree is twice as intense as a temperate one, one in the second degree is four times so, one in the third degree is eight times so, and, finally, one in the fourth degree is sixteen times so.

This is the distinguishing feature of al-Kindī's theory, and it is the chief target of Ibn Rushd's criticism. It is the viewpoint associated with al-Kindī in the relatively meager medieval literature which carries on their debate.[4] This summary suffices for our purpose, and we shall not go into any more details. Nevertheless, before moving on to Ibn Rushd, we should like to call attention to these features of al-Kindī's treatise:

1. Originality. Al-Kindī is quite assertive concerning the original nature of his investigations, saying that they constitute an advance in scientific knowledge.[5] Although earlier authorities spoke about the four degrees as applied to simples, none explained how this is to be done with regard to compounds. Yet the determination of these degrees for compounds is a more urgent matter than it is for simples. Al-Kindī writes, "I see that the attainment of knowledge concerning the strengths of compound medications would [yield] tremendous benefits." As we shall see, the Andalusians too are quite cognizant of the fact that they are undertaking new research.

2. Nicomachus. Near the beginning of his treatise, al-Kindī describes five different mathematical series; from among these he chooses the geometric series noted above, since it takes "natural precedence" over the rest.[6] This notion is taken directly from the *Introduction to Arithmetic* of Nicomachus of Gerasa. At the end of part one, chapter XVII, Nicomachus lists five species of "the greater"; and, at the beginning of the next chapter, he asserts that "the multiple is the species of the greater first and most original by nature."[7] However, it seems that this point is stated a bit more emphatically in al-Kindī's version of that treatise, which, we must note, displays throughout a very different text than the Greek; it is not just a question of variants, but of a substantial reworking of the entire treatise. Al-Kindī's version is preserved, along with notes to his lectures on the book, only in the Hebrew translation of Qalonymos ben Qalonymos.[8] At this particular juncture of the *Introduction to Arithmetic*, there is a long explanatory note by al-Kindī, which discusses the five series in cosmological context, furnishing additional evidence of his interest in the topic.[9] A full account of al-Kindī's pharmacology, especially his attempt at mathematization, would pay close attention to the various manifestations of his interest in Pythagoreanism.[10]

3. Galen. Al-Kindī cites some Galenic texts, and refers to some difference of opinion he has with other scholars concerning their interpretation.[11] All in all, however, the correct interpretation of Galen's pronouncements seems to have been a

relatively minor issue (and even less important to Ibn Rushd, though not entirely neglected by him; see below) in this particular debate.

II Ibn Rushd

Ibn Rushd criticizes al-Kindī and his teachings towards the end of the fifth book of *al-Kulliyyāt, Kitāb al-Adwiya wa-l-Aghdhiya (The Book of Drugs and Foodstuffs)*, in a chapter (the last in the fifth books) entitled "*Al-Qawl fī Qawānīn al-Tarkīb*," "The Chapter on the Rules [or: Laws] of Composition," that is, of making compound medicines. We shall examine this chapter, following Ibn Rushd's exposition of the principles and laws that govern this branch of pharmacology, paying close attention in particular to the critique of al-Kindī with which it culminates. As we shall see, though, the chapter was never really finished. Several variants, especially at the end, testify to Ibn Rushd's shifts of thought concerning the issues at hand. His mature views are expressed not in *al-Kulliyyāt*, but in a separate monograph devoted to one of the most famous compound drugs, the theriac.[12] That essay is tightly organized and systematically developed. Significantly, Ibn Rushd makes no mention there at all of pharmacological calculations. As it seems to me, he came to realize that mathematics was not his forte, and he concentrated instead on an analysis in terms of natural philosophy. Our concern here, then, is with a relatively youthful venture on the part of the great philosopher, an excursion into a subdiscipline that he later abandoned. Nonetheless, this episode is instructive concerning developments in Andalusian science and as well for the intellectual biography of Ibn Rushd.

Ibn Rushd begins the chapter with an explanation of the "necessity" (*ḍarūra*) of mixing simples to form compounds—that is to say, why the physician, despite his preference for simples, may have no choice but to prescribe a compound. According to Ibn Rushd, three "things" (*ashyāʾ*) force the physician to have recourse to compounds: a simple of the required properties (*quwā*) is not available; there is available a simple that possesses the requisite properties, but not in the right quantities; the simple that is otherwise appropriate has an additional, undesired property that must be neutralized. Having established these three general classes, Ibn Rushd moves on to a detailed exposition of each one, including many examples. In the course of this exposition, he refers several times to the *qānūn* which is relevant to the case at hand. That term refers most often to rules that must be kept in mind when preparing medications; however, these rules are based upon affirmative statements concerning the behavior of natural substances, the human body, or the interaction between the two—statements that often take on the appearance of the natural laws.

Although a satisfactory analysis of the issue is beyond the purview of this chapter, I would suggest that the term *qānūn*—speaking of how the term was used, not necessarily how it was defined—originally referred to manipulative, procedural rules, i.e. general instructions as to how to act, but evolved to mean natural law, i.e. general observations concerning the natural world, which form the basis or justification for the procedures accepted by the medical profession.

One reason for compounding drugs would be to reduce a substance hot in the third degree to one hot in the second. In order to achieve this purpose we have two options; we can either compound it with one *cold* in the first degree, or with one *hot* in the first degree. Ibn Rushd appears to realize that the second option is more problematical.[13] We are compounding two substances of identical qualities, but the heat, rather than intensifying, reaches an equilibrium between the degrees of the two simples. Ibn Rushd feels that this calls for an explanation:

> This law (*al-qānūn*), I mean that a drug having less heat will reduce the heat of [a drug possessing] more [heat] was confirmed (*yuṣaḥḥiḥuhu*) by Galen. He took as evidence [the case of] hot and tepid water. When the hot is mixed with the tepid, its heat is necessarily reduced.[14]

Since a hot drug, when mixed with a substance hotter than it, will result in a net reduction of the heat, one may suppose that a patient with a hot disease, say a burning fever, could be treated with a hot drug of a lower degree than his illness; theoretically, this should reduce the patient's fever. Experience shows, however, that a hot drug will clearly harm the patient. For example, a person suffering from a burning fever who is given honey to drink will suffer greatly. In order to justify the law of combination given in the preceding paragraph, Ibn Rushd first resorts to a computation. Pepper is hotter than nard. (We learn elsewhere in the *Kulliyyāt* that the former is hot in third degree, the latter in the second.) This means that it has a greater proportion of heat, "as if you were to say (*ka-ʾannaka qulta*)," a ratio of heat to cold that is 5:1, whereas the corresponding ratio in nard is 2:1. "Thus, necessarily, when we mix a dirham of pepper with a dirham of nard, the ratio of cold to hot in the combination is greater than its ratio in pepper. If you contemplate this, it is evident."[15] The computation would seem to be the simple addition of the parts, yielding a ratio of 2:7, which is indeed greater than 1:5.

It is noteworthy that Ibn Rushd does not help the reader along by explaining anything at all about the mathematical operations employed in pharmaceutical computations. Some details relating to his method are broached only at the end of the chapter, in the course of the critique of al-Kindī, and in a manner that is not wholly satisfactory. In particular, Ibn Rushd does not exploit his computational apparatus in order to solve the vexing problem of the way simples of identical qualities but different degrees react; he does not develop any computus that would show precisely by how much a simple, say, hot in the first degree will reduce another hot in the second. As we shall see, Ibn Rushd has only the most general idea how to approach the problem—though this does not stop him from castigating rival approaches. A Jewish writer, Mordecai ben Joshua, also known as Viola de Rhodes (Provence, 14th? century), attempted to fill in the gaps in Ibn Rushd's exposition. In fact, though, the computus that he devises owes much to the theory of al-Kindī.[16]

The simple example of nard and pepper clearly does not in and of itself provide a sufficient explanation, and in the following paragraph Ibn Rushd elaborates. Although his tone is not defensive, it seems that he is trying to anticipate objections to the rules that he has just stated. Thus he begins by explaining that the "parts" of hot and cold spoken about above occupy an intermediate state of being; "they do not exist in pure actuality (*bi-l-fiʿl al-maḥḍ*), but rather in some sort of intermediate state (*bi-ḍarb min al-tawassuṭ*) between potentiality and actuality."[17] He adds that, with the aid of the concept of "intermediate state," we can understand how homogeneous bodies can nonetheless be analyzed into unequal numbers of parts of hot and cold. Another, unstated advantage of this characterization of the "parts" is that it avoids any confusion between the quantity of the substances and the analysis of their qualities. In the example given above, one dirham of pepper has a total of six parts, a dirham of nard has three, and a dirham of the mixture has nine—but all weigh one dirham. The division into parts is purely an analytical device for determining the relative strengths of qualities. This matter too will be clarified presently.

Because these "parts" may also be viewed as potentialities, they represent the readiness of the drug to be acted upon by the human body, more specifically by the body's innate heat. This fact solves the puzzle alluded to earlier: why a simple with a relatively low degree of heat will harm a patient suffering from fever, i.e., increase his fever, even though it will reduce the heat of a hotter simple when the two are mixed together. Even though (to return to the example given above) honey has a relatively low degree of heat, it is all the same "hot," and this heat will, under the agency of the innate heat, become "fiery substance" (*jawhar nārī*). We may conclude, then, that there is an essential difference between the way the "parts" react: when mixed with another simple, they recombine according to a simple arithmetical procedure (as yet unspecified), but when consumed by the human, they are activated (though this technical term is not employed here) by the innate heat.

Ibn Rushd speaks again of a *qānūn* in connection with the third cause for compounding medicines, i.e., in order to mask one of the qualities inherent in a simple. This case is of particular interest, for two reasons. First, Ibn Rushd emphasizes the generality and importance of the *qānūn*. "It is as we have said a general law (*qānūn jāmiʿ*)."[18] Second, he gives credit to the Banū Zuhr for clarifying and sharpening a *qānūn* that is only implicit in the writings of the ancients. "It is found in the compounds of the ancients even though they did not refer to it explicitly (*bi-l-qawl*), nor did they call attention to it. The people who [provided] the best information about this are none other than the Banū Zuhr. Upon my life, they have many merits (*maḥāsin*) in this art."[19] This is one of several major accomplishments in the field of medicine that stand to their credit. Ibn Rushd may have hoped that his own work, together with that of the Banū Zuhr, could produce an alternative to Ibn Sīnā's monumental *al-Qānūn fī al-Ṭibb*. We shall return to this in the second part of our chapter.

The rule describes two synergetic effects:

1. The combination of simples may produce in the resultant compound properties that were unpredictable on the basis of the constituent ingredients. Ibn Rushd observes that this particular rule applies only to secondary and tertiary qualities: "On the whole (*bi-l-jumlati*) this rule applies only (*innamā*) to secondary and tertiary powers."[20]
2. The reaction of the human body to medication cannot be fully predicted on the basis of an analysis of the ingredients that make up the compound. Ibn Rushd writes: "The actions of drugs upon [human] bodies are only a relative matter (*amr iḍāfī*). In truth, this is not something that is consequent upon the parts of the drug itself. It may happen that a drug that is itself less hot will be, relative to the human body, hotter than a drug that itself possesses greater heat."[21]

The reaction of the human body to a given drug—be it simple or compound—is a strong variable in pharmacology, and one which can be established only *post facto*, through experience, experiment, and observation. This fact was common knowledge, and it underlies, as far as I can tell, the disinterest (if not disdain) which some authorities, most notably, Ibn Sīnā, display concerning pharmacological theory. It is almost paradoxical that Ibn Rushd, who invested more effort than most in the theoretical analysis of the action of medication should also have so strongly emphasized the essentially unpredictable nature of compound medications. His discussion of the issues leads eventually to the sweeping conclusion that any given combination of drugs may have an effect on the human body that is stronger than we might have predicted on the basis of an analysis of the qualities of the individual ingredients.

To return to the application of this rule in the chapter under discussion here. According to Ibn Rushd, "this *qānūn* is an important (*muhimm*) *qānūn* in medicine. Indeed, if a person would only bear it in mind as he ought to, he would hardly ever administer a cure with a simple drug. By my life, it is found [implicitly] in the compounds of the ancients. . . ."[22] Ibn Rushd is going against the grain of a medical tradition, stated by Maimonides among others, that the physician should administer a compound only when there is no appropriate simple; and we have seen that, in the opening paragraph to the chapter, Ibn Rushd appears to confirm this tradition. This apparent inconsistency is one of several features which, taken together, indicate that this chapter displays work in progress, rather than Ibn Rushd's mature views on the issue.

The next *qānūn* addresses the matter of the *quantity* of simples that are to be used in the preparation of compounds. This *qānūn* as well has several facets (*awjuh*). For the most part they are general guidelines, almost self-evident. Unlike the preceding, they do not foreshadow any laws of nature; nor do they involve any mathematization of pharmacology. One rule states that one employs a smaller quantity of a strong drug and a larger quantity of a weak drug—the exact quantities, of course, depend upon the desired result of the final product. Another rule concerns compounds in which one particular element is overwhelmingly dominant, the other ingredients being ancillary to it. The exact quantities of all of the ingredients will be determined by the ultimate strength that the medication should have, or other considerations, such as the distance which the drug must traverse through the body in order to reach the diseased organ. Ibn Rushd then speaks specifically of laxatives, offering again a trivial example: if four drugs are required for a potion, then the physician mixes a quarter dosage of each and has the patient drink the combination. He then concludes:

> These are all the rules and laws (*jamīʿ al-dustūrāt wa-l-qawānīn*) which are employed with regard to quantity. However, since the most important thing for the doctor to know when compounding medicines is the degree of the primary, secondary, and tertiary powers, if this is possible, we must say something about it. We state: when someone wishes to determine the rank (*martaba*), as far as primary qualities are concerned, of a compound drug, the way to do this is to take into consideration the degrees of the primary drugs which are in it.[23]

With that statement, Ibn Rushd takes up the computus. He begins with some trivial examples which, however, are important for establishing the rules that will be employed in more complex cases: "When you know the simple law (*al-qānūn al-basīṭ*), you will necessarily know the complex law (*al-qānūn al-murakkab*) by means of an investigative procedure (*bi-wajh al-naẓar*)."[24]

Finally Ibn Rushd is ready to reveal his most complex computus, that which applies to drugs compounded from simples of opposing qualities and varying degrees. From the preceding, however, the procedure is quite clear:

it is all a matter of simple arithmetic, provided that we deal with comparable units of power, not of weight: "I mean by their equivalence, not the equivalence in weight, but rather the equivalence in power."[25] The cold ingredient will reduce the hot by the amount of its degree; if it is cold in the first degree, it will reduce the hot drug by one degree. Drugs of opposing qualities will reduce one another in accordance with the number of degrees assigned to them. This is in fact an extension of the second class; it is exactly the same computus by which it has already been determined that drugs of opposing qualities but equal degrees neutralize each other's effect.

The computus of degrees can be combined with the notion of specific quantitative units, yielding a more complete computus. For example, two units of a drug cold in the first degree, when mixed with one unit of a drug hot in the third degree, will reduce it by two degrees, not one (the sum of the degree computus alone). Since the result will be proportional not only to the differences in degrees (which are always integers, when speaking of simples) but their relative quantities (in terms of specific units of each substance), the final computation may lead to fractional degrees. Ibn Rushd illustrates with one case in which the final result is "an amount in the middle between the third and the second,"[26] but he does not go into any further detail. The treatise of Viola de Rhodes alluded to above works out these fractions in great detail. Before ending his discussion on this matter, Ibn Rushd hints that this very same principle explains why a double dosage of a drug can be fatal.[27] We may readily supply the full explanation: a larger quantity of a drug functions (more precisely, multiple unit dosages function) as unit dosages of a higher degree.

The last and most troublesome class of drugs is comprised of those that are compounded of simples of identical qualities but varying degrees. Ibn Rushd has no complete solution to this problem. He does however, have some strong opinions about solutions that others have proposed, Indeed, it is by way of this particular issue that Ibn Rushd takes on al-Kindī and his computus. The problem, as Ibn Rushd intimates, is simple, if not "self-evident (*bayyin bi-nafsihi*)."[28] Since a cold drug will reduce a hot drug by the arithmetical difference of their respective degrees—for example, a drug cold in the first will reduce a drug hot in the second by one degree, yielding a compound hot in the first—it stands to reason that a drug of the same quality, e.g. one hot in the first degree, will reduce the same drug (hot in the second) by less; but by how much less? Ibn Rushd can as yet provide no answer. He argues for a scheme in which drugs of the opposite quality will reduce by the greatest amount, temperate drugs will reduce by less, and drugs of identical qualities by even less; he speaks of the "proportion" (*nisba*) by which these types of substance will reduce. One might have expected that his earlier remarks on the number of "parts" in certain substances, as well as the clear cut method discussed in the previous section would have presented some opportunities. But he offers no clear formulae for computing the resultant powers of the drug.[29]

He does, however, have some definite things to say about *computi* that other physicians have proposed. It is here that Ibn Rushd first mentions al-Kindī by name, placing the blame for introducing confusion and irrelevancies into the field of medical pharmacology squarely upon his shoulders:

> The person who first plunged them into this matter is none other than the man known as al-Kindī. That is because this man wrote a treatise in which he sought to speak about the rules (*al-qawānīn*) by which the nature of a compound drug may be known. But he went astray (*kharaja*) in speaking about the art of numbers and the art of music, in the matter of someone who looks into something only incidentally.[30] This man adduced in that book senseless and hideous things.[31]

After this blistering introduction, Ibn Rushd is ready to display some specific points of criticism. Let us summarize the arguments which he raises:

1. Al-Kindī's computus contradicts the reasoning which underlies the system by which drugs are graded by degrees. Ibn Rushd does not here say who designed this system, preferring instead to refer to the inventors anonymously, in the third person; in a later passage, however, at least according to the reading of one of the manuscripts, the system is associated with Galen. A drug that causes sensible heat—apparently, the most barely sensible heat is intended—is assigned to the first degree. Then, drugs "whose distance from that [first degree] is [the same as] that [first drug's] distance from temperance" are classified in the second degree. "Without any doubt, it is double the first"—not four times, as in the theory of al-Kindī. Similarly, drugs whose distance from the second degree is the same as the distance of the second from the first are assigned to the third degree, "and likewise for the fourth."
2. What could ever have forced the medical profession to adopt the "double ratio" advocated by al-Kindī? It is not entirely clear whether Ibn Rushd rejects al-Kindī's assertion (drawn as we have seen from Nicomachus) that the "double ratio" is the most natural, or whether he is simply unaware of it.
3. According to the system of al-Kindī, third-degree drugs would be already fatal; that is, if we follow Ibn Rushd's interpretation, whereby any drug four times as intense as the first degree is fatal. What, then, are we to do with fourth degree drugs?
4. The increments between the degrees are unequal. "What greater disorder (*ikhtilāl*) could happen to the art?"[32] Ibn Rushd elaborates, somewhat unclearly: al-Kindī has defeated his own purpose. (Here again Ibn Rushd uses the third person, and it is possible, though in my view unlikely, that he means to say that

al-Kindī has missed the mark with regard to 'his' [Galen's] purpose.) He wished to preserve the [uniform] increase in degrees, but by doing so geometrically, rather than arithmetically, he has caused the increments between each degree to increase successively. "So if there were a fifth degree, it would be thirty-two 'parts'. . . This is all delusion and drivel, and a discussion of things which have no reality."[33]

The last sentence, in which Ibn Rushd reinforces his earlier denunciation of al-Kindī, is a fitting finale to his tirade. However, the manuscript tradition indicates that Ibn Rushd continued to reflect upon the problem; he was especially concerned with the proper understanding of one of Galen's pronouncements. His later thoughts are contained in this paragraph, which is found in one of the Arabic manuscripts of *al-Kulliyyāt* (St. Petersburg 124) and reproduced in some of the Hebrew and Latin translations:

> The way by which al-Kindī came to err is that he made the first degree double the temperate with regard to the quality, hot or cold. This then required him to maintain the double ratio. It may be said to him that Galen intended by the first degree that which adds one part in ten to the temperate. In this way, if the double ratio were to be compounded for [each] increase in the degrees, it would not entail that the drug which is in the fourth degree would be sixteen times the temperate. That it is arranged (*tarattaba*) in this way [is indicated by] Galen's saying:[34] "By the first degree I mean that which is evident to the sense, when first there appears a change in the body." Had he meant by the first degree double the temperate, then the change which is evident in the body [upon the application of a substance in the first degree] would not be the first change. Contemplate this; it is quite clear. However, when a well-known person commits an error, people by habit follow him, since the power of imitation (*quwwat al-taqlīd*) overwhelms their natures.[35]

A somewhat abbreviated and slightly different version of this paragraph is found in the Hebrew translation of Shlomo ben Avraham: "But what Galen said about this, when he stated, 'I intend by the first degree, that change in the body which is first apparent for a drug.' Had he meant by the first degree double the temperate, then the change which is evident in the body upon [the application of] the drug would not be the first change. Contemplate this; it is quite clear. However, it is customary for people to follow the claims of a famous person, since they accept his opinion."[36] Yet a third version of this paragraph is found in the Latin *translatio antiqua*.[37]

In his later reflections upon the problem, Ibn Rushd has arrived at three additional arguments:

(P1) A quasi-mathematical argument, whose meaning is not entirely clear to me, which implies a contradiction between a decimal system planned by Galen and al-Kindī's geometrical series.

(P2) A textual argument, based upon a direct quotation from Galen. According to Ibn Rushd, since the first degree signifies the minimum sensual excitation, it cannot possibly be double the temperate (which does not stimulate the senses at all).

(P3) A third argument for the supposed popularity of al-Kindī (though it could equally be directed against Galen), namely people's ingrained habit of accepting without question the pronouncements of "famous people."

Argument P2 leads us to raise the question of Ibn Rushd's familiarity with the entire Galenic corpus; and, even more to the point, it forces us to wonder, as Gauthier did, whether Ibn Rushd read al-Kindī's treatise in its entirety. For al-Kindī cites two Galenic texts, neither of which is referred to by Ibn Rushd: *Tarkīb al-Adwiya* (*On Compound Drugs*), in ten chapters, better known in Arabic as *Kitāb al-Mayāmir*, and the book on drugs written for Andromachus, apparently a reference to the work known in Arabic as *Kitāb al-Tiryāq ilā Bīsūn* or *Kitāb al-Tiryāq ilā Fīṣun*.[38] In addition, al-Kindī cites these sources as proof that Galen had unambiguously rejected the very computus advocated by Ibn Rushd: "Moreover, Galen has already refuted the school which maintains that the strength (*quwwa*) of the fourth degree is four times the first, and the third three [times the first]. . .".[39] Ibn Rushd, on the one hand, gives no indication that he knows of these passages, nor, in particular, that they were cited by al-Kindī—and it seems almost inconceivable that he could simply ignore them, if he wishes to make a convincing critique. Yet, on the other hand, he quotes a different Galenic dictum (not cited by al-Kindī, at least not in this context) whose misinterpretation, so he claims, is the source of al-Kindī's error.

Argument P3 is interesting for other reasons. As we have already suggested—and we shall return to this point in the second part of this chapter—al-Kindī's treatise seems to have generated very little interest. How, then, are we to understand Ibn Rushd's complaint concerning the uncritical acceptance which al-Kindī's theory enjoys? It may very well be the case that, despite the lack of written evidence, al-Kindī's theory—or, at least, the basic idea that with each higher degree, the intensity of the drug doubles—did have wide currency. Alternatively, the appeal to habit as a source of error may have become a cliché; at the very least, Ibn Rushd's contemporary, Moses Maimonides, offers the very same explanation for persisting errors in both a medical and a philosophical context.[40]

Ibn Rushd has now completed his critique of al-Kindī. The two final paragraphs of the chapter do help to fill us in on the context of the critique, and we shall look at them briefly. Ibn Rushd briefly discusses the concept of khāṣṣa ("specific").[41] This leads him to note a point of disagreement with Ibn Sīnā. According to the latter, most of the active properties of the theriac are *khawāṣṣ*, specifics whose cause cannot be determined (*la yumkin taʿlīluhu*). For that reason, Ibn Sīnā did not want to make any changes at all in the traditional recipe of Andromachus. "As for me," continues Ibn Rushd, "I see it fit to add many drugs to the theriac."[42] These are substances which may or may have not been known to the ancients, such as aloe (*ʿūd*), ambergis (*ʿanbar*), and cloves (*qaranful*). Ibn Sīnā holds the view that the properties of the theriac are occult, and, therefore, the physician should not tamper with the formula of the ancients, whose efficacy has been proven empirically over the centuries. By contrast, Ibn Rushd, though not denying a role for empiricism, feels that the properties of the theriac, or some of them at least, are explicable in terms of the predictable reactions of the constituent simples; therefore, there is no reason why a physician ought not to experiment with different formulations. Ibn Rushd thus closes the chapter with a strong, personal statement of confidence in rational pharmacology and in the advance of science.

The next few lines are meant to introduce the discussion of some specific compounds and their properties. This is in line with the plan of this section of *al-Kulliyyāt* (*Kitāb al-Adwiya wa-l-Aghdhiya*), which is outlined in the opening paragraph. There Ibn Rushd writes: "Afterwards we shall move on to the rules for the composition [of drugs; *qawānīn al-tarkīb*]. We shall mention the best known compound drugs, and we shall make known their natures, in accordance with what the rules necessitate in that matter.[43] When that is completed, the purpose of this section will have been achieved."[44] The list itself is not found in the edition of the Arabic text, nor in any of the Hebrew manuscripts which I have examined.

The very last sentence of this chapter is relevant, in a strange way, to the critique of al-Kindī. Justifying the particular organization which he has chosen for his materials, Ibn Rushd writes, "Just like the authority on music (*ṣāḥib al-mūsīqā*) will speak of the well-known instruments only after he has presented the elements of melodies and the ways in which they are combined, in order that training should take place in that way (*li-yaqaʿa bi-dhālika al-irtiyāḍ*), so also is the case here."[45] Ibn Rushd could not have forgotten that, only a few pages earlier, he had criticized al-Kindī for appealing to music in order to justify a decision made in pharmacological theory. Music, however, is not at all a major factor in al-Kindī's treatise, and Ibn Rushd's citation of that particular point—while ignoring many other highly relevant arguments adduced by al-Kindī—seems to be purely polemical. Why Ibn Rushd should then have chosen to make a similar analogy in his own treatise is truly puzzling.

The Contexts

Ibn Rushd's participation in the controversy surrounding models employed by the astronomers of his day has been quite adequately contextualized.[46] There is, first of all, evidence of dissatisfaction with the Ptolemaic models on the part of a number of prominent thinkers, from all parts of the Islamic world: Thābit ibn Qurra and Ibn al-Haytham in the east, and Ibn Bājja and Ibn Ṭufayl in the west. The westerners, so it seems, reexamined celestial physics as part of a wide-ranging reappraisal of Aristotelian natural philosophy. Within this Andalusian trend, Ibn Rushd stands out for his resolute commitment to Aristotle—not to Aristotelianism, by which I mean the body of doctrines, including various accretions and commentaries, not all of them mutually compatible, built up around Aristotle's teachings, but rather to a purified, strict reading of Aristotle's own writings. Thus, within the framework of the astronomical controversy, Ibn Rushd's attitude differs from that of Ibn Bājja, whose dynamics, intended or not, contain a radical reworking of Aristotle, and from that of al-Biṭrūjī, who, it has been shown, preferred a number of teachings, especially the theory of impetus and the identification of *tashawwuq* ("desire," "yearning") as the driving force of the cosmos, which he may have discovered in the writings of the maverick thinker Abu 'l-Barakāt.[47] In short, Ibn Rushd's interest in the problem situates itself into both the wider Islamic and the narrower Maghribian-Andalusian contexts; in addition, there is a distinct twist to his views, which can only be explained by taking into account his own personal inclinations and interests.

What about medicine? To be sure, Galen's writings evoked widespread criticism throughout the Islamic cultural orbit. Abū Bakr al-Rāzī wrote a book length refutation of Galen.[48] Ibn Sīnā's *al-Qānūn* is replete with critical remarks, including the standard accusation that, in attempting to philosophize, Galen intruded into a field in which he did not belong.[49] Moses Maimonides, a native of Cordova and Ibn Rushd's contemporary, devoted the twenty-fifth section of his *Fuṣūl Mūsā* to a stinging attack on Galen's teachings in both medicine and philosophy.[50] Ibn Rushd as well offers numerous strictures on Galenic doctrines.[51] His critical posture, therefore, conforms to a general trend among Arabic medical writers.

This point, however, is only of limited value in approaching our particular problem. Ibn Rushd's critique is directed at al-Kindī, not Galen. To be sure, there is evidence that the proper interpretation of some Galenic texts was one factor in the dispute between Ibn Rushd and al-Kindī. Moreover, it is noteworthy that Abu 'l-ʿAlāʾ ibn Zuhr, patriarch of the Banū Zuhr family of physicians, defended Galen against the strictures of al-Rāzī.[52] As we have seen, Ibn Rushd's work in medicine, particularly in pharmacology, is deeply indebted to that of the Banū Zuhr. If Ibn Rushd's attitude towards Aristotle can be characterized as unswerving admiration, his attitude towards Galen was more ambivalent; his attack upon al-Kindī's fits neither into the critique of Galen nor into his defense.

Two general observations are in order at this juncture, built upon remarks of Ibn Rushd concerning scientific activity going on in his day. First, an important distinction between the work undertaken in pharmacology and in astronomy must be drawn. Research into pharmacology is part of a comprehensive reinvestigation into the science, one whose aim is to attain *new* information. By contrast, the program in astronomy was undertaken in order to *recover* the true astronomy of Aristotle; but pharmacology has no supreme authority, nor any lost knowledge of the ancients. True, the ancients had by experience arrived at some effective medical prescriptions, but even then, they did not elaborate the underlying theory. This important task was assumed only by later generations; and it is only one of several areas in pharmacology where the moderns can make new and important contributions. Second, however, there exists a significant similarity between the works in pharmacology and astronomy. Credit for advancing the new research programs belongs to the Andalusians, in particular, the medical clan of the Banū Zuhr.

Let us return to the specific issue of the critique of al-Kindī. As far as I can tell from the extant sources, at the time that Ibn Rushd launched his attack, al-Kindī's monograph seems to have engendered hardly any interest at all. Indeed, one searches in vain for references to al-Kindī's computus in the two most important medical encyclopedias, Ibn Sīnā's *al-Qanūn* and al-Majūsī's *al-Kāmil*. Al-Kindī is cited about a dozen times in the *Qānūn*, invariably in connection with this or that particular remedy, but his computus is not mentioned. Indeed, Ibn Sīnā does seem to have held much stock in pharmacological computations at all. In the introductory "scientific tract" (*al-maqāla al-ʿilmiyya*) to book five of the *Qānūn*, Ibn Sīnā surveys the various reasons why it may be necessary to employ compound medications: to combine or intensify various secondary qualities, to delay the digestion of the medication until it reaches its destination, and so forth. Within this list we find a simple computus, not that of al-Kindī, described as follows:

> Perhaps we may need a drug which heats four parts (*ajzāʾ*), but we can only find one which heats three parts, and another which heats five parts. We then combine them, in the hope that the result of the combination will heat four parts.[53]

Of the various reasons which Ibn Sīnā gives for compounding medicines, it is only in connection with this computation that he adds the disclaimer, "in the hope that" (*rājīn an*) the desired result will be obtained. Al-Majūsī also limits himself to a single trivial example. If the physician requires a drug hot in the second degree, but no such simple is available, he may compound two other simples, one hot in the third degree and the other hot in the first. The resulting compound will be hot in the second degree.[54]

Even among the Andalusians, who, as we shall see, were especially interested in pharmacology, the impact of al-Kindī's book seems to have been minimal. In the sources available to me, I have found only one clear trace. Ibn Buklārish (fl. Saragossa, eleventh century), author of *al-Mustaʿīnī*, exhibits a computus which is surely based upon al-Kindī, though the latter is not named.[55] Significantly, the only tract which directly addresses the theory of al-Kindī was written by the same Abu 'l-ʿAlāʾ ibn Zuhr. Unfortunately, his *Treatise in the Explication of the Letter of Yaʿqūb bin Isḥāq al-Kindī on Compounding Drugs* (*Maqāla fī basṭihi li-risālati Yaʿqūb ibn Isḥāq al-Kindī fī Tarkīb al-Adwiya*) is not extant.[56]

Why was there so little interest in al-Kindī's book? For all practical purposes, recipes for compound drugs were passed on by tradition. Medical practitioners surely experimented with different formulas. Ibn Rushd, as we have seen, is especially confident of his ability to improve existing formulae by the application of rational laws. On the whole, though, Ibn Sīnā's approach (criticized by Ibn Rushd) was the norm. Traditional recipes were passed on from generation to generation; even when practitioners experimented with old or entirely new formulae, their work was not guided by mathematical rules, such as those worked out in the treatise of al-Kindī. Pharmacological theory was of little practical use, both before and after Ibn Rushd. This accounts both for the weak interest in al-Kindī's treatise in the period preceding Ibn Rushd, as well as the relatively low level of interest in the controversy, once Ibn Rushd published his critique.[57] Ibn Rushd's work in pharmacology built upon earlier advances of the Banū Zuhr and, presumably, he was familiar with the exposition of al-Kindī's theory by Abu 'l-ʿAlā. However, this very limited tradition is only a small part of our story.

For a fuller appreciation of the historical context of this episode, it must be considered against the background of two larger developments. The first of these is the conspicuous concern which Andalusian and Maghribi scientists evinced in pharmacology; one ought to include as well related disciplines such as botany and botanical lexicography. This special interest has often been connected with the gift of a beautiful copy of Dioscorides' *Materia Medica* by the Byzantine emperor Constantine VII to the Umayyad caliph ʿAbd al-Raḥmān III. This event, which took place in 948/9, certainly had an immediate impact. An impressive group of Andalusian savants set about to study the text, and the Byzantine emperor was persuaded to send a monk in order to assist the Andalusians in understanding the Greek text. Three centuries later Ibn al-Bayṭār still found cause to write a commentary on *Kitāb Diyāsqūrīdūs*. All in all, the corpus of pharmacological writings produced in Spain and Morocco is impressive indeed.[58]

It seems especially noteworthy that, among the Andalusians who displayed a keen interest in medical pharmacology, we find two savants who also figure prominently in the reexamination of astronomy: Ibn Bājja and Ibn Ṭufayl. The former wrote (in addition to some notes to part of *De plantiis*) a *Discourse on Galen's Book of Simples* and, in collaboration with Abu'l-Ḥasan Sufyān, a book recording experiences with the drugs described by Ibn Wāfid.[59] Moreover, none other than *al-shaykh al-akbar*—the noted mystic Muḥyī al-Dīn ibn ʿArabī—informs us about a debate or contest or sorts between Ibn Bājja and Ibn Zuhr regarding medicinal plants.[60]

As for Ibn Ṭufayl: Ibn Rushd records correspondence which he had with Ibn Ṭufayl concerning the proper definitions of the terms "food" (*ghidhāʾ*) and "drug" (*dawāʾ*), and especially the distinction between the two. Their disagreement seems to have hinged upon the interpretation of a statement made by Galen.[61] As we have already stressed several times, Ibn Rushd's most important collaboration was that carried out with the Banū Zuhr. In an earlier study, we took note of the lack of communication between some of the major players in the Andalusian revolt against Ptolemaic astronomy.[62] Although scholars can discern a community of interest between Ibn Bājja, Ibn Ṭufayl, Ibn Rushd, and al-Biṭrūjī, there is very little evidence of any joint effort or even discussion between these indivduals. By contrast, there is abundant evidence of lively and fecund exchanges in pharmacology. This may indicate that, of the two sciences, astronomy and pharmacology, it was the latter which aroused more intense interest on the part of the Andalusians.

The second context, and the one which promises to be of greater significance for the history of thought in Islamic culture, is the attempt—perhaps one ought to call it a program—of the Andalusians to construct an alternative to the syntheses which were produced in the East. The most important target of this enterprise, in connection with both medicine and philosophy, was the work of Ibn Sīnā. Abu'l-ʿAlāʾ ibn Zuhr's negative opinion of Ibn Sīnā's *al-Qānūn* has been recorded by a number of authors: "Previously, he [Ibn Zuhr] had not encountered this book [*al-Qānūn*], but when he examined it he condemned it and discarded it, and did not include it in his private library. He kept tearing off the margins of its leaves which he used for writing prescriptions for his own patients."[63]

In the final paragraph of *al-Kulliyyāt*, which as the title implies, deals with generalities, Ibn Rushd notes that his book must be supplemented by another treatise of the type known as *kanānīsh* (singular: *kunnāsh*), containing instructions for the treatment of specific diseases. He hopes to have the opportunity to compose a work of this sort himself. In the meantime, however, he warmly recommends "the work known as *al-Taysīr*, which was written in our own time by Abū Marwān ibn Zuhr. It was I who requested of him [to write] this book. I made a copy, and this was the cause of its publication (*khurūj*)."[64] In connection with this passage, A.Z. Iskandar has observed: "The *Kulliyyāt* of ibn Rushd and the *Taysīr* of ibn Zuhr were meant to serve as a complete work on medicine, possibly to serve instead of *K. al-Qānūn* which did not appeal to Abu 'l-ʿAlāʾ Zuhr b. ʿAbd al-Malik b. Marwān b. Zuhr and, very likely, his son and disciple Ibn Zuhr, also the latter's pupil, ibn Rushd."[65] Whether the combined efforts of Ibn Zuhr and Ibn Rushd were undertaken in response to the work of Ibn Sīnā, al-Majūsī, or al-Rāzī, it does seem to be the case that they were striving to produce an alternative to the great compilations emanating from the eastern reaches of Islam. Ibn Rushd's critique of al-Kindī may be viewed as one small contribution to this program.

The point which we would like to stress is that the purpose of this venture was not *tahāfut*, the "destruction" of noxious ideas, but rather the construction of an alternative. Moreover, the effort to compose a comprehensive work on medicine was part of a far-reaching program to create alternatives in other fields of science and philosophy. Earlier writers, especially Ibn Bājja, had prepared the groundwork. Ibn Bājja was certainly an insightful and penetrating thinker, but he does not seem to have been interested in or capable of producing a systematic work. Two relatively short works of Ibn Ṭufayl are generically identical—they even bear the same title—as some treatises of Ibn Sīnā: *Ḥayy ibn Yaqẓān* and *al-Urjūza fī al-Ṭibb*.[66] These may, then, be viewed as Andalusian alternatives to Avicennian writings. The central tasks of this enterprise were taken up by Ibn Rushd. We have already noted his collaboration with Abū Marwān Ibn Zuhr in the field of medicine. Perhaps his series of commentaries to Aristotle were intended to serve as the Andalusian response to Ibn Sīnā's *al-Shifāʿ*. In philosophy, however, Ibn Rushd, had no collaborators; he did it all himself.[67]

This leads us to our final observation. Although the broad program of pharmacological research as well as many of the specific statements made by Ibn Rushd do find their place in the contexts which we have suggested above, they do not exhaust the story. Ibn Rushd's personality certainly stands out in his bold confidence in the power of rational inquiry. This, however, is just one aspect of the strong sense of dissatisfaction with the work of his colleagues—or most of them I should say, excluding in particular the Banū Zuhr—which informs and infects Ibn Rushd's medical writings. Even after all of the contexts have been listed and described, one must take into account Ibn Rushd' strong individuality, his personality, and his idiosyncrasies, if one wishes to obtain a satisfactory understanding of the work of this great thinker.

NOTES

1. S. Shaybān and U. al-Ṭālibī, eds., *Al-Kulliyyāt fī al-Ṭibb li-Ibn Rushd* (Cairo, 1989), p. 19. All of my references are to this edition; in my view, it supersedes the edition prepared by J. M. Fórneas Besteiro and C. Alvarez de Morales, 2 vols. (Madrid, 1987).

2. L. Gauthier, *Antécédents Gréco-Arabes de la Psychophysique* (Beirut, 1938).

3. I have borrowed the term computus from Latin astronomy, where it refers to a system for carrying out certain computations, most notably for reckoning the date of Easter.

4. The relatively low level of interest in this issue will be discussed in the final section of this chapter. One writer who does express interest in the debate is Joshua Lorki, in his (as yet unedited) *Gerem ha-Maʿalot* (see, e.g., Mss. Paris, BN héb 1143, ff. 146a–b; St Petersburg, RNL, Heb I 334, ff. 90b–91a). He notes that the "later Christian physicians, such as Arnald de Villnova and Bernard de Gordon" have decided in favor of al-Kindī, and so shall he. Arnald's *Aphorismi de gradibus* has been edited by Michael McVaugh (*Arnaldi de Villanova Medica Omnia*, vol. 2, 1975).

5. See Gauthier, pp. 1–2 of the Arabic treatise.

6. Ibid., 5.

7. M. L. D'Ooge, F. E. Robbins, and L. C. Karpinski, *Nicomachus of Gerasa, Introduction to Arithmetic* (London, 1926), pp. 213–214.

8. I discuss the Hebrew manuscripts, giving as well preliminary evidence for the nature of al-Kindī's version, in "Studies in Medieval Hebrew Pythagoreanism: Translations and Notes to Nicomachos; Arithmological Texts," to appear in *Micrologus*. See also M. Steinschneider, *Die hebraeischen Uebersetzungen des Mittelalters* (Berlin, 1892), 517–519, and S. Brentjes, "Untersuchungen zum Nicomachus Arabus," *Centaurus* 30 (1987), 212–239. Note that the version of Thābit ibn Qurra (published by Wilhem Kutsch, Beirut, 1958), while in general following the Greek quite closely, also differs significantly in a number of places; these are not noted by Kutsch, despite the inclusion of occasional Greek terms in his notes to the text.

9. See, e.g., ms. Paris, BN héb 1029, ff. 10b–11a.

10. Further hints at al-Kindī's familiarity with doctrines ascribed to Pythagoras are evident in some of the newly discovered manuscripts of his *Risāla fī ʾl-qawl fī ʾl-nafs*, discussed by Charles Genequand, "Platonism and Hermetism in al-Kindī's *Fī al-Nafs*," *Zeitschrift für Geschichte der Arabisch-Islamischen Wissenschaften* 4 (1987/88), 1–18, esp. pp. 5–8; but see also p. 11 n. 40.

11. Gauthier, pp. 24–25 of the Arabic text.

12. Ibn Rushd's *Kitāb al-Tiryāq* was published by G. C. Anawati and S. Zayed in *Rasāʾil Ibn Rushd al-Ṭibbiyya* (Cairo, 1987), pp. 389–422. The Hebrew translations are discussed by Steinschneider, *Hebraeischen Uebersetzungen*, p. 676, to which should now be added Ms. St Petersburg, Academy of Sciences C67. A Spanish translation by María Concepción Vázquez de Benito is included in the volume *Averroes: Obra Médica*, published as part of VIII Centenario Averroes (Seville and Malaga, 1998), pp. 261–277. I am currently preparing an annotated English translation.

13. The problematics emerge later on (see below, p. 9ff.); but Ibn Rushd offers no solution.

14. *al-Kulliyyāt*, p. 304.

15. Ibid., 305.

16. I have studied the copy of his treatise found in Ms Moscow, Ginzburg 462, ff. 135b-139b; I hope to discuss it in detail in a future publication. See M. Steinschneider, *Die hebraischen Uebersetzungen des Mittelaters* (Berlin,1892), p. 799, no. 215.

17. *al-Kulliyāt*, p. 305.

18. Ibid., 308.

19. Ibid., 308.

20. Ibid., 308. These terms are distinguished as follows (*al-Kulliyyāt*, pp. 218–219). The four elemental qualities (hot, cold, wet, dry) are called primary. Secondary properties are quasi-elemental effects, e.g., hardening or softening, which are induced throughout the entire body. Tertiary properties are limited to one particular organ, for example, a diuretic.

21. Ibid., 307.

22. Ibid., 308.

23. Ibid., 309.

24. Ibid., 309. In this context, *basīṭ* and *murakkab* can only refer to the simplicity or complexity of the computation.

25. Ibid., 310.

26. Ibid., 310.

27. Ibid., 311.

28. Ibid., 311.

29. Here too Viola de Rhodes attempts to work out the details according to the system of Ibn Rushd.

30. Literally: according to what happens to the person who looks into the thing with an incidental investigation (*ʿalā jihati mā yaʿriḍ li-man yanẓur fī ʾl-shayʾ al-naẓara alladhī bi-l-ʿaraḍ*).

31. *al-Kulliyyāt*, p. 312.

32. Ibid., 313.

33. Ibid., 313.

34. The Hebrew translations (see following notes) exhibit: "That which Galen said shows this [i.e., that this is the correct interpretation]." So also the Latin *Antiqua translatio*, as translated by Gauthier, p. 97. Both of these versions (which may not be independent of one another) clearly read *tara* instead of *tarattaba*.

35. *al-Kulliyyāt*, p. 313 n.1794. The St Petersburg manuscript is described by the editors on p. 14, and, like the Madrid manuscript, it exhibits Ibn Rushd's own later version of the book. This passage is found in the anonymous Hebrew translation, mss. Munich 29, f. 179b, and St Petersburg, Academy B288, f. 157a.

36. Ms. St Petersburg, Academy C81, f. 165a. Concerning the two Hebrew translations, see Steinschneider, *Hebraeischen Uebersetzungen*, pp. 671–675. The activity of Shlomon ben Avraham can now be dated to the early decades of the thirteenth century; see Y. T. Langermann, "Some New Medical Manuscripts from St Petersburg," forthcoming in *Korot*.

37. It is translated by Gauthier, pp. 96–97.

38. This work was the subject of the unpublished doctoral dissertation of Lutz Richter-Bernburg, *Eine arabische Version der peudogalenischen Schrift De Theriaca ad Pisonem* (Göttingen, 1969).

39. Gauthier, Arabic text, p. 24.

40. *Guide of the Perplexed*, part I, chapter 31 (trans. S. Pines, Chicago, 1963, vol. 1, p.67), and *al-Maqāla fī ʾl-Rabw* (*Treatise on Asthma*), soon to be available in an edition and translation by Gerrit Bos.

41. There is special section on the *khāṣṣa* (or *khāṣṣiyya*) in *al-Kulliyyāt*, 233ff. It is odd that Ibn Rushd does not refer to it here. Cf. Y. T. Langermann, "Gersonides on the Magnet and the Heat of the Sun," in G. Freudenthal (ed.), *Studies on Gersonides, a Fourteenth Century Jewish Philosopher-Scientist* (Leiden,1992), 267–284, esp. pp. 273–274 concerning Ibn Rushd's claim that all physical properties can be explained by the proportionalities of the four qualities, in effect denying the possibility of any appeal to specific, occult properties.

42. *al-Kulliyyāt*, p. 314.

43. Clearly by this statement Ibn Rushd intends the properties that are rationally learned by the systematic application of the *qawānīn*.

44. *al-Kulliyyāt*, p. 215.

45. *al-Kulliyyāt*, p. 314.

46. A. I. Sabra, "The Andalusian Revolt against Ptolemaic Astronomy: Averroes and al-Biṭrūjī," in E. Mendelsohn (ed.), *Transformation and Tradition in the Sciences* (Cam-

bridge, 1984), 133–153; rprt., A. I. Sabra, *Optics, Astronomy, and Logic. Studies in Arabic Science and Philosophy* (Variorum, 1994). Cf. Y. T. Langermann, "Arabic Cosmology," *Early Science and Medicine* 2 (1997), 185–213, esp. pp. 202–206.

47. See now the important study of J. Samsó, "On al-Biṭrūjī and the *Hay'a* Tradition in al-Andalus," in *idem*, *Islamic Astronomy and Medieval Spain* (Variorum, 1994), essay XII. Note that Ibn Bājja wrote a short treatise *Fī māhiyyat al-shawq al-ṭabī'iyy* (*On natural desire and its quiddity*), published by Jamāl al-Dīn al-'Alawī, *Rasā'il Falsafiyya li-Abī Bakr bin Bājja* (Beirut, 1983), pp.97–102. The treatise reported by Ibn Abī Uṣaybi'a, *'Uyūn al-Anbā' fī Ṭabaqāt al-Aṭibbā'*, ed. N. Riḍā (Beirut, n.d.) (hereafter: IAU), p. 516, has *tashawwuq* instead of *shawq* in the title. See further S. Pines, "La Dynamique d'Ibn Bâjja," *Mélanges Alexandre Koyré* (1964), pp. 442–468; rprt., *The Collected Works of Shlomo Pines*, vol. 2 (Jerusalem and Leiden, 1986), pp. 440–466. For a very recent and thorough discussion of the history of the problem, see A. I. Sabra, "Configuring the Universe: Aporetic, Problem Solving, and Kinematic Modeling as Themes of Arabic Astronomy," *Perspectives on Science* 6 (1998), 288–330.

48. *Kitāb al-Shukûk 'alâ Jâlînûs*, ed. M. Mohaghegh (Tehran, 1993); cf. S. Pines, "Razi Critique de Galien," *Actes du VIIIe Congrès International d'Histoire des Sciences* (Paris, 1953), pp. 480–487; rprt., *Collected Works*, vol. 2, pp. 256–263.

49. See, e.g, the end of the very first *faṣl* (ed. *Būlāq*, vol. 1, p. 5): "When Galen tried to construct a proof for the first division, he did not wish to attempt that as a physician, but rather insofar as he wished to be a philosopher speaking about natural science."

50. The philosophical critique (but not the medical critique) was published by J. Schacht and M. Meyerhof, "Maimonides against Galen, on Philosophy and Cosmogony," *Bulletin of the Faculty of Arts*, vol. 5, part 1 (Cairo, 1939). Some medical critiques are discussed by Y. T. Langermann, "Maimonides on the Synochous Fever," *Israel Oriental Studies* 13 (1993), pp. 175–198, esp. pp. 186–188.

51. J. C. Bürgel, "Averroes 'contra Galenum'," *Nachrichten der Akademie der Wissenschaften in Göttingen*, Philologisch-Historische Klasse, no. 9 (1967), pp. 263–340.

52. His *Kitāb Ḥall Shukūk al-Rāzī 'alā Kutub Jālīnūs* is listed by IAU, p. 519. Mohaghghegh, *al-Shukūk*, p. 112n40, reports the existence of a manuscript in Mashhad.

53. *al-Qānūn fī 'l-Ṭibb*, ed. Būlāq, vol. 3, p. 309.

54. *al-Kāmil min al-Ṣinā'a 'l-ṭibbiyya*, second section (*'amalī*), book ten, chapter ten. I do not have access to the Būlāq edition, and consulted instead the ms. Bethesda, National Library of Medicine WZ 225 A398k 1736. The same trivial example (but substituting cold for hot) is reproduced by Ibn Rushd in the course of his detailed and extensive analysis of compound medications; see *al-Kulliyyāt*, p. 304. Although both authorities employ the same simple arithmetical rule, there are some significant differences between their examples.

55. This section of *al-Musta'īnī* is available in M. al-Khattabi, *Pharmacopée et régimes alimentaires dans oeuvre (sic!) des auteurs hispano-musulmans* (Beirut,1990), pp. 312–314. I have recently identified a partial copy of *al-Musta'īnī*, Arabic in Hebrew letters, in ms. Vatican Ebr. 530/1.

56. It is listed by IAU, p. 519.

57. See M. McVaugh, "Quantified Medical Theory and Practice at Fourteenth-Century Montpellier," *Bulletin of the History of Medicine* 43 (1969), 397–413, esp. p. 405: "The course of medical practice is, for him, still determined by the results of the established tradition. And in this he is followed by all later writers at Montpellier, none of whom shows any signs of using mathematical theory as other than a descriptive tool." McVaugh's remarks about Bernard de Gordon and the other physicians of Montpellier would, on the face of it, seem to apply to most writers in the Arabic tradition as well.

58. M. Meyerhof, "Esquisse d'histoire de la Pharmacologie chez les Musulmans d'Espagne," *al-Andalus* 3 (1935), 1–41; M. al-Khattabi, *Al-Aghdhiya wa-l-adwiya ʿinda muʾallifī ʾl-gharb al-islāmī: madkhal wa-nuṣūṣ* [French title: *Pharmacopé et régimes alimentaires dans oeuvre (!) des auteurs hispano-musulmans*] (Beirut, 1990).

59. Ibn Bājja's treatises are listed by IAU, 516–517. M. Asín Palacios, "Avempace Botánico," *Al-Andalus* 5 (1940), 255–299, includes a text and translation of Ibn Bājja's *Kalām fī ʾl-Nabāt* (*Discourse on Plants*). Ibn Bājja's essay on the *nīlūfar*, a heliotropic plant, was published by al-ʿAlawī, *Rasāʾil*, pp. 106–107.

60. Asín Palacios, "Avempace Botánico," pp. 257–258.

61. *al-Kulliyyāt*, p. 215, n. 16. Cf. IAU

62. Langermann, "Arabic Cosmology," pp. 202–206.

63. A. Z. Iskandar, *A Catalogue of the Arabic Manuscripts on Medicine and Science in the Wellcome Historical Medical Library* (London,1967), p. 36, translating from a work Ibn Jumayʿ.

64. *al-Kulliyyāt*, p. 422.

65. Iskandar, *Wellcome Catalogue*, p. 37. Iskandar adds, however, that Ibn Rushd did write a commentary on Ibn Sīnā's *Urjuza*.

66. For the work of Ibn Ṭufayl see the excellent collection of essays edited by Lawrence Conrad, *The World of Ibn Ṭufayl: Interdisciplinary Perspectives on Ḥayy bin Yaqẓān* (Leiden, 1996). Concerning *Ḥayy* see further D. Gutas, "Ibn Ṭufayl on Ibn Sīnā's Eastern Philosophy," *Oriens* 34 (1994), 222–241. Portions of the *Urjūza* were published by M. Muḥammad, "Qirāʿa fī Urjūzat Ibn Ṭufayl fī ʾl-Ṭibb," *Majallat Maʿhad al-Makhṭūṭāt al-ʿArabiyya* 30 (1986), 47–82.

67. According to the indications of some Hebrew manuscripts, Ibn Rushd may have collaborated with his son, Abū Muḥammad ʿAbd Allāh, in writing a few short philosophical monographs; this point still requires clarification. See Steinschneider, *Hebraeischen Uebersetzungen*, pp. 198–199. On Ibn Rushd's acquaintances, see J. Puig Montada, "Materials on Averroes's Circle," *Journal of Near Eastern Studies* 51 (1992), 241–260.

Index